Electronic Devices, Circuits, and Systems

Third Edition

Electronic Devices, Circuits, and Systems

Michael M. Cirovic
California State Polytechnic University
San Luis Obispo, California

James H. Harter
Mesa College
Phoenix, Arizona

PRENTICE-HALL, INC., Englewood Cliffs, New Jersey 07632

Library of Congress Cataloging-in-Publication Data

Cirovic, Michael M.
  Electronic devices, circuits, and systems.

   Previous ed. published under title: Basic
electronics. Reston, Va. : Reston Pub. Co., c1979.
   Includes index.
   1. Electronics. I. Harter, James H. II. Cirovic,
Michael M. Basic electronics. III. Title.
TK7816.C53   1987   621.381   86-22564
ISBN 0-13-250655-6

Editorial/production supervision and
  interior design: Linda Zuk, Wordcrafters Editorial Services, Inc.
Cover design: Wanda Lubelska Design
Manufacturing buyer: Carol Bystrom

© 1987, 1979, 1974 by Prentice-Hall, Inc.
A division of Simon & Schuster
Englewood Cliffs, New Jersey 07632

Previously published as *Basic Electronics*, second edition,
by Cirovic, Reston Publishing Company, Inc., 1979.
Reprinted by permission of Prentice-Hall, Inc.

*All rights reserved. No part of this book may be
reproduced, in any form or by any means,
without permission in writing from the publisher.*

Printed in the United States of America

10 9 8 7 6 5 4 3 2 1

ISBN 0-13-250655-6 025

Prentice-Hall International (UK) Limited, *London*
Prentice-Hall of Australia Pty. Limited, *Sydney*
Prentice-Hall Canada Inc., *Toronto*
Prentice-Hall Hispanoamericana, S.A., *Mexico*
Prentice-Hall of India Private Limited, *New Delhi*
Prentice-Hall of Japan, Inc., *Tokyo*
Prentice-Hall of Southeast Asia Pte. Ltd., *Singapore*
Editora Prentice-Hall do Brasil, Ltda., *Rio de Janeiro*

Dedicated to my mother
Smilja
to whom I am grateful
for much more than my existence

Michael M. Cirovic

Contents

Preface   xiii

PART I   DEVICES   1

Chapter 1   Semiconductor Physics   3

1.1 Classifying Matter, 3
1.2 Semiconductors, 6
1.3 Doped Semiconductors, 11
1.4 Conduction in Semiconductors, 13
Review Questions, 14

Chapter 2   Semiconductor Diodes   16

2.1 Depletion Region and Contact Potential, 18
2.2 Diode with Bias, 19
2.3 Diode Characteristics, 21
2.4 Diode Equivalent Circuit, 24
2.5 Special Purpose Diodes, 28
Review Questions, 30
Problems, 31

Chapter 3   Bipolar Junction Transistors   34

3.1 Currents in the BJT, 35
3.2 BJT Static Characteristics, 41
3.3 Transistor Ratings, 46
Review Questions, 48
Problems, 49

Chapter 4 BJT Biasing, Bias Stability, and Small-Signal Models 51

4.1 Biasing, 51
4.2 BJT Low-Frequency Model, 63
4.3 BJT High-Frequency Model, 76
Review Questions, 80
Problems, 80

Chapter 5 Field-Effect Transistor Operation, Biasing, and Models 82

5.1 Junction Field-Effect Transistor, 82
5.2 Biasing the JFET, 87
5.3 MOSFET, 94
5.4 Biasing the MOSFET, 101
5.5 Small-Signal FET Model, 102
5.6 FET High-Frequency Model, 105
Review Questions, 105
Problems, 106

Chapter 6 Thyristors and Related Devices 108

6.1 UJT and PUT Transistors, 108
6.2 Multilayered Diodes, 112
6.3 Thyristors, 119
Review Questions, 126

Chapter 7 Photoelectric Devices 128

7.1 Photoconductive Cells, 128
7.2 Photovoltaic Cells, 131
7.3 Photodetectors, 132
7.4 Photoemitters, 137
7.5 Optical Couplers/Isolators, 139
7.6 Fiber Optic Emitters and Detectors, 139
Review Questions, 141

PART II CIRCUITS 143

Chapter 8 Rectifiers and Filters 145

8.1 Half-Wave Rectifier, 145
8.2 Full-Wave Rectifier, 148
8.3 Rectifier Filters, 153
Review Questions, 158
Problems, 159

Chapter 9 Amplifier Fundamentals 160

9.1 Gain Calculations—Systematic Analysis, 161
9.2 Single-Stage BJT Amplifier, 163
9.3 Single-Stage FET Amplifier, 166
9.4 Frequency Response, 169
Review Questions, 175
Problems, 176

Chapter 10 Practical Amplifier Considerations 178

10.1 Input and Output Impedance, 178
10.2 Real and Apparent Gain, 179
10.3 Amplifier Loading, 182
10.4 Impedance Matching, 184
10.5 Cascading of Amplifiers, 186
Review Questions, 190
Problems, 191

Chapter 11 Tuned Amplifiers 192

11.1 Single-Tuned Amplifiers, 192
11.2 Coupling of Tuned Amplifiers, 201
11.3 Double-Tuned Amplifiers, 202
Review Questions, 206
Problems, 207

Chapter 12 Power Amplifiers 209

12.1 Classes of Power Amplifiers, 209
12.2 Series-Fed Class-A Amplifiers, 209
12.3 Power Efficiency and Dissipation, 214
12.4 Harmonic Distortion, 220
12.5 Single-Ended Class-A Amplifiers, 223
12.6 Transformer-Coupled Push-Pull Amplifiers, 225
12.7 Other Push-Pull Amplifiers, 231
12.8 Complementary-Symmetry Amplifiers, 233
12.9 IC Power Amplifiers, 235
12.10 Summary, 238
Review Questions, 239
Problems, 240

Chapter 13 Negative Feedback Amplifiers 242

13.1 General Feedback Concepts, 242
13.2 Voltage-Feedback Amplifiers, 243
13.3 Current-Feedback Amplifiers, 247
13.4 Effect of Feedback on Frequency Response, 251
13.5 Series- and Shunt-Feedback Amplifiers, 253
13.6 Effect of Feedback on Nonlinear Distortion and Noise, 258
Review Questions, 259
Problems, 259

Chapter 14 Differential and Operational Amplifiers 261

14.1 Emitter-Follower Circuits, 261
14.2 Differential Amplifiers, 264
14.3 Integrated Circuit DIFF AMPs, 271
14.4 Monolithic Operational Amplifiers, 272
14.5 Basic Amplifier Configurations, 276
Review Questions, 280
Problems, 281

Chapter 15 Sinusoidal Oscillators 282

15.1 Criteria for Oscillation, 282
15.2 Hartley Oscillators, 285
15.3 Colpitts Oscillators, 286
15.4 RC Phase-Shift Oscillators, 288
15.5 Tuned-Output Oscillators, 291
15.6 Twin-T Oscillators, 292
15.7 Wien-Bridge Oscillators, 293
15.8 Amplitude-Stabilized Oscillators, 294
15.9 Crystal Oscillators, 295
Review Questions, 296
Problems, 297

Chapter 16 Clipping, Clamping, and Wave-Shaping Circuits 298

16.1 Single-Level Clipping Circuits, 298
16.2 Two-Level Clipping Circuits, 301
16.3 Clamping Circuits, 303
16.4 Wave-Shaping Circuits, 308
Review Questions, 315
Problems, 315

PART III SYSTEMS 317

Chapter 17 Regulated Power Supplies 319

17.1 Regulators, 319
17.2 Current-Limiting Circuits, 332
17.3 IC Regulators, 334
17.4 Complete Power Supply, 339
Review Questions, 344
Problems, 345

Chapter 18 Power Control Systems 346

18.1 Principles of Power Control, 346
18.2 Triac Light-Intensity Control, 352
18.3 SCS Alarm Circuit, 353
18.4 SCR Universal Motor Speed and Direction Control, 354
18.5 12-Volt Battery Charger, 355
18.6 Electronic Crowbar, 356
Review Questions, 357

Chapter 19 Analog Systems 359

19.1 Principles of Analog Computation, 359
19.2 Assorted Analog Circuits, 368
19.3 Application of Analog Systems, 370
Review Questions, 374

Appendix A Manufacturers' Data Sheets 375

Appendix B Answers to Selected Problems 428

Index 435

Preface

The age in which we live has been termed the *Age of Technology*. Electronics has played a major role in this technology: It has provided us with the hardware to improve communciation, to enable us to explore the universe, to help physicians provide us with better health care, and to process huge amounts of information and with data to liberate us from the more mundane tasks. The list of accomplishments, as well as future possibilities, is unending. However, in some cases, the hardware of the technological revolution has created problems that are not easily solved. The future technologist, unlike his or her predecessor, must be more aware of and concerned with the consequences of technology on people and their environment.

The purpose of this book is threefold: first, to introduce a variety of semiconductor devices (integrated circuits and discrete devices), their basic operation, and their characteristics; second, to illustrate how these devices are used in simple electronic circuits, as well as how these circuits are analyzed and designed; third, to present complex electronic systems as simple extensions and examples of the use of devices and simple circuits. The prerequisites for understanding the material are basic college mathematics (algebra) and a basic course in electronic circuits.

Part I presents the basic physics and physical principles that make understanding the operation of electronic devices possible. This is a brief description, not a mathematical discussion, leading to the terminal characteristics of devices. The terminal characteristics directly lead to and suggest biasing schemes that follow. With the devices properly biased, terminal characteristics under signal conditions are presented, leading to the use of models and equivalent circuits in the systematic analysis of circuits containing devices.

Part II deals with applying the devices introduced in Part I in simple circuits. Methods of analysis stressing approximations and practical considerations are used, and some design problems are illustrated.

In Part III more complex electronic circuits and systems are described. In some cases, actual circuits are examined; in other cases, block diagrams are used.

By using this three-tiered building-block approach, the reader is able to move from simple, basic concepts to complex systems and is able to see how even the most complex electronic system is a logical extension of very simple circuits.

The third edition of *Electronic Devices, Circuits, and Systems* (formerly titled *Basic Electronics*) has undergone an extensive content revision. Many linear integrated circuits have been added, as have several power MOSFET devices. New to this edition are topics on fiber optic devices, heat sinks, electric static discharge (ESD), operational amplifiers, and IC thyristor triggers.

In the 13 years since the first edition was published, most electronics curricula have created separate courses in digital and communication circuits and systems. In response to the evolution of the electronics curriculum, topics dealing with communication circuits as well as digital circuits have been deleted from the text.

We would be remiss in not acknowledging six reviewers whose valued suggestions have shaped the third edition of the text. Our thanks to Russell Puckett, P.E., Texas A & M University; William Campas, John Tyler Community College; Roger Scheunemann, Group III Electronics; Conrad Zalace, GM Hughes Electron Dynamics Division; Stephen Cheshier, Ph.D., Southern Technical College; and Gary Lyon, Mesa Community College. We are also grateful to the manufacturers who provided technical data sheets and illustrative materials.

Electronic Devices, Circuits, and Systems

Part I

Devices

Electronic circuits use many different components. Besides resistors, capacitors, and inductors, there is a large group of active components called *electronic devices*, or simply devices. In circuits, these devices exist in two different forms: either as a separate component, called a *discrete component*, or as a unit with many components in one package, called an *integrated circuit*. In either form, the electronic device is usually the most important and, at the same time, the most complicated part of a circuit. You will be able to understand and fully use electronic circuits only if you first study each device in the circuit by itself and then as it relates to the whole circuit. Once you understand how the device works, predicting the operation of a circuit containing the device becomes relatively easy and straightforward.

Part I studies various semiconductor devices so that a knowledge of their terminal characteristics is gained along with an understanding of these characteristics.

Chapter 1

Semiconductor Physics

The rapid advances in semiconductor technology from the 1950s to present have revolutionized electronics. In replacing vacuum tubes, semiconductor devices brought smaller size, increased lifetime, lower power consumption, lower operating temperatures, and lower costs. To understand how these results were possible, we have to examine the physics of semiconductors. Our treatment will be descriptive rather than mathematical. The physical principles involved will be emphasized; not their formal mathematical descriptions.

1.1 CLASSIFYING MATTER

The ability of a material to conduct electricity or to sustain an electric current will be used as a means of classifying matter. Conduction takes place as a result of the motion of charged particles, usually *free electrons*. Thus, the ability of any material to conduct is directly proportional to the number of charged particles that can be set in motion within the material. Materials that have relatively large numbers of free electrons (such as metals) are very capable of sustaining an electric current. These materials are called *conductors*. Other materials that do not readily sustain an electric current under normal conditions are called *insulators*. Insulators have very few or no free electrons. It should be realized that the terms *conductor* and *insulator* are not absolute; that is, some conductors do not conduct as well as other conductors, while some insulators do not insulate as well as other insulators.

1.1.1 Shell Structure

The electrical conductivity of a material depends on the structure of the individual atoms, as well as the manner in which the atoms are arranged within the material. All matter is made up of atoms in assorted configurations. Each atom has its electrons arranged in shells or orbits around the nucleus. In the electrically neutral atom, the positive charge of the nucleus is balanced by an equal negative charge

TABLE 1.1 SUMMARY OF SHELL STRUCTURE

Shell number (letter)	Subshell number (letter)	Maximum number of electrons allowed	Total possible in shell
1(L)	1(1s)	2	2
2(M)	1(2s)	2	8
	2(2p)	6	
3(N)	1(3s)	2	
	2(3p)	6	18
	3(3d)	10	

of the electrons in orbit around the nucleus. The distinguishing factor among atoms of different materials is the size of the nucleus and the number of electrons in orbit around the nucleus. As an example, an oxygen atom has a large nucleus with 16 electrons in orbit; in contrast, a hydrogen atom has a small nucleus with only one electron in orbit.

In atomic structure, there is a specific scheme, called *shells*, for the arrangement of electrons into orbits. Furthermore, there is a prescribed maximum number of electrons that can be sustained in any one shell or orbit at any given time. It should be noted that the shells are divided into subshells (suborbits), each of which can sustain a maximum allowed number of electrons. This structure, together with the number of electrons allowed for the first three shells, is given in Table 1.1. Perhaps the easiest way to visualize electrons in a shell is to think of each electron as requiring its own little space, with only a limited number of spaces in each shell.

The shells and subshells closest to the nucleus are filled first, until all the electrons for that particular atom are accommodated. For example, in a hydrogen atom, the one and only electron is found in the L shell (the first shell). In an oxygen atom, its 16 electrons are distributed as follows: 2 in the first shell (the L shell); 8 in the second shell (2 in the 2s subshell, 6 in the 2p subshell); and 6 in the third shell (2 in the 3s subshell, 4 in the 3p subshell). Thus, the pattern is clear. The innermost shells and subshells are filled completely before any of the other shells. The outermost shell containing electrons is called the *valence shell*, and it plays the important role of determining the electrical as well as the chemical properties of elements.

Using aluminum, with a total of 13 electrons as an example, the shell structure is: $1s^2$, $2s^2$, $2p^6$, $3s^2$, $3p^1$ (where the superscript numbers indicate the number of electrons in that particular subshell). The valence shell is incomplete and contains three electrons. For the valence shell to be complete, it needs either to gain five electrons or to give up three electrons. In either case, the aluminum atom becomes *ionized*. (As previously noted, an atom is electrically neutral but when it gains or loses electrons, it develops a net charge and is said to be ionized.)

Aluminum is known to be trivalent; that is, it gives up three electrons when reacting with other elements. The reason for giving up three and not acquiring five electrons is that less energy is involved in liberating three electrons simply because of the lower number of electrons involved.

Aluminum is a good electrical conductor because of the three loosely bonded valence electrons. The energy binding the three valence electrons to the nucleus of the aluminum atom is weak and only a small amount of energy is needed to liberate the three valence electrons. The energy present at room temperature is sufficient to free the valence electrons for electrical conduction. A bar of aluminum is made up of literally billions of atoms and at room temperature countless free electrons are available for conduction.

A good insulator results from elements that have filled or completed valence shells. In these cases, no electrons are freed at room temperature because of the strong binding forces between the electrons in the filled shells and the nucleus. A material made up of such elements is called *inert*; it does not provide any free electrons that could take part in conduction.

1.1.2 Energy Bands

The difference in the binding energy of valence electrons in the valence shell is used as the basis for another means of classifying materials. An electron in the valence shell is said to have energy corresponding to the valence band of energy or, simply, *valence band*.* However, as a result of acquiring a specific amount of additional energy, an electron in the valence shell becomes free of the nucleus; with its new energy, an electron is characterized as being in the conduction band of energy or, simply, *conduction band*. Differentiation among materials can be made on the basis of the amount of energy needed to liberate a single valence electron from the influence of the nucleus. The amount of energy between the highest energy in the valence band, labeled E_v, and the lowest energy in the conduction band, labeled E_c, is a characteristic of the material and is called the *energy gap*, labeled E_g. From these definitions, we can write

$$E_g = E_c - E_v \qquad (1.1)$$

In a metal or other good electrical conductor at room temperature, there is an overlap between the conduction and valence bands, as shown in Figure 1.1. Consequently, in a conductor, many electrons are free to take part in conduction and very little energy is needed to sustain a sizable electric current.

In an insulator, the energy gap is large; that is, the conduction and valence bands are far apart, as shown in Figure 1.1. As a result, a large amount of energy is required to liberate even a small number of electrons that could then contribute to conduction.

---

* Extremely large numbers of electrons are involved in even small samples, and each electron has a discrete amount of energy slightly different from any other electron. The range of energy possessed by all the electrons in all the valence shells constitutes a dense set of energy values called a *band*—in this case, a valence band.

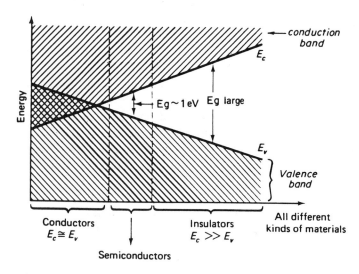

Figure 1.1 Classification of matter on the basis of conductivity.

1.2 SEMICONDUCTORS

From Figure 1.1 we see that there is no sharp dividing line between conductors and insulators; there are materials that are neither good conductors nor good insulators. These materials are called *semiconductors* and are characterized by energy gaps on the order of 1 electron volt (1 eV), as shown in Figure 1.1. The principle semiconductor material used in electronics is silicon (Si) with germanium (Ge) as a secondary material.

An atom of silicon has 14 electrons, and its electron-shell configuration is $1s^2, 2s^2, 2p^6, 3s^2, 3p^2$. Because the third shell is incomplete, it is the valence shell. In order for the third shell to be complete, it must either acquire four electrons in its 3p subshell or lose the four electrons that it already has in the 3s and 3p subshells. In either gaining or losing electrons, exactly the same number of electrons is involved; therefore, exactly the same energy is involved and neither process is more likely to occur. Thus, silicon neither acquires nor gives up electrons. Instead, each silicon atom enters into a unique sharing of its four valence electrons, called *covalent bonding*.

1.2.1 Covalent Bonding

In covalent bonding, each silicon atom shares two electrons with each of its four nearest neighbors. As shown in Figure 1.2, each silicon atom in the *crystal lattice* is connected by means of a single *covalent bond* (2 shared electrons) to each of four other atoms located at the corners of a regular tetrahedron. This basic structure is repeated millions of times in a crystal and is illustrated schematically in Figure 1.3.

The crystal structure for germanium is similar to that of silicon. Its valence

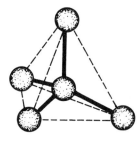

Figure 1.2 Diamond lattice construction of silicon (regular tetrahedron).

shell has four electrons with space for four more. Thus, germanium also forms covalent bonds and has a crystalline structure similar to that of silicon.

The picture of a semiconductor drawn thus far is somewhat oversimplified. In reality, atoms in a crystal are not stationary but are instead three-dimensional, vibrating in random fashion. The temperature of the crystal is a quantitative measure of how rapid and random this motion actually is. You can think of the positions of atoms depicted in Figures 1.2 and 1.3 as average positions since none of the atoms is ever completely stationary.

Electrons are also in constant motion, but their path cannot be predicted

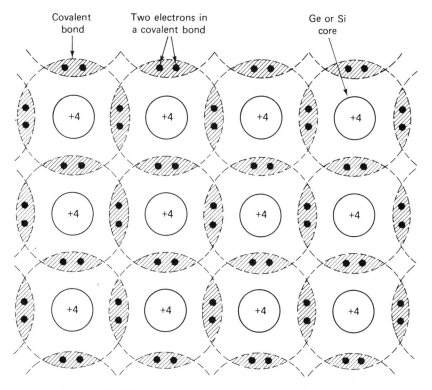

Figure 1.3 Schematic representation of a semiconductor crystal.

with certainty. As the temperature of the crystal increases, the vibration of the atoms becomes more violent; that is, excursions away from the average position become longer. As we shall see later, this increased vibration inside a crystal at elevated temperature limits the usefulness of semiconductors.

1.2.2 Free Electrons

The binding forces on electrons exerted by the covalent bonds are quite strong, much stronger than the ionic binding forces on the electrons exerted by the nucleus. However, even at room temperature, the vibration of the atoms is enough to place a serious strain on the covalent bonds. An analogy can be drawn between the covalent bond and its nearest atoms with an elastic pod (the bond) containing two peas (the two electrons in the bond). The elastic pod is attached to two large balls (the two atoms between which the bond exists). The balls are in constant motion, vibrating about some average position. As the balls move apart, the elastic pod stretches and shrinks to accommodate their motion. Because the vibration is random, the two balls may move directly away from each other at some instant. The pod may not be able to stretch that much and may break, releasing one of its peas. Exactly the same process can happen in a semiconductor crystal. The vibration causes some of the covalent bonds to break, liberating electrons. These liberated electrons are called *free electrons*.

Each atom, before engaging in covalent bonding, started out being electrically neutral; that is, it had exactly the same positive charge in its nucleus as the negative charge carried by its electrons in orbit around the nucleus. The crystal, as a whole, is also electrically neutral because it contains only atoms that are neutral. In the schematic presentation shown in Figure 1.3, we have conceptually combined the nucleus and the inner complete shells of each atom into one unit called the *core*. The core for both silicon and germanium has a net charge of $+4$ (the magnitude of the charge of one electron being used as a unit). The core together with four valence electrons, each one of which appears in each of the four surrounding covalent bonds, constitute a neutral atom.

Upon the breakup of a covalent bond, an electron has gained enough energy to become free. While in the bond, it had an amount of energy corresponding to the valence band; once the bond is broken, the electron energy corresponds to the conduction band. Each electron that is liberated by the breakup of a covalent bond has gained energy in the amount of the energy gap for that material.

1.2.3 Electron-Hole Pair

Returning to our analogy, when the pod ruptures to release a pea, a hole is left behind in the pod where the electron used to be. In semiconductor terminology, the absence of an electron in a covalent bond is also termed a *hole*. Every time a covalent bond is broken, an *electron-hole pair* is formed. (From now on, *electron* denotes a free electron unless otherwise stated.) The free electron roams around randomly and carries with it a negative charge. It, therefore, leaves behind a net positive charge near the broken bond. The positive charge comes from the core of one of the atoms near the broken bond, but there is no way to identify which

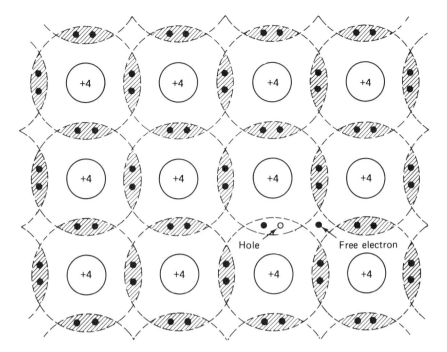

Figure 1.4 Crystal lattice showing a broken covalent bond.

core. We can associate the positive charge with the broken bond, specifically with the absence of an electron we have called a hole. Thus, the breakup of a covalent bond results in a free electron of negative charge and a hole of positive charge. In a pure (also called *intrinsic*) semiconductor, the number of free electrons is exactly equal to the number of holes, as illustrated schematically in Figure 1.4. On the average, at room temperature, about 1 out of 10^{13} bonds is broken in silicon crystals, and about 1 out of 10^{10} bonds in germanium crystals. The fewer broken bonds in silicon are a reflection of its larger energy gap.

1.2.4 Charge Carriers

Let us examine what happens inside an intrinsic semiconductor if we apply a voltage to it. The negatively charged free electrons are attracted to the positive terminal, and an electric current results. Inside the semiconductor, the electric field that is set up by the applied voltage also acts on the positively charged hole. As we noted earlier, the positive charge is actually in a nucleus of one of the atoms and cannot move as such. However, the electric field acting in the vicinity of the hole can cause a valence electron, called a *bound* electron, from an adjacent bond to move over and occupy the hole. This shift, in effect, accomplishes transport of a hole together with a positive charge, as shown in Figure 1.5. If the electric field is directed from left to right, the hole will successively move to the right.

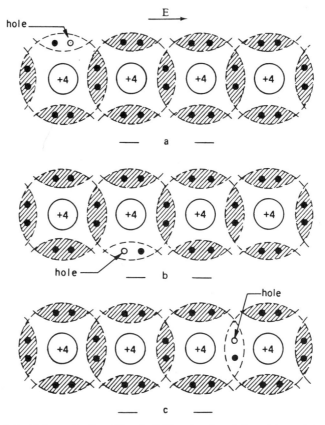

Figure 1.5 Hole conduction: Electric field acting from left to right and the successive motion of a hole, as shown in (a), (b), and (c).

The concept of the motion of a hole is at first hard to accept, for it may seem silly to say that "nothing" can move. We can again return to our analogy with pods and peas and imagine many large balls (cores) connected by many pods (bonds). Let us say that the pods are also interconnected so that peas can be interchanged among them at will. When an attempted interchange occurs between one pod containing two peas and another pod containing one pea and one hole, the pod that had two peas may wind up with only one pea and a hole. In other words, it traded a pea for a hole.

In reality, when we say that a hole has moved, it is actually a series of different bound electrons that have moved from one bond to another. The key here is that the electrons contributing to hole conduction are bound electrons; that is, they start in one bound state and wind up in another bound state. Note that it is incorrect to say that the motion of holes is the result of free electron movement in the opposite direction. We now see that current flow in a semiconductor results from the motion of two distinct charge carriers: negatively charged electrons and positively charged holes.

There are two inherent drawbacks to the use of intrinsic (i.e., pure) semiconductors. First, the number of charge carriers available to sustain an electric current is relatively low; hence, the name semiconductor. The second problem concerns *recombination*. Free electrons may be captured by vacancies in covalent bonds (holes), and in the process useful charge carriers are lost.

1.3 DOPED SEMICONDUCTORS

Conduction in semiconductor materials is greatly enhanced by the addition of carefully chosen impurities in precisely controlled amounts. Adding impurities to intrinsic semiconductors is called *doping*.

By adding appropriate impurities, two types of semiconductors can be formed. In one kind, called N-*type*, conduction results primarily from electron flow. In the second type, called P-*type*, the predominant conduction is by means of holes.

1.3.1 N-type Semiconductors

N-type semiconductors are formed by adding small amounts (typically, one part in a million) of *pentavalent* impurities to the intrinsic semiconductor crystal. Typically, such impurities are chemical elements of phosphorus, arsenic, or anti-

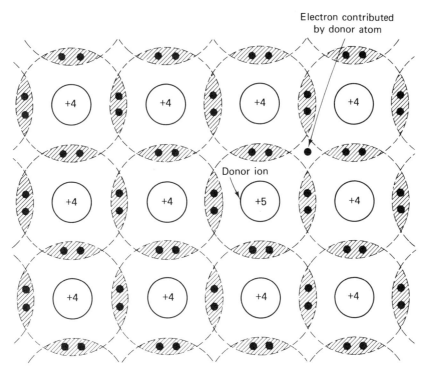

Figure 1.6 N-type semiconductor.

mony. All of these elements have valence shells with five electrons. The pentavalent impurity atoms take the place of silicon atoms in the crystal structure and, like silicon atoms, the impurity atoms form covalent bonds with neighboring silicon atoms. Figure 1.6 pictures *N*-type semiconductor material. Each impurity atom contributes one electron to each of four covalent bonds, and each has one *excess* electron that is not taking part in a covalent bond. This excess electron is only weakly bound to the core and, at room temperature, usually has enough energy to be considered a free electron. The addition of pentavalent atoms, called *donor atoms* or just *donors*, increases the free electron population in a semiconductor without increasing the hole population. Therefore, the essential characteristic of an *N*-type semiconductor is that electrons are more (and usually much more) plentiful than holes.

Because current in an *N*-type semiconductor depends on electrons, they are called the *majority* (charge) *carriers*, and holes are referred to as the *minority* (charge) *carriers*.

1.3.2 P-Type Semiconductors

P-type semiconductors are formed by introducing *trivalent* impurities into the intrinsic semiconductor. Typically, such impurities are chemical elements of boron, gallium, and indium. These trivalent impurity atoms all have three valence

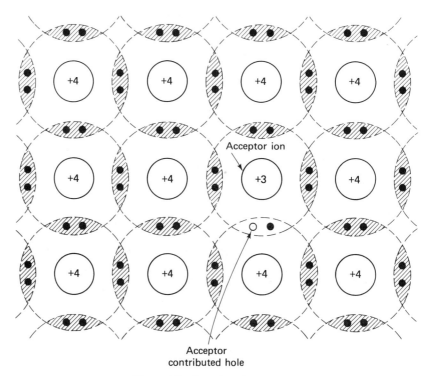

Figure 1.7 *P*-type semiconductor.

electrons. Upon replacing silicon atoms in the crystal, the trivalent atoms cannot satisfy all four covalent bonds around them. As shown in Figure 1.7, one covalent bond near each of the *P*-type impurity atoms is incomplete in that it contains a hole. *P*-type impurity atoms are called *acceptors*, because each can accept one electron into its incomplete bond. The effect of adding acceptor atoms to a semiconductor is to increase the number of holes without an increase in the number of electrons.

Because current in a *P*-type semiconductor is mainly the result of holes, they are termed *majority carriers*, and electrons in *P*-type semiconductors are called the *minority carriers*. We must emphasize that the terms *majority* and *minority* carriers only have significance and specific meaning when the type of semiconductor, *N* or *P*, is specified.

1.4 CONDUCTION IN SEMICONDUCTORS

1.4.1 Drift

Two processes in semiconductor materials account for the flow of current. The first process is called *drift*. For example, suppose a potential difference is applied across the ends of a piece of *N*-type semiconductor. Because of the electric field that is established inside the semiconductor by the applied voltage, both majority carriers (in this case, electrons) and minority carriers (holes) will move in opposite directions to constitute a net current. The motion of carriers under the influence of an electric field is defined as *drift*.

For any material, we can measure the current resulting from a known applied voltage. The ratio of the voltage to the current is the *resistance*. Thus,

$$\frac{V}{I} = R = \rho \frac{L}{A} \tag{1.2}$$

where ρ is the *resistivity* of the material (in Ohm-m), L is the length of the sample (in m), and A is the cross-sectional area of the sample (in m^2). Resistance is a function of the shape, size, and nature of the material; resistivity is a function of the nature of the material itself. For example, one ton of copper has the same resistivity as one ounce of copper. However, the resistance of the two quantities of copper can be drastically different. The lower the resistivity of a material, the better it is able to conduct. In semiconductor materials, the higher the level of doping, the higher the number of free charge carriers and the lower the resistivity. Thus, the effect of doping is to lower the resistivity. From Equation 1.2, we note that drift current is proportional to the applied voltage ($I \propto V$). For the same magnitude of voltage applied to two samples of silicon, one intrinsic and the other doped, the current in the doped semiconductor is larger than the current in the intrinsic semiconductor.

1.4.2 Diffusion

The second process in semiconductors contributing to conduction is known as *diffusion*. Diffusion is frequently encountered in nature. A drop of ink in a glass of water diffuses through the water until all the water is evenly discolored by the

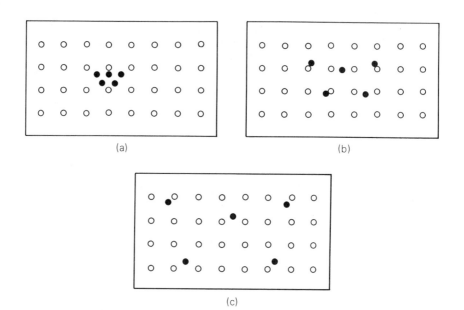

Figure 1.8 Diffusion: (a) large electron concentration in a localized area; (b) and (c) electrons diffuse away from the area of highest density.

ink. In semiconductors the same process occurs. For example, suppose a sample of *P*-type silicon has an excess number of electrons introduced into a small region of the sample. These electrons will diffuse through the sample, resulting in an even distribution throughout the sample. This motion of minority carriers is strictly a statistical phenomenon and not the result of any electrostatic attraction or repulsion. Diffusion takes place in a direction away from the region of highest density of the minority carriers. For our example, electrons diffuse away from the localized region into which they were introduced. This phenomenon is pictured in Figure 1.8. Diffusion is important because it is a process that describes the motion of excess minority carriers in a semiconductor; that is, electrons in *P*-type and holes in *N*-type.

As a sidelight to diffusion, note that when a few minority carriers are diffusing, they encounter a tremendously large number of majority carriers. There is always a danger that the two will recombine; thus, majority carriers may be lost as far as conduction is concerned.

REVIEW QUESTIONS

1. What are the pertinent properties of conductors?
2. Give examples of materials that are good conductors and state the reasons why they are good conductors.
3. What are the pertinent properties of insulators?

4. Give examples of materials that are good insulators and state why they are good insulators.
5. What is the mechanism by which conduction takes place inside a semiconductor?
6. What are some of the properties of an element that can be deduced from its electron-shell and subshell structure?
7. What are the ways in which atoms of unlike elements differ? In what ways are they similar?
8. What is the importance of the valence shell and valence electrons?
9. What is energy gap?
10. Characterize conductors, insulators, and semiconductors on the basis of energy gap.
11. Which material conducts electricity better, one with a small energy gap or one with a large energy gap? Why?
12. What is a covalent bond?
13. Under what conditions do atoms form covalent bonds?
14. What are some of the distinguishing characteristics of semiconductors as compared with conductors and insulators?
15. Give examples of semiconductors. Specifically, why are these materials classified as semiconductors?
16. What is the characteristic structure of a semiconductor?
17. What is an intrinsic semiconductor?
18. What is doping?
19. What is the name for an impurity atom that makes a semiconductor N-type?
20. Give examples of commonly used N-type impurities.
21. What is the characteristic that makes an impurity an N-type impurity?
22. What are the majority carriers in N-type semiconductors? Why?
23. What are the minority carriers in N-type semiconductors? Why?
24. What is the name for an impurity atom that makes a semiconductor P-type?
25. Give examples of commonly used P-type impurities.
26. What is the characteristic that makes an impurity P-type?
27. What are the majority carriers in P-type semiconductors? Why?
28. What are the minority carriers in P-type semiconductors? Why?
29. What are the two conduction processes in semiconductors? Give the conditions for each.
30. Describe conduction in an N-type semiconductor. Which carriers are responsible for most of the current? Why?
31. Describe conduction in a P-type semiconductor. Which carriers are responsible for most of the current? Why?
32. Describe the action as a result of the presence of an excess localized minority carrier concentration.

Chapter 2

Semiconductor Diodes

The study of semiconductor devices starts with an introduction to the semiconductor junction diode. Once you understand the basic physics of semiconductors, you will be able to easily understand and even predict the properties and behavior of semiconductor devices.

A junction diode results when a *P*-type semiconductor is brought into physical and electrical contact with an *N*-type semiconductor. A junction between the *N*-type and *P*-type semiconductors is formed, and, as we shall see, this *PN* junction is the essential building block for a variety of devices.

To manufacture a junction diode, a crystal of semiconductor material is grown that is usually greater than 3 inches in diameter and may be 6 inches in diameter. The basic process for growing a silicon crystal is depicted in Figure 2.1. To initiate the process, a "seed" crystal is inserted into the molten silicon and slowly drawn upwards. The molten semiconductor crystallizes and "grows" on the seed. A small, controlled amount of an impurity may be dissolved in the molten semiconductor if *doped* (either *N*- or *P*-type) crystals are desired.

Using a diamond saw, the crystal is then cut into thin slices called *wafers*. The wafers, once polished, are ready for further processing into diodes, transistors, or integrated circuits. From each wafer, literally hundreds of individual devices can be fabricated. The actual yield depends on the type of device and the number of imperfections in the wafer.

In actual practice, diffusion and ion implantation are the methods used in processing diodes, transistors, and integrated circuits. To illustrate some of the basic principles involved, we shall confine our discussion to the older diffusion method. The following description assumes that wafers cut from an *N*-type silicon crystal are being processed.

The first step is to form a thin layer of silicon-oxide (SiO_2) on the surface of the wafer by passing hot steam over the wafer. A thin layer of oxide (SiO_2) forms, as shown in Figure 2.2(a). Next, the oxide is coated with a *photoresist*, a substance that reacts to light (usually ultraviolet light). A *mask* is made, and the photoresist is exposed through the mask. Using a suitable chemical agent, the

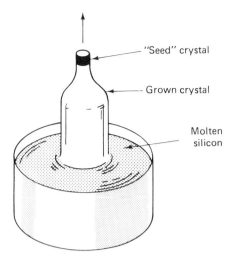

Figure 2.1 Growing a silicon crystal.

desired region of oxide is etched away. The etchant is chosen so that it will react only with the oxide that has been exposed, that is, the region where the photoresist has been exposed and removed. The resultant window in the oxide is shown in Figure 2.2(b). A suitable trivalent acceptor impurity (for example, boron gas) is passed over the window and diffuses into the lightly doped *N* region to form the *P* region, as shown in Figure 2.2(c).

Successive oxidation, masking, and etching processes expose a small part of the *N* and *P* regions onto which aluminum layers are deposited to form the metallic contacts for the connection of leads at a later time. Each of the many diodes thus formed on the single wafer is tested. Those that do not pass the test are marked with ink so they can be identified and later discarded. The wafer is then cut into small pieces called *die*; each die contains a single device, in this

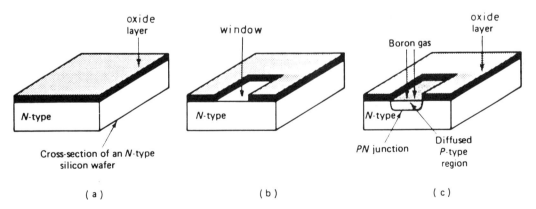

Figure 2.2 Formation of a junction *PN* diode: (a) an oxide film of silicon dioxide (SiO$_2$) is deposited; (b) the film is etched; (c) the *P*-type region is formed by diffusing boron into the *N*-type material.

Chap. 2   Semiconductor Diodes

case a diode. Leads are attached. Finally, the devices are packaged in protective jackets of epoxy or other suitable material.

2.1 DEPLETION REGION AND CONTACT POTENTIAL

Imagine that we have separate P-type and N-type samples of material and that each has just been brought into contact with the other to form a diode. Unfortunately, this method will not yield a diode; however, it can be used to see what actually does happen when a PN junction is formed. At the junction that is formed, one side (the N side) has a large number of electrons, whereas the other side (the P side) has a large number of holes. Conditions are quite favorable for diffusion to take place: Electrons from the N side diffuse across the junction into the P side, and holes from the P side diffuse across the junction into the N side, as shown in Figure 2.3(a). Before they were brought into contact, both sides were electrically neutral; that is, in the N region, for each electron with a negative charge, there is a donor ion in the crystal with a positive charge; in the P region, for each hole with its positive charge, we can associate an acceptor ion with a negative charge. As a result of the diffusion, the N side develops a net positive charge and the P side develops a net negative charge. This difference is due to the fact that the N side lost electrons and gained holes at the same time that the P side lost holes and gained electrons. The diode, as a whole, is still electrically neutral. It has not lost or gained any charge, but there is a localized imbalance of charge. Since the diode, as a whole, is still electrically neutral, the net positive charge in the N side must be exactly equal to the net negative charge in the P side.

The charges thus developed, negative in the P side and positive in the N side, are not mobile or free to move. They are *stationary* because they are caused by the ions in the crystal. The net negative charge in the P side is in the form of electrons that have fallen into holes near acceptor atoms; the net positive charge in the N side comes from the donor nuclei that have lost their fifth electrons.

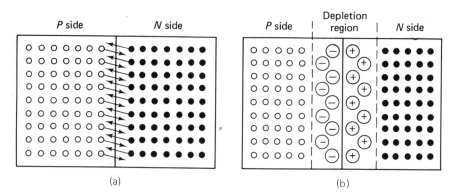

Figure 2.3 Formation of depletion region: (a) carriers diffuse across the junction until (b) equilibrium is reached.

Diffusion of carriers across the junction stops and equilibrium is reached when no additional carriers have enough energy to overcome the *electric field* that has built up at the junction as a result of the redistribution of charge. This situation is depicted in Figure 2.3(b). The positively charged donor ions in the N side repel holes, while negatively charged acceptor ions in the P side repel electrons.

The region on either side of the junction where the stationary charges appear is called the *depletion region*. The name refers to the fact that this region has been depleted of mobile charge carriers (holes and free electrons). Other names often used for this region are transition region and space-charge region.

The electric field that is set up across the depletion region acts from + to −, or from the N side to the P side. A field acting over the width of the depletion region causes a potential difference between the N and P regions. This difference in potential is called the *contact potential* or *barrier potential* and it is the equilibrium potential difference across a PN junction. By "equilibrium," we mean that neither holes nor electrons are crossing the junction, and the electric field across the depletion region is constant and not changing. The magnitude of the contact potential is a few tenths of a volt (0.3- to 0.7-V), depending on the doping concentrations in the N and P regions as well as on the type of semiconductor (germanium or silicon) used.

2.2 DIODE WITH BIAS

In Section 2.1, the diode was described under equilibrium conditions with no externally applied voltage, as shown in Figure 2.4(a). (The diode circuit symbol is also shown.) We will now consider the action of the diode when an external voltage is applied, or when the diode is *biased*.

When the positive end of a battery is connected to the P side (called the *anode*) and the negative end of the battery to the N side (called the *cathode*), a fairly high current is observed. This current is called the forward current, and the diode is said to be *forward biased*, as shown in Figure 2.4(b). If we reverse the polarity of the battery and connect the negative end to the anode and the positive end to the cathode, a minute current is observed. This current is called the reverse current, and the diode is said to be *reverse biased*, as shown in Figure 2.4(c).

2.2.1 Forward Bias

Under conditions of forward bias, as pictured in Figure 2.4(b), the externally applied voltage acts in opposition to the contact potential, thus lowering the effective or net potential at the junction. Physically, electrons enter the N side through the cathode lead, and as majority carriers in the N side, the electrons *drift* toward the junction. Because the potential barrier at the junction is lowered by the forward bias, electrons cross the junction and, as minority carriers in the P side, *diffuse* toward the anode lead. At the same time, holes, created at the anode lead by the liberation of bound electrons, *drift* through the P side, cross the junction, and, as minority carriers in the N side, *diffuse* toward the cathode

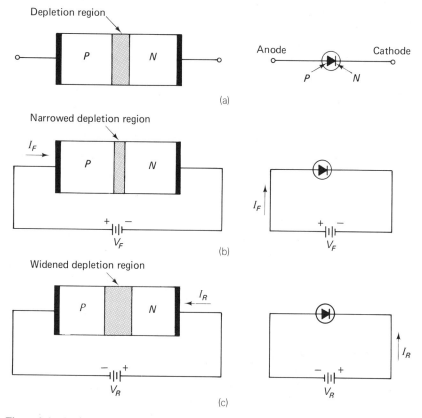

Figure 2.4 *PN* junction diode: (a) no bias; (b) when forward biased, the depletion region narrows; (c) when reverse biased, the depletion region widens.

lead. At the cathode, these holes are filled by electrons from the external wiring. At any instant of time, the same number of electrons enter the N side as leave the P side. The rather high forward current (I_F), is due to the motion of majority carriers, which, by definition, are large in number. A large number of charge carriers cause a high current.

Another consequence of applying a forward bias to the diode is the *narrowing* of the depletion region, as illustrated in Figure 2.4(b).

2.2.2 Reverse Bias

When the diode is reverse biased, the externally applied voltage acts in the same direction as the contact potential, causing an increase in the effective potential barrier across the depletion region, and a *widening* of the depletion region. The external voltage causes holes in the N side and electrons in the P side to move in the direction of the junction. However, holes in N type and electrons in P type are both *minority* carriers. The resulting current is, of necessity, very low because minority carriers are by definition extremely few in number.

2.3 DIODE CHARACTERISTICS

The *curve tracer* shown in Figure 2.5 is used to display the terminal characteristics of diodes as well as other semiconductor devices. With the curve tracer properly set up, an instant display of the pertinent $V - I$ relationships is seen on the screen. A camera may be attached to the display face of a curve tracer so that a permanent record of the display may be taken.

The forward characteristic of typical silicon and germanium diodes, obtained on a curve tracer, are shown in Figure 2.6. Notice the nonlinear shape of the $V - I$ characteristic. Also notice the basic forward and reverse bias $V - I$ characteristic of any diode, shown in Figure 2.7. As shown, in the forward direction even small voltages result in appreciable currents, whereas in the reverse direction, the current is almost negligible until a certain voltage, labeled V_B on the characteristic, is reached. For voltages more positive than V_B, a good mathematical model for the junction diode characteristic is given by the *diode equation*

$$I = I_o(e^{V/V_t} - 1) \tag{2.1}$$

where I is the diode current resulting from an externally applied voltage V, I_o is a constant (whose value depends on the particular diode under test) with units of current, V_t is the voltage equivalent of temperature (whose value equals 25 mV at room temperature), and e is the base of natural logarithms.

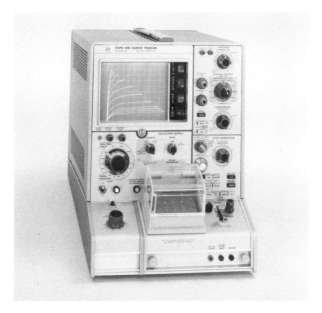

Figure 2.5 One of several models of curve tracer commercially available. (*Courtesy of Tektronix*)

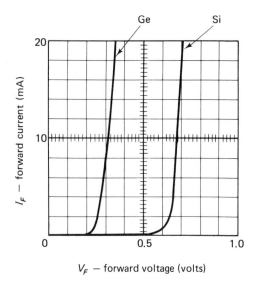

Figure 2.6 Comparison of silicon (Si) and germanium (Ge) junction diode characteristics.

2.3.1 Reverse Saturation Current

There is agreement between the actual diode current measured for a given voltage and that predicted by the diode equation with the constant I_o properly evaluated. To see the significance of this constant, we can evaluate the diode current for some value of applied voltage, for example, -0.25 V. At room temperature, this applied voltage in the diode equation yields a current of approximately $-I_o$. In fact, we get this answer for the current for any reverse voltage between V_B and about -0.2 V. In other words, the current *saturates* or levels off at a fixed value $(-I_o)$ and is not a function of the applied voltage. The negative sign indicates that the current actually flows in the reverse direction. The name attached to I_o now becomes evident; it is called the *reverse saturation current*. As we stated

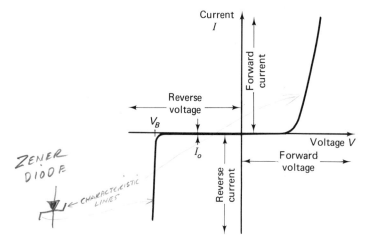

Figure 2.7 Typical junction diode characteristics.

earlier, the reverse saturation current is different for different diodes. The factors that determine the value of I_o at room temperature for a particular diode are (1) the type of material (silicon or germanium), (2) the doping levels of the P and N regions, and (3) the geometry of the junction.

Because of the differences in energy gap and the mobility of carriers in silicon and germanium semiconductors, diodes made of these materials exhibit different reverse saturation currents (as well as the difference in the forward characteristics already seen in Figure 2.6). In general, germanium diodes have reverse saturation currents in the microampere (10^{-6} A) range, while silicon diodes show reverse saturation currents in the nanoampere (10^{-9} A) range.

Temperature plays an important part in the magnitude of the reverse saturation current. When temperature is increased, a larger number of covalent bonds are broken; thus, a larger number of charge carriers become available for conduction. For example, a 10°C rise in the temperature of the semiconductor causes the reverse saturation current to almost *double* in value. Obviously, a large number of devices are extremely temperature sensitive. A visual example of this phenomenon will be shown in the discussion of the bipolar junction transistor characteristics in Chapter 3.

2.3.2 Breakdown Voltage

As previously stated, the reverse current quickly saturates as the voltage is advanced in the reverse bias direction. The current becomes essentially insensitive to additional changes in the reverse voltage. However, the voltage cannot be advanced in the reverse direction indefinitely. At some critical reverse voltage, labeled V_B in Figure 2.7, a large change in the reverse current results from a minute change in the reverse voltage. Under this condition, the diode is voltage saturated and is said to be in *breakdown*. The voltage at which this occurs, V_B, is called the *breakdown voltage*. There are two mechanisms that account for the sudden increase in reverse current.

The first mechanism is called *Zener breakdown*. It occurs when the applied reverse voltage sets up an electric field across the depletion region, which is high enough to cause the breaking of covalent bonds. As a result, mobile carriers, far in excess of those that had set up the reverse saturation current, are present. The result is a sudden increase in the current because of a sudden increase in the number of minority charge carriers.

The second mechanism is called *avalanche breakdown*. Here the highly accelerated carriers collide with the stationary ions inside the depletion region. A high enough field exists in the depletion region so that the carriers are moving fast enough (with enough energy) to break covalent bonds upon impact. This collision results in the liberation of many more mobile charge carriers, which are in turn accelerated in such a way that they now have the capability of producing other free charge carriers upon collision. The effect is the same as that of an avalanche. A few free carriers liberate many other free carriers, which in turn liberate others, and so on. The effect is cumulative. As a result, the abrupt creation of many extra free charge carriers causes an abrupt increase in the reverse current.

Diodes exhibiting reverse breakdowns anywhere from about 1 V to hundreds of volts are available commercially. Breakdowns below 5 V are usually explained by the Zener mechanism, and those above 8 V by the avalanche mechanism. However, there is nothing to preclude the possibility of both mechanisms occurring simultaneously.

Note that if protective circuitry is used to limit the current, a diode may be operated under breakdown conditions safely and without damage. In fact, as we shall see in later sections, special diodes are constructed for the purpose of exhibiting breakdown at a certain voltage and are used specifically for this property.

2.4 DIODE EQUIVALENT CIRCUIT

Thus far the essential property of the diode has been examined; that is, the diode passes current in the forward direction while it blocks current in the reverse direction. We will now look at the diode from an equivalent point of view where it is seen as an appropriate combination of elements and sources. Upon examining the diode $V - I$ characteristic in Figure 2.7, it may be concluded that a single *equivalent circuit* for the diode would be extremely complicated and cumbersome to use. To avoid this complexity, the diode characteristic is divided into three distinct regions, as shown in Figure 2.8. Each of the three regions uses a different *linear model* of the diode.

2.4.1 Region 1

In region 1, the diode is forward biased and a straight tangent line is fit to the characteristic, as shown in Figure 2.8. The tangent line intersects the voltage axis at the *cut-in voltage,* V_A. Thus, for forward voltages above V_A, the diode characteristic can be approximated by a straight line. For example, the point of the characteristic labeled P_1, where the current is I_1 and the voltage is V_1, can be represented by the cut-in voltage, V_A, and a fixed resistance r (the *dynamic resistance*). The dynamic resistance is characterized by the slope of the tangent line

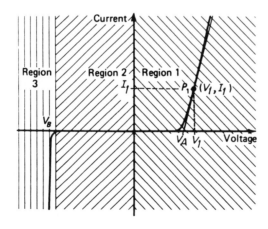

Figure 2.8 Definition of three regions of the diode characteristic for the purpose of obtaining piecewise linear diode models.

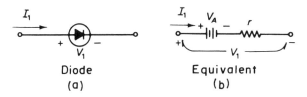

Figure 2.9 Diode under forward bias: (a) diode symbol and (b) the equivalent circuit used to replace the symbol where $r = (V_1 - V_A)/I_1$.

in Figure 2.8 and is defined as

$$r = \frac{\Delta V}{\Delta I} = \frac{\text{change in voltage}}{\text{change in current}} = \frac{1}{\text{slope of tangent}} \quad (2.2)$$

The representation of the *equivalent circuit* is pictured in Figure 2.9.

We can evaluate r from the inverse of the slope of the tangent line in Figure 2.8, noting that ΔV is $V_1 - V_A$ and ΔI is $I_1 - 0$ or just I_1. Thus,

$$r = \frac{V_1 - V_A}{I_1} \quad (2.3)$$

To check the equivalence between Figures 2.9(a) and 2.9(b), we can calculate that the voltage drop between the terminals in Figure 2.9(b) is $V_A + I_1 r$. Substituting for r from Equation 2.3, we obtain this voltage drop of $V_A + V_1 - V_A$, or simply V_1. This result is identical to the voltage drop in Figure 2.9(a). Therefore, the equivalent circuit in Figure 2.9(b) can be used to replace the diode symbol once the cut-in voltage and dynamic resistance are evaluated. This procedure is illustrated by Example 2.1.

Example 2.1.

Determine the cut-in voltage (V_A) and dynamic resistance (r) for the silicon diode whose forward characteristics are indicated in Figure 2.6.

Solution. A straight line drawn tangent to the characteristic intercepts the voltage axis at about 0.65 V. Therefore, the cut-in voltage $V_A \approx 0.65$ V. To find the slope, take ΔI from 0 to 8 mA, which yields a ΔV from 0.65 to 0.68 V. The dynamic resistance is then calculated using Eq. 2.2.

$$r = \frac{\Delta V}{\Delta I} \quad r \approx \frac{0.68 - 0.65}{8 - 0} \frac{V}{mA} = 3.75 \, \Omega$$

We must make one important qualification to this procedure: Both the cut-in voltage and the dynamic resistance values depend on the range of current and voltage from which they are evaluated. If we take the same silicon diode as in Example 2.1 and display its characteristics in the range between 0 and 1 mA, we would calculate a higher dynamic resistance with a lower cut-in voltage. On the other hand, if we looked at the range from, say, 0 to 100 mA, we would obtain a lower dynamic resistance and a slightly higher cut-in voltage. Nevertheless, the use of the piecewise linear equivalent circuit is valid over the selected range of current and voltage.

2.4.2 Region 2

In region 2, the current is essentially equal to the reverse saturation current, I_o, and it is just about independent of the applied voltage. The characteristic curve does have a slight slope in region 2, but it is so small that it can be ignored without fear of a significant error being introduced. The equivalent circuit for the diode in region 2 is then just a current generator of magnitude I_o, as shown in Figure 2.10(a). We are assuming that the dynamic resistance in region 2 is essentially infinite or is an open circuit.

2.4.3 Region 3

In region 3, the diode is in breakdown. There, the diode maintains a constant voltage, V_B, while the current varies over a wide range. The characteristic is almost perpendicular to the voltage axis in this region. Thus, the dynamic resistance is very small and so may be assumed to be zero. The diode equivalent circuit in breakdown (region 3) is just a voltage source with a voltage equal to the breakdown voltage, V_B, as shown in Figure 2.10(b).

2.4.4 Capacitive Effects

Let us now briefly examine the capacitive effects in a diode. The preceding equivalent circuits are insufficient to predict the behavior of diodes at very high frequencies. At very high frequencies, these models may be modified by adding a capacitor in parallel with each equivalent circuit. To understand why the parallel capacitor is necessary in the equivalent, we must study the internal operation of the diode.

When the diode is forward biased, large numbers of holes are being injected into the N side and, simultaneously, large numbers of electrons are being injected into the P side. These processes cause *excess* concentrations of charge in the two

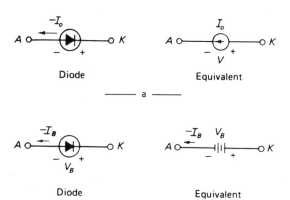

Figure 2.10 Diode 1 under reverse bias: (a) reverse voltage less than the breakdown voltage and (b) at breakdown voltage.

regions. If the amplitude of the applied voltage is changing rapidly, as is the case at very high frequencies, these excess concentrations of charge cannot increase or decrease instantaneously. Thus, there is a time lag between the change in voltage and the resulting change in current. This lag is a capacitive effect. The equivalent capacitance present in a diode under forward bias conditions is called *diffusion capacitance* and may be a few hundred picofarads (pF) in magnitude.

Under reverse bias conditions, the depletion region is widened so that there are more uncovered stationary charges on either side of the junction. These stored charges, positive in the *N* side and negative in the *P* side, give rise to a capacitance similar to a parallel-plate capacitance. The equivalent capacitance under reverse bias conditions is called *depletion-region capacitance* (also known as *space-charge capacitance*, *transition-region capacitance*, or *barrier capacitance*) and is typically a few picofarads in magnitude.

At high frequencies, we have two separate capacitive effects associated with a *PN* junction: depletion-region capacitance when the junction is reverse biased and diffusion capacitance when the junction is forward biased. Both effects can be neglected at low frequencies.

2.4.5 Recovery Time

During forward conduction, there are a great number of minority carriers (electrons in the *P*-material and holes in the *N*-material) moving through the diode to make up the forward current. At the moment the bias polarity is changed and the diode is reverse biased, the forward current stops and the reverse current is, at that moment, large, as noted in Figure 2.11. The large reverse current is due to the great number of minority carriers available for conduction in the reverse direction. Once the minority carriers move back across the junction (electrons to

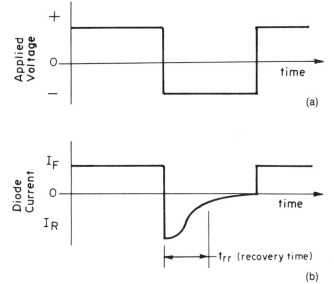

Figure 2.11 The current in a diode (b) is the result of the applied square-wave voltage (a).

the *N*-material and holes to the *P*-material), the reverse current falls to the typical value of reverse saturation current.

The time that it takes for the minority carriers to clear out of their respective *P*- and *N*-materials is called the *reverse recovery time, t_{rr}*. To be specific, the reverse recovery time is the time between the application of the reverse bias to the time the reverse current recovers to 10% of its maximum value. The reverse recovery time is directly related to the magnitude of the forward current just prior to the application of the reverse bias voltage, that is, a large I_F will result in a long t_{rr}.

Although the reverse recovery time varies from diode to diode and from circuit to circuit, you can expect a typical range of values for this parameter to be from a few nanoseconds to several microseconds.

In closing out this discussion of the junction diode and its properties, we invite you to look over the 1N4001 through 1N4007 data sheet in Appendix A.

2.5 SPECIAL PURPOSE DIODES

2.5.1 Zener Diode

Some of the general diode characteristics already mentioned are so useful that special diodes are designed to emphasize a specific property. Such is the case with the *Zener diode*.

Zener or breakdown diodes are much like ordinary diodes except that they are manufactured to exhibit breakdown in the reverse direction at a specific voltage. These diodes, which are operated in the reverse bias condition, are used because of their breakdown characteristics. The Zener diode circuit symbol is shown in Figure 2.12(b). Zener diodes can exhibit breakdowns at a very wide range of voltages, as noted in the 1N5221 through 1N5272 data sheet in the Appendix. Here you will see listed fifty-two different Zener voltages. When looking at this data sheet, you will notice that the nominal Zener voltage (V_Z) is specified at a particular test current (I_{ZT}). For the 6.8 V Zener in Figure 2.13, the test current is 20 mA or about 25% of the maximum current ($I_{ZM} = 70$ mA).

2.5.2 Varactor Diode

Another common type of diode that is also operated in the reverse biased condition is the voltage-variable capacitance diode or *varactor diode* (also called a *varicap*). When reverse biased, the depletion region of the varactor widens with an increase in the negative bias voltage. If you think of the depletion region as a parallel-plate capacitor, then the varactor is effectively a voltage-controlled capacitor. In the varactor, the voltage varies the depletion-region width, which is analogous to the mechanical rotation of a variable capacitor, which varies the separation between the plates of a capacitor. The circuit symbol for a varactor is given in Figure 2.12(c). A typical varactor diode (voltage-variable capacitance diode) data sheet may be found in Appendix A.

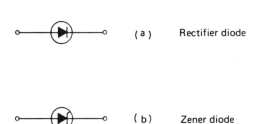

(a) Rectifier diode

(b) Zener diode

(c) Varactor diode

(d) Tunnel diode

(e) Schottky-barrier diode

Figure 2.12 Various diode symbols.

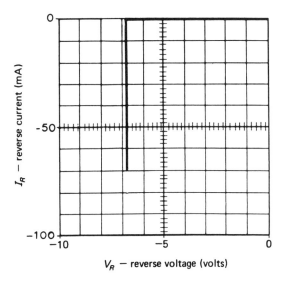

Figure 2.13 Curve-tracer display of a 6.8 V Zener diode reverse characteristic curve.

Sec. 2.5   Special Purpose Diodes

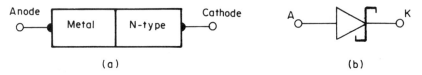

Figure 2.14 Schottky-barrier diode: (a) structural representation; (b) schematic symbol.

2.5.3 Schottky-Barrier Diode

The *Schottky-barrier* or *hot-carrier* diode has several unique properties. Because of the metal to silicon junction (gold to *N*-type silicon), as noted in Figure 2.14(a), no minority carriers are present in the metal anode of the Schottky-barrier diode during forward conduction. Since electrons (majority carriers in metal) are the only charge carriers, then virtually no *charge storage* is present in the metal anode of the junction. When the bias voltage is reversed, the forward current stops and the diode instantly recovers without a transient. Since the reverse recovery time is essentially zero, Schottky-barrier diodes find application in high-speed switching. Low-power Schottky-barrier diodes can switch from ON to OFF in times approaching one nanosecond (1 GHz), while high-power Schottky-barrier diodes can switch tens of amperes at frequencies approaching 100 kHz (10 μs) without the troublesome forward and reverse recovery transients experienced with junction diodes.

Besides majority carrier conduction, the Schottky-barrier diode has a low forward voltage drop (V_F) as shown in the data sheet of the MBD501 low-power Schottky-barrier diode (see Appendix A). Schottky-barrier power rectifiers have a typical V_F of 0.7 V at I_F of 50 A. This specification may be compared to a typical junction power rectifier with a rating of 1.5 V at 50 A.

REVIEW QUESTIONS

1. What is the role of a *seed* in crystal growing?
2. What are some of the aspects of diode manufacture?
3. Enumerate and briefly describe the steps used in the diffusion process of diode manufacture.
4. Describe the dynamics of the formation of the *depletion region*.
5. In a diode under equilibrium, what causes the *contact potential*?
6. When a diode is *forward biased*, which terminal is positive and which is negative?
7. When a diode is *reverse biased*, which terminal is positive and which is negative?
8. Does a voltage in the forward bias direction add to or subtract from the contact potential? Why?
9. Describe the motion of charge carriers inside a forward-biased diode.

10. What is the effect of forward bias on the depletion region in a diode?
11. Does a reverse voltage add to or subtract from the contact potential? Why?
12. What is the effect of reverse bias on the depletion region in a diode?
13. Describe the motion of charge carriers inside a reverse-biased diode.
14. In the diode characteristics, what is the significance of the *reverse saturation current*?
15. What factors affect the reverse saturation current in a diode?
16. What order of magnitude is the reverse saturation current at room temperature in silicon? In germanium?
17. What effect does temperature have on the reverse saturation current? Why?
18. Describe the conditions inside a diode under *Zener breakdown*.
19. Describe the conditions inside a diode under *avalanche breakdown*.
20. What precautions must be taken to protect a diode from permanent damage when it is operating under reverse breakdown conditions?
21. What is *cut-in voltage*?
22. What is *dynamic resistance*?
23. What order of magnitude (ohms, kilohms, or megohms) is the diode dynamic resistance in region 1? (Refer to Fig. 2.8.)
24. What order of magnitude is the diode dynamic resistance in region 2? (Refer to Fig. 2.8.)
25. What order of magnitude is the diode dynamic resistance in region 3? (Refer to Fig. 2.8.)
26. What is meant by a *piece-wise linear model* in the case of the diode?
27. What is the capacitance in a forward-biased diode called? What is its cause? When must you account for it?
28. What is the capacitance in a reverse-biased diode called? What is its cause? When must you account for it?
29. Which gives rise to a larger capacitance in a diode, forward or reverse bias? Why?
30. What are the unique characteristics of a *Zener diode*?
31. What are the unique characteristics of a *varactor diode*?
32. Draw the circuit symbols for the following: rectifier diode, Zener diode, varactor diode, and Schottky-barrier diode.

PROBLEMS

1. Make a plot of the diode equation [Eq. (2.1)] for a silicon diode with $I_o = 1$ nA. Use 0.02 V increments in the voltage between -0.1 and 0.7 V. Use 0.1 V increments between -2 and -0.1 V. (Assume the breakdown voltage to be below -2 V.)
2. Repeat Problem 1 for a germanium diode with $I_o = 1$ μA, and go up to 0.3 V in the forward direction.

3. Plot the two diode curves obtained in Problems 1 and 2 on the same graph paper and compare them, stating pertinent differences.

*4. From the characteristics in Fig. 2.6 and the diode equation, make a reasonable prediction of the saturation current for the germanium diode.

5. Repeat Problem 4 for the silicon diode.

*6. Determine the cut-in voltage and dynamic resistance for the germanium diode (whose characteristics are given in Fig. 2.6) that would produce a good piece-wise linear model in the vicinity of 0.3 V.

7. Repeat Problem 6 in the vicinity of 0.5 V.

8. Repeat Problem 6 for the silicon diode in the vicinity of 0.55 V.

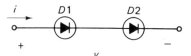

Figure 2.15 Circuit for Problem 9.

9. Draw the $V - I$ characteristics for the circuit consisting of two diodes in series, as shown in Fig. 2.15, if the two diodes are identical and have reverse saturation currents of 10 μA.

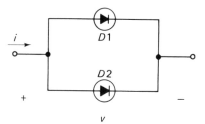

Figure 2.16 Circuit for Problem 10.

*10. Draw the $V - I$ characteristics for the diode circuit shown in Fig. 2.16 if the two diodes are identical and have reverse saturation currents of 5 μA.

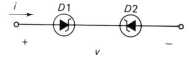

Figure 2.17 Circuit for Problem 11.

11. The two Zener diodes in Fig. 2.17 are identical and have the reverse characteristics indicated in Fig. 2.13. Sketch the resulting $V - I$ characteristics for the combination shown in Fig. 2.17.

---

* Answers to starred problems are in Appendix B.

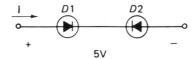

Figure 2.18 Circuit for Problem 12.

12. The diodes in Fig. 2.18 have 5 V applied as shown. If the diodes are identical, with $I_o = 100$ nA, determine (a) the current, I; (b) the voltage, V_{D1}, across $D1$; (c) the voltage, V_{D2}, across $D2$. (Hint: The voltages across the two diodes are *not* equal, but the current is the same through both diodes.)

Chap. 2   Problems   33

Chapter 3

Bipolar Junction Transistors

The bipolar junction transistor (BJT) is a three-terminal device that contains two *PN* diodes sandwiched back to back. The three terminals are connected to three regions of semiconductor material that are named for the function each performs in the transistor. The three regions are named *emitter, base*, and *collector*. The emitter emits or dispatches charge carriers into the base, where control over the carriers is exercised; eventually the carriers are gathered in the collector region. The name *base* stems from the historical development of transistors. Originally, a transistor was made by alloying the emitter and collector regions to a relatively large sample of doped semiconductor. This sample served as the base for the manufacture of the transistor; the corresponding region took on that name.

BJT structure. The base region, which is situated between the emitter and collector regions, is always doped opposite to the emitter and collector regions. The emitter and collector are always formed by the same type of doped material (i.e., *N*-type or *P*-type); consequently, there are two possible structures for the BJT. The first structure, which has an *N*-type base and a *P*-type emitter and collector, is classed as a *PNP* transistor. The second structure, which has a *P*-type base and an *N*-type emitter and collector, is classed as an *NPN* transistor. In these two classes of bipolar junction transistors (*PNP* and *NPN*), the type of doping used in the base region is specified by the middle letter.

Fabrication processes. Bipolar junction transistors are manufactured using the same methods previously described for the junction diode. Since the *diffusion* process was previously described in the manufacture of junction diodes, it will once again be used to describe the manufacture of BJT. The following discussion will detail the fabrication of an *NPN* transistor. In the manufacture of BJTs, the first three steps of the process, illustrated in Figure 2.2, are used along with additional steps including masking, insulating (oxidation), etching, and diffusing, as shown in Figure 3.1.

The geometry in Figure 3.1(d), although not to scale, allows the most efficient

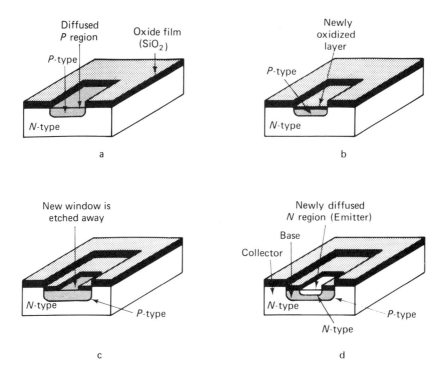

Figure 3.1 Formation of a planar *NPN* transistor by the diffusion process: (a) end result of processes shown in Figure 2.2; (b) oxidation; (c) new window is masked and etched away; (d) an *N* region is diffused to complete the *NPN* structure.

operation of the transistor. Here we see a *diffused planar* transistor wherein the emitter region is surrounded by the base; the base in turn is surrounded by the collector.

Another method of BJT manufacture is by *epitaxial growth*. In this process, an epitaxial layer is first grown upon the collector substrate and then the *PNP* and *NPN* structure is formed by growing additional regions one on top of another. The resulting transistor is called an *epitaxial mesa* transistor. The name *mesa* comes from its geographical mesa-like appearance when viewed through a microscope.

Although a number of fabrication processes and geometries may be used in manufacturing bipolar junction transistors, the operation of all BJTs is based on the same principles.

3.1 CURRENTS IN THE BJT

The *NPN* and *PNP* types of BJTs are shown structurally in Figure 3.2, together with their schematic symbols. An easy way to remember the convention for the schematic symbols is to note the direction of the arrow on the emitter terminal.

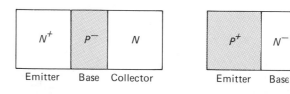
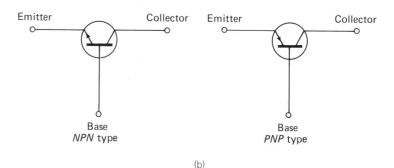

Figure 3.2 *NPN* and *PNP* transistors: (a) structure diagram and (b) schematic symbols. Note: + superscript = heavy doping; − superscript = light doping.

If the arrow points in, it indicates a *PNP* transistor; if it points out, it indicates an *NPN* transistor.

3.1.1 BJT Operation

A fairly high current is caused to flow through the BJT. This current is controlled at the base and may be either increased or decreased. Our discussion will first use the *NPN* transistor as an example and then it will be expanded to include the *PNP* transistor.

In an *NPN* transistor, electrons are the majority carriers inside the emitter. In order to cause these electrons to cross the base-emitter junction and enter the base region, an external voltage is applied between the base and emitter terminals. The base and emitter regions constitute a diode that must be *forward biased* in order to cause majority carriers—electrons in the *N*-type emitter—to cross the junction. Once electrons from the emitter enter the *P*-type base, they diffuse through the base. Some electrons find their way to the base terminal and flow out; however, most of the electrons reach the collector-base junction. In order for electrons to cross the junction from the *P*-type base into the *N*-type collector, a *reverse-biased* voltage must be applied between collector and base terminals. In this manner, the electrons that started in the emitter and wound up in the collector constitute the main current, whereas those electrons that flowed out of the base make up the small controlling current.

However, we have not accounted for all the currents in the BJT. As a result of the forward bias on the base-emitter junction, we would expect not only the injection of electrons into the base but also holes from the base to enter the emitter. Although holes do enter the base, their effect is minimal on the net current flow in the transistor because the P^- base material is very lightly doped, while the N^+ emitter material is very heavily doped. (Note that the "$+$" superscript = heavy doping; "$-$" = light doping.) The BJT is purposely made that way. Thus, when the junction is forward biased, the heavily doped N^+-type emitter offers extremely large numbers of electrons for conduction, but the lightly doped P^--type base has only a small number of holes to offer. The current across the base-emitter junction within an *NPN* transistor is essentially electron flow.

At the collector-base junction, the reverse bias transports electrons, which started in the emitter, from the base into the collector. Besides this action, the reverse bias causes another current to flow across the collector base junction. This current is made up of the *minority* charge carriers in both the base and collector; that is, electrons from the *P*-type base and holes from the *N*-type collector. The current component resulting from this minority charge flow across the collector-base junction is quite small. It is called the *collector-base reverse saturation current*, or *collector cutoff current*, and is usually labeled I_{CBO} or simply I_{CO}. As a reverse saturation current, it is very sensitive to temperature.

Obviously, there is no way that we can actually measure the separate components of the currents inside the transistor. We simply note that electrons supplied to the emitter and flowing toward the base correspond to a *conventional* current flow out of the emitter. We label the net conventional terminal current I_E for the emitter current. Similarly, electrons flowing out of the base and collector correspond to net conventional terminal currents, I_B and I_C, into the base and collector respectively. The three conventional terminal currents, together with the internal motion of charge carriers, are indicated in Figure 3.3. The circuit diagram for the normal operation of an *NPN* transistor, with the base-emitter diode forward biased by V_{EE} and the collector-base diode reverse biased by V_{CC}, is shown in Figure 3.4.

As a result of the externally applied voltages, the depletion region at the base-emitter junction narrows and the depletion region at the collector-base junction widens, as shown in Figure 3.5.

3.1.2 Current Relationships

The net current entering the transistor in Figure 3.4 is the sum of I_B and I_C and the net current leaving the transistor is I_E. Because electric charge is neither piling up nor depleting inside the transistor, the net current into the transistor must be equal to the net current out of the transistor. Thus,

$$I_E = I_C + I_B \tag{3.1}$$

Assuming that the emitter current remains constant in Equation 3.1, the smaller the base current, the larger the collector current. In order to make the

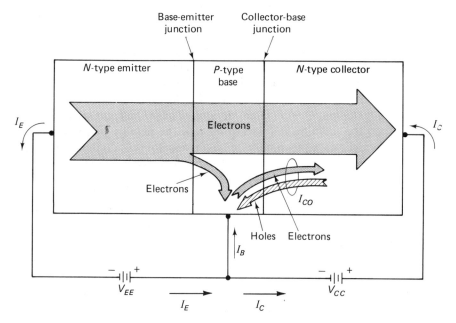

Figure 3.3 Motion of charge carriers (electrons) within an *NPN* transistor, as well as the directions of the resulting conventional terminal currents: I_E, I_B, and I_C.

collector current as large as possible, the base current must be minimized. We saw that in an *NPN* transistor electrons from the emitter diffuse through the P^--type base and eventually reach the collector. Some of these electrons are lost in the base due to recombination. To minimize this recombination, the base is only a lightly doped P^--type. Furthermore, the physical width of the base (distance from the edge of the emitter to the edge of the collector) is made small (typically 1/10,000 in.), which assures that almost all the electrons reach the collector.

The collector current is made up of the current in the collector resulting from electrons that started in the emitter, labeled I_{NC}, together with the reverse saturation current I_{CO}. Thus,

$$I_C = I_{NC} + I_{CO} \tag{3.2}$$

The ratio of the electron current in the collector to the total electron current

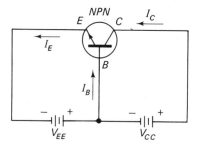

Figure 3.4 Bias for normal operation of an *NPN* transistor, showing actual directions of conventional currents. Notice that the base-emitter junction is forward biased (V_{EE} + to P, − to N) and the collector-base is reverse biased (V_{CC} + to N, − to P).

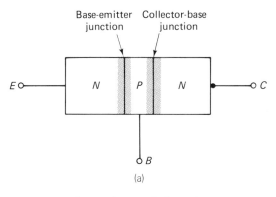

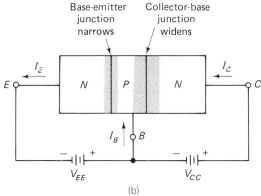

Figure 3.5 Depletion regions inside an *NPN* transistor: (a) no external bias; (b) with normal bias, i.e., emitter-base forward biased, collector-base reverse biased.

in the emitter, labeled α (alpha), is an important parameter for the BJT. In terms of an equation, α is defined as

$$\alpha = \frac{I_{NC}}{I_E} \qquad (3.3)$$

Note that no matter how few of the electrons that started in the emitter are lost in the base, I_{NC} is always less than I_E. This is due to the simple fact that we can never collect a larger number of carriers in the collector than the number that started in the emitter; thus, α is always less than 1. It is, however, usually very close to 1, typically between 0.98 and 0.9995. The significance of α is that it is the *dc short-circuit current gain in the common-base configuration*. We shall discuss this gain shortly.

If we use Equation (3.2) in the defining equation for α, we have

$$\alpha = \frac{I_C - I_{CO}}{I_E} \qquad (3.4)$$

Solving this equation for I_C gives us an important current relationship for the BJT:

$$I_C = \alpha I_E + I_{CO} \qquad (3.5)$$

We must digress briefly to consider the *PNP* transistor. Its operation is completely analogous to that of the *NPN*. The majority carriers (in this case, holes) from the *P*-type emitter are injected into the base by forward biasing the base-emitter diode. Holes are further transported through the base, and some of them are gathered by the collector with the aid of a reverse bias on the collector-base diode. In fact, all of the preceding discussion for the *NPN* transistor holds for the *PNP* transistor if the appropriate substitutions (*N* instead of *P*, *P* instead of *N*, and hole instead of electron) are made.

Some circuit conditions are the same for both *NPN* and *PNP* transistors. For normal operation of both *NPN* and *PNP* transistors, the base-emitter diode is forward biased and the collector-base diode is reverse biased. In the case of the *PNP* transistor, the main current is caused by hole motion; therefore, the conventional current is in the same direction as the flow of holes. Thus, the emitter current flows into a *PNP* transistor; the base and collector currents flow out, as shown in Figure 3.6. (Note that the battery polarity is reversed from what it was in Figure 3.4.) Taking the algebraic sum of the terminal currents in a *PNP* transistor yields the same result as it did for the *NPN* transistor, given in Equation (3.1).

It is always good practice to indicate by arrows the directions of the terminal currents. Under normal operation, we can be certain of indicating the *actual* conventional current direction if we use the arrow on the emitter of the transistor symbol to get the actual direction of the emitter current and if we remember from Equation (3.1) that both I_C and I_B are in a direction opposite to that of I_E. For example, the arrow on a *PNP* transistor points into the transistor. Therefore, I_E flows into the transistor, and both I_C and I_B flow out.

With our digression out of the way, we can make a substitution in Equation (3.5) for I_E, using Equation (3.1) where $I_E = I_C + I_B$. Thus,

$$I_C = \alpha(I_C + I_B) + I_{CO} \tag{3.6}$$

When Equation (3.6) is solved for I_C, the following results:

$$I_C = \frac{\alpha}{1-\alpha} I_B + \frac{1}{1-\alpha} I_{CO} \tag{3.7}$$

In order to simplify Equation (3.7), we must first define β (beta), the *dc*

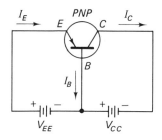

Figure 3.6 Bias for normal operation of a *PNP* transistor, showing actual directions of conventional current flow.

short-circuit current gain in the common-emitter configuration:

$$\beta = \frac{\alpha}{1 - \alpha} \tag{3.8}$$

Now Equation (3.7) can be rewritten in a more useful form:

$$I_C = \beta I_B + (\beta + 1)I_{CO} \tag{3.9}$$

The most important and useful transistor equations are those given in Equations (3.1), (3.5), and (3.9).

3.2 BJT STATIC CHARACTERISTICS

When the transistor is operated as an amplifier, we apply the input between two terminals and obtain the amplified output at another pair of terminals. Since the transistor has only three terminals, one terminal must be designated as being common to both input and output. Because of the versatility of the BJT, it can operate with any one of its three terminals as the common terminal. The amplifier configuration is usually named for the common terminal, that is, *common-base* (*CB*), *common-emitter* (*CE*), and *common-collector* (*CC*) or *emitter-follower*.

The static characteristics for the BJT in any of the three possible configurations are contained in two curves. One curve, the *input characteristics*, gives the voltage-current relationship (*V-I* relationship) at the input terminals for different values of either output current or output voltage. The second curve, the *output characteristics*, gives the *V-I* relationship at the output terminals, with either the input current or voltage as the parameter.

3.2.1 Common-Base Configuration

In the common-base (*CB*) configuration, the input is applied between emitter and base, while the output is taken between collector and base (as illustrated in Figure 3.7 for an *NPN* transistor).

The *CB* characteristics, shown in Figure 3.8(a), relate the emitter current to the emitter-base voltage and the collector-base voltage. For normal operation, the base-emitter junction is forward biased, so the general appearance of the *CB* input characteristics resembles that of a forward-biased diode. From the tight grouping of the curves, we learn that the effect of the output voltage V_{CB} is not

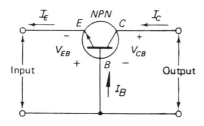

Figure 3.7 *CB* configuration using an *NPN* transistor. For normal operation, V_{EB} is negative, V_{CB} is positive, and conventional currents are in the directions shown.

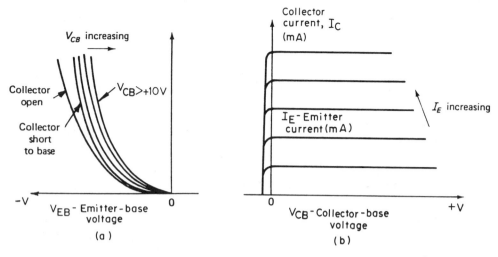

Figure 3.8 CB characteristics for an NPN transistor: (a) input characteristics with base-emitter junction forward biased, V_{EB} from 0 to $-V$; (b) output characteristics with collector-base junction reverse biased, V_{CB} from 0 to $+V$.

great. For fixed values of V_{EB}, as V_{CB} is increased, a larger emitter current results. For silicon transistors, the approximate emitter-base cut-in voltage is about 0.6 V for all values of collector-base voltage ($V_{EB} \approx 0.6$ V).

The CB output characteristics, shown in Figure 3.8(b), relate the collector current to the collector-base voltage and emitter current. For normal operation, the collector-base junction is *always* reverse biased. Under these conditions, the collector current saturates to a value determined by the emitter current and is essentially independent of the collector-base voltage. Note that collector current flows even for some small forward bias ($V_{CB} = -0.2$ V) at the collector-base junction, because carriers injected from the emitter into the base have enough energy to cross the collector junction. Some small forward bias at the collector is required to prevent the carriers from crossing into the collector. For silicon transistors, this voltage is typically $+0.5$ V for PNP and -0.5 V for NPN.

An actual CB output characteristic curve from a curve tracer for a silicon NPN transistor is shown in Figure 3.9. By using Equation (3.5) and assuming that I_{CO} is small enough to be negligible, we can write an equation to evaluate α from Figure 3.9. Thus:

$$I_C \approx \alpha I_E \quad (3.10)$$

Using Equation (3.10), evaluate α from Figure 3.9. For any value of V_{CB} when $I_E = 0.3$ mA, I_C is approximately 0.29 mA. Therefore, $\alpha \approx I_C/I_E = 0.29/0.3 = 0.97$.

3.2.2 Common-Emitter Configuration

In the common-emitter (CE) configuration, the input is applied between base and emitter, while the output is taken between collector and emitter, as shown in Figure 3.10 for an NPN transistor.

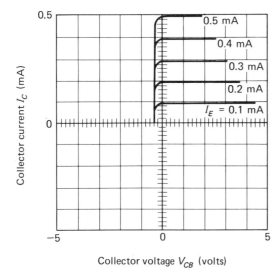

Figure 3.9 CB output characteristics as displayed on a curve tracer.

The *CE* input characteristics of Figure 3.11(a) relate the base current to the base-emitter voltage and the collector-emitter voltage. As can be seen, V_{CE} does not have much control on the input characteristics; and for values of V_{CE} larger than 0.5 V, all the input curves coincide with one another. The shape of the curves, as might be expected, is that of a forward-biased diode with a base-emitted cut-in voltage of about 0.6 V ($V_{BE} \approx 0.6$ V) for silicon transistors.

The *CE* output characteristics shown in Figure 3.11(b) relate the collector current to the collector-emitter voltage and the base current. For all values of V_{CE} larger than about 0.2 to 0.5 V, I_C saturates to a value essentially independent of V_{CE} and I_C is only controlled by the magnitude of the base current, I_B.

When the base is open-circuited (that is, $I_B = 0$), the collector current, labeled I_{CEO}, is given by

$$I_C \approx I_{CEO} = (\beta + 1)I_{CO} \qquad (3.11)$$

I_{CEO} is the collector leakage current in the *CE* configuration with the base disconnected, as shown in Figure 3.11(b). Although appreciably larger than I_{CO}, it is usually very small ($\leqslant 100$ μA) and may be neglected in the determination of β. Actual *CE* characteristics for an *NPN* silicon transistor, illustrated in Figure 3.12,

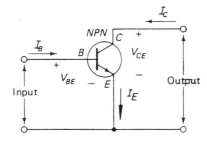

Figure 3.10 *CE* configuration using an *NPN* transistor. For normal operation, both V_{BE} and V_{CE} are positive, with $V_{CE} \gg V_{BE}$. This condition results in a positive V_{CB} (the correct reverse bias on the collector-base junction).

Sec. 3.2 BJT Static Characteristics

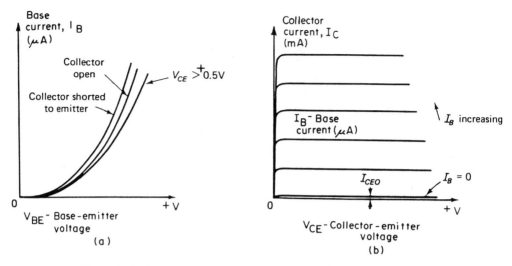

Figure 3.11 *CE* characteristics for an *NPN* transistor: (a) input characteristics with the base-collector junction forward biased, V_{BE} from 0 to +V; (b) output characteristics with the collector-base junction reverse biased.

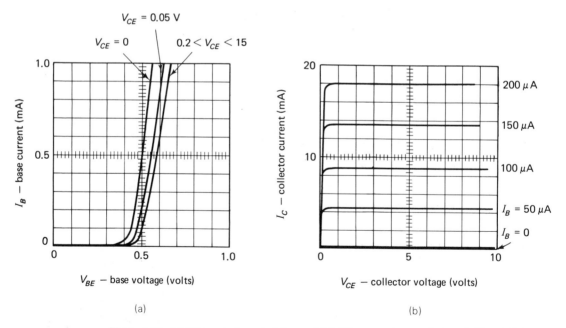

Figure 3.12 (a) *CE* input characteristics and (b) *CE* output characteristics of silicon *NPN* transistor (2N2369).

reveal I_{CEO} to be much less than 0.1 mA. Therefore, Equation (3.9) can be rewritten without the I_{CEO} term as

$$I_C \approx \beta I_B \tag{3.12}$$

The value of β can be evaluated from the *CE* output characteristics of the type shown in Figure 3.12(b). For example, at V_{CE} = 5 V, with I_B = 100 μA, we read I_C to be about 8.8 mA. Substituting and solving for β in Equation (3.12) results in: $\beta = I_C/I_B = 8.8/0.1 = 88$.

Note here that the symbol h_{FE} (or H_{FE}) is often used in data sheets instead of β. However, because of the similarity between the dc and ac symbols, h_{FE} and h_{fe}, and the possibility of confusing the two, β will be used to denote the dc short-circuit current gain in the *CE* configuration.

As you might expect, β is a function of the operating point. In the *CB* configuration, as the number of carriers injected from the emitter into the base is increased, so is the number of carriers reaching the collector. However, these two increases are not directly proportional; the number of carriers reaching the collector increases by a smaller amount than the increase in the carriers leaving the emitter. Consequently, there is a decrease in the transistor α, which is also a decrease in β. The same is true at very low injection levels. Not only does the β vary among transistors of the same type number, but it also varies for the same transistor under different operating conditions, i.e., different voltage and current. Figure 3.13 demonstrates the typical variation of β as a function of collector current.

Another property of the *CE* output characteristics that should not be overlooked is the extreme temperature sensitivity of these characteristics. Figure 3.14 shows a double exposure of the *CE* output characteristics as displayed on a curve tracer. The first exposure was taken at room temperature (T_1); the second exposure was taken after the transistor case was heated by placing a match near it. For the same base current, at the higher temperature a larger collector current is observed. This difference is caused by the increase in the collector leakage current I_{CO}, as well as a slight increase in β. The collector leakage current I_{CO} approximately doubles in value for every 10°C rise in temperature. The *CE* output characteristics depend on $(\beta + 1)I_{CO}$ [see Equation (3.9)]. Therefore, the most noticeable change as a result of temperature increase is in the *CE* characteristics.

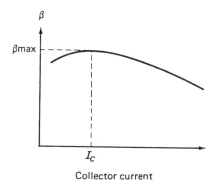

Figure 3.13 Variation of β as a function of quiescent collector current.

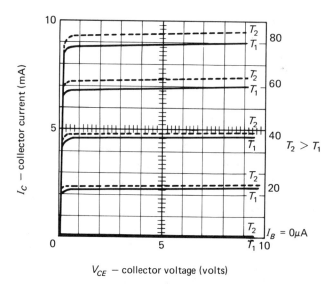

Figure 3.14 Double exposure of a curve-tracer display of CE output characteristics showing the effect of increased ambient temperature.

In general, we can say that transistor characteristics and parameters are very sensitive to temperature. When we design a circuit, we must allow for heat dissipation and for the temperature sensitivity of transistors.

3.2.3 Common-Collector Configuration

In the common-collector (CC) or emitter follower configuration, the input is applied to the base, while the output is taken from the emitter. The emitter follower (*CC* configuration) may be treated as a special case of the *CE* configuration.

3.3 TRANSISTOR RATINGS

Literally thousands of different BJTs are made, including both *PNP* and *NPN* as well as many others with different types of construction. Some examples are given in Figure 3.15. Many types of transistors exist mainly because the applications in which they are used are so varied.

In order to help customers select the proper transistor for a particular application as well as provide transistor specifications, manufacturers publish data sheets, manuals, and application notes. Examples of typical data sheets for diodes, BJTs, FETs, ICs, and assorted other devices and circuits are given in Appendix A. Among other quantities, the manufacturer usually specifies the maximum permissible V_{CE}, I_C, and P_D (device dissipation). Also specified is the maximum allowable collector junction temperature, T_J. In operation, exceeding any one of these maximum values may result in permanent damage to the transistor.

For example, a Texas Instruments *NPN* silicon power transistor TIP29A is listed by the manufacturer as having a maximum continuous I_C of 1 A, maximum V_{CEO} of 60 V, and maximum continuous device dissipation, P_D, of 30 W (see Appendix A). These three maximum values define the permissible region of op-

Figure 3.15 Several types of semiconductor devices. (Courtesy of Texas Instruments)

eration for this particular transistor. The total continuous device dissipation can be approximated by the power dissipated at the collector junction, because the power dissipated in the forward-biased emitter junction is usually a few orders of magnitude smaller. A reasonable approximation for the power dissipated at the collector is the product of the collector voltage and current:

$$P_D = I_C V_{CE} \tag{3.13}$$

where P_D is the permissible power dissipation; I_C, the quiescent collector current; and V_{CE}, the operating point collector-emitter voltage. The permissible region of operation may be shown graphically on the CE output characteristics. It has five distinct boundaries. Two boundaries are provided by the transistor characteristics; one being the cutoff region, the other the saturation region, as shown in Figure 3.16. The other three boundaries are provided by the maximum values specified by the manufacturer for the collector current, collector-emitter voltage, and total power. The maximum power curve is obtained by choosing values for I_C lower than $I_{C\max}$ and by calculating from Equation (3.13) values of V_{CE} to give the specified maximum permissible power.

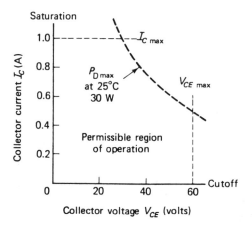

Figure 3.16 Permissible region of transistor operation.

In their data sheets, manufacturers also specify different breakdown voltages (usually denoted by a capital B before the symbol: e.g., BV_{CEO}, BV_{CBO}). These symbols use triple subscript notation. The first two subscripts denote the two terminals between which the voltage is to exist; the last subscript O denotes that the test is made with the unmentioned terminal held open. Thus, BV_{CBO} is the collector-base breakdown voltage with the emitter terminal open-circuited. The last subscript may also be R (specific resistance is placed in the unmentioned terminal circuit) or X (unmentioned terminal is shorted). Other aspects of manufaturers' data sheets will be treated as the need arises.

REVIEW QUESTIONS

1. Discuss the role of the emitter region in the operation of a BJT.
2. Discuss the role of the base region in the operation of a BJT. Why is the width of the base usually very small?
3. Discuss the role of the collector region in the operation of a BJT.
4. Name some of the common methods used in the fabrication of transistors.
5. Briefly describe one method used for the manufacture of transistors.
6. Identify the doping type for each of the regions inside a *PNP* transistor. Repeat for an *NPN* transistor.
7. For normal operation, how is the base-emitter diode biased? How is the collector-base diode biased?
8. Draw a block diagram of an *NPN* transistor and include batteries with the proper polarity for normal operation.
9. Repeat Question 8 for a *PNP* transistor.
10. What is α? How is it related to the transistor terminal currents?
11. What typical range of values can you expect for α?
12. What is β? How is it related to the transistor terminal currents?
13. What typical range of values can you expect for β?

14. How are the transistor α and β related?
15. What is the collector reverse-saturation current and what is it caused by?
16. What carrier makes up the largest component of the current inside the transistor: *PNP* and *NPN*?
17. What are the three possible transistor configurations?
18. In the *CB* input characteristics, what are the functions that are plotted? What is the parameter?
19. In the *CB* output characteristics, what are the functions that are plotted? What is the parameter?
20. Repeat Question 18 for the *CE* input characteristics.
21. Repeat Question 19 for the *CE* output characteristics.
22. What is the meaning of the symbol I_{CEO}?
23. What is the meaning of the symbol BV_{CBO}? Under what conditions is it measured?
24. What is the meaning of the symbol BV_{CEO}? Under what conditions is it measured?
25. What is the meaning of the symbol BV_{CER}? Under what conditions is it measured?
26. From the data sheets for a 2N2222 *NPN* transistor (see Appendix A), determine the following: (a) $V_{CEO\text{max}}$, (b) $I_{C\text{max}}$, (c) $P_{D\text{max}}$ at $T_A = 25°C$.
27. What are the five boundaries that determine the permissible region of operation for a BJT?
28. What are the consequences of exceeding the manufacturer's maximum specifications?
29. How does the transistor β vary as a function of the operating point (I_C)? Support your answer with an example from a manufacturer's data sheet (see Appendix A).
30. What effect does an increase in temperature have on transistor characteristics?

PROBLEMS

*1. In an *NPN* transistor, the collector and emitter currents are measured to be 2.0 and 2.01 mA, respectively. Determine the base current and the transistor's α and β.

2. The α of a certain transistor is determined to be 0.99 with an uncertainty in the last digit of ±1. What is the β for this transistor and what is the uncertainty in β?

3. For the *CB* output characteristics in Fig. 3.9, determine the transistor α and β when $V_{CB} = 2$ V and $I_E = 0.1$ mA.

4. Repeat Problem 3 for I_E of 0.2, 0.3, 0.4, and 0.5 mA. Make a plot of α as a function of I_E. Also make a plot of β as a function of I_E.

5. On the same graph, sketch the *CB* output characteristics for two transistors, one of which has $\alpha \cong 1$ and the other $\alpha = 0.8$. State the difference between their characteristics in words.
6. Refer to Fig. 3.12. Determine the collector current when the transistor voltages are: $V_{BE} = 0.6$ V and $V_{CE} = 5$ V. [Hint: First determine I_B from Fig. 3.12(a).]
*7 Determine β for the transistor with the *CE* output characteristics in Fig. 3.12(b) if $I_B = 200$ μA, and $V_{CE} = 7$ V.
8. Determine the β for the transistor with the characteristics in Fig. 3.12 when $I_C = 10$ mA and $V_{CE} = 9$ V. (Hint: First determine β for the nearest I_B curve.)
9. From the data sheet in Appendix A, determine the base current for a 2N2222 transistor when the collector current is 0.5, 1, and 10 mA. Make a sketch of I_B as a function of I_C using log-log paper. Use curve 2 of Figure 7 and assume $\beta = h_{fe}$.
10. Sketch the permissible region of operation for a 2N2222 transistor (see Question 26).

Chapter 4

BJT Biasing, Bias Stability, and Small-Signal Models

The basic operation of a BJT was introduced in Chapter 3. In this chapter, we will consider the transistor as an amplifier; but first, we must learn how to select and establish the proper operating point for the transistor. Several basic *biasing* schemes will be discussed from both an analytical and a graphical standpoint.

Once the transistor is properly biased (operating point set), it is next considered as a two-port device and its two-port characteristics are analyzed. In this analysis, it is assumed that the transistor operates in a reasonably linear fashion. The analysis concludes with a discussion of small-signal ac models that can be used to determine the performance of transistors under small-signal operation.

4.1 BIASING

In the following analysis, we shall use approximations. You may be somewhat confused to see the same quantity neglected in one case and kept in the analysis in another case. The key to when and why something may be ignored lies in a comparison of the quantity in question with some other quantity. For example, in a series circuit, a voltage drop of 0.5 V cannot be neglected if the total voltage drop across the elements of interest is 2 or 3 V. However, the same voltage drop of 0.5 V may be ignored in another circuit where the total voltage drop under consideration is in excess of, say, 10 V. Follow the rule of thumb that when one quantity is at least one order of magnitude (a factor of 10) larger than another quantity, the smaller quantity may be justifiably neglected, with little loss in accuracy, since $\pm 10\%$ is typically the resistor tolerance.*

---

* A comparison of unlike quantities (e.g., voltage and current, resistance and conductance) should not be attempted. Only like quantities may be compared (e.g., two voltages, two currents, two resistors) with the aim of neglecting one of these.

4.1.1 Calculation of the Operating Point

Fixed-bias circuit. To establish proper operation of the BJT, the base-emitter junction must be forward biased and the collector-base junction must be reverse biased. The circuit in Figure 4.1, called the *fixed-bias circuit,* provides the transistor with proper bias for normal operation. The ac input is applied between the base and emitter terminals shown in Figure 4.1, with the output taken between the collector and emitter terminals.

A fairly common notation for the indication of the connection of a dc power supply (or battery) is shown. We use the convention that the point labeled $+V_{CC}$ in the diagram has the dc source connected to it. The other terminal of the source is understood to be connected to the common return point (ground).

The starting point in the analysis is to define voltage and current polarities and directions in the circuit. For the *NPN* transistor used, both the collector and base current flow *into* the transistor. Although we shall analyze *NPN* transistors here, the procedure applies equally well to *PNP* transistors with the appropriate reversal of the dc supplies, voltage, and current polarities.

In the analysis that follows, we are interested in determining the operating, or *quiescent* (Q), point. This point is specified by the values of the base and collector currents as well as the collector-emitter voltage set up by the specific values of the dc supply V_{CC} and bias resistors R_B and R_C. We shall denote the transistor currents and voltages at the operating point by adding a Q subscript to the symbol. Thus, I_{BQ} would denote the quiescent base current. We shall assume that the supply voltage V_{CC} and the bias resistors are all known. In addition, we shall assume that the necessary transistor parameters are also known.

Taking the summation of voltages from the source V_{CC}, through R_B, across the base-emitter junction to ground, and setting it equal to zero, we obtain $V_{CC} - I_B R_B - V_{BE} = 0$ which yields:

$$V_{CC} = I_B R_B + V_{BE} \qquad (4.1)$$

We now come to the first approximation. The base-emitter voltage drop is a forward-biased diode drop. Typically for silicon transistors, this drop is 0.5 to 0.6 V; for germanium, between 0.2 and 0.3 V. In Equation (4.1) when V_{BE} of this

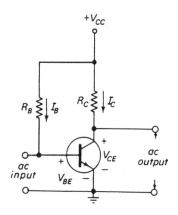

Figure 4.1 Fixed-bias circuit.

magnitude is compared to V_{CC}, which is typically larger than 10 V, V_{BE} may be justifiably neglected. Thus, we have:

$$V_{CC} \cong I_B R_B \tag{4.2}$$

This is the *bias curve equation*. Because both V_{CC} and R_B are fixed in value, Equation (4.2) determines the quiescent base current. Therefore,

$$I_{BQ} \cong \frac{V_{CC}}{R_B} \tag{4.3}$$

In a similar manner, we take the summation of voltages in the collector circuit and set it equal to zero to obtain $V_{CC} - I_C R_C - V_{CE} = 0$ which yields:

$$V_{CC} = I_C R_C + V_{CE} \tag{4.4}$$

This is the *load line equation*. We shall discuss the significance of this name when we examine graphical techniques for determining the operating point.

If the β of the transistor is known, the quiescent collector current is calculated from the relationship:

$$I_{CQ} = \beta I_{BQ} \tag{4.5}$$

where I_{BQ} has already been determined in Equation (4.3). Using the value of I_{CQ} thus calculated in Equation (4.4), we determine the quiescent collector-emitter voltage from the relationship:

$$V_{CEQ} = V_{CC} - I_{CQ} R_C \tag{4.6}$$

So we see that if the values of the dc supply voltage, the bias resistors, and the transistor β are known, the dc operating (quiescent) point for the transistor in Figure 4.1 may be determined in the straightforward manner just indicated. Example 4.1 illustrates this procedure.

Example 4.1

The circuit values in Fig. 4.1 are $R_B = 200$ kΩ, $R_C = 2$ kΩ and $V_{CC} = 20$ V. A silicon *NPN* transistor with β = 50 is used. We want to determine the *Q*-point.

Solution. We first check to make certain that V_{BE} may indeed be neglected. For silicon transistors, $V_{BE} \cong 0.6$ V, which is indeed negligible with the V_{CC} of 20 V. Thus from Eq. (4.3):

$$I_{BQ} \cong \frac{20}{200} \text{ mA} \cong 0.1 \text{ mA}$$

The collector current is calculated next using Eq. (4.5):

$$I_{CQ} \cong 50(0.1 \text{ mA}) \cong 5 \text{ mA}$$

The collector voltage is then determined from Eq. (4.6):

$$V_{CEQ} = 20 - (5)(2) \cong 10 \text{ V}$$

The operating point, therefore, is seen to depend on the dc supply voltage, the bias resistors, the transistor used, and, as we shall see, temperature.

We must now turn our attention to a practical problem. The manufacturer usually specifies a range for β rather than a single specific value. In Appendix A, for a type 2N2222A, the minimum β is specified as 50, while the maximum value specified is 300. In other words, a specific transistor of type 2N2222A may have a β anywhere within the specified range; it might be 50 or 300 or any value in between. Large β spreads in transistors of the same type number are by no means unusual; instead, they are the rule. Even if the α of a transistor is controlled within fairly tight tolerances, the corresponding β has a wide range of values.

From the preceding example, we can readily see that any variation in β causes the operating point in the circuit shown in Figure 4.1 to vary drastically. To see this effect, note that if the transistor used in the example had a β of 100 or more, the transistor would have a *$V_{CEQ} \cong 0$, and could not be used to amplify any signal.

Thermal stability. Another problem is related to the temperature sensitivity of transistor terminal currents. Recall that the relationship between the collector and base currents also contained a term involving the collector reverse-saturation current. The equation is repeated here for convenience:

$$I_C = \beta I_B + (\beta + 1)I_{CO} \qquad (4.7)$$

The collector reverse-saturation current about doubles for every 10°C rise in junction temperature. Moreover, this increase is multiplied by β + 1. Therefore, the collector current is also very much a function of the junction operating temperature.

A condition known as *thermal runaway* may occur in the BJT with the increase in collector current. Such a condition can cause the transistor to saturate. Assume that the collector junction is at room temperature. As soon as any collector current is caused to flow, the junction temperature increases because of the power being dissipated at the junction. This increase in temperature is reflected as an increase in I_{CO}. From Equation (4.7) we see that any increase in I_{CO} also causes I_C to increase. As a result of the increase in I_C, the power dissipated at the collector junction increases, causing the junction temperature to rise accordingly. This cycle may be repeated until the maximum junction temperature is exceeded, in which case the transistor may be damaged permanently, or until the collector current increases sufficiently to cause the transistor to be saturated. Either of these possibilities must be avoided.

A numerical measure, called the *thermal stability factor S,* has been defined to indicate how well thermal runaway has been controlled. The thermal stability

---

* When V_{CEQ} is small (<0.6 V) the transistor is said to be *saturated*.

factor is defined as:

$$S = \frac{\text{change in } I_C}{\text{change in } I_{CO}} \tag{4.8}$$

From the definition we see that the lower the numerical value of S, the less likely it is for thermal runaway to occur. The thermal stability factor is determined by the bias circuit configuration. For the circuit in Figure 4.1, S is given by

$$S = \beta + 1 \tag{4.9}$$

The expression just given for S indicates that for the fixed-bias circuit (Figure 4.1), the thermal stability factor is a constant that depends only on the β of the transistor used. Moreover, because β is typically quite large, so is S. Consequently, the fixed-bias circuit is not very well protected from thermal runaway.

4.1.2 Self-Bias Circuit

The self-bias arrangement, shown in Figure 4.2, offers a number of advantages over the fixed-bias circuit. It provides us with the ability to design for a specific operating point with almost complete immunity to the β spread. We can also design for a specified stability factor.

Analysis of the self-bias circuit is similar to that used in the fixed-bias circuit. First, we recognize that the two resistors (R_1 and R_2) in the base circuit provide a voltage divider. The circuit is redrawn for more convenience in Figure 4.3(a). We may simplify it by finding the Thevenin equivalent circuit to the left of points A and N. If we label the equivalent resistance by R_B and the equivalent open-

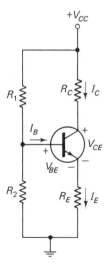

Figure 4.2 Self-bias circuit.

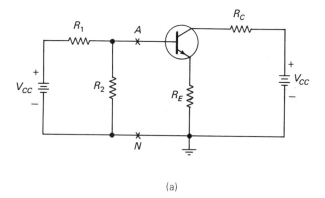

(a)

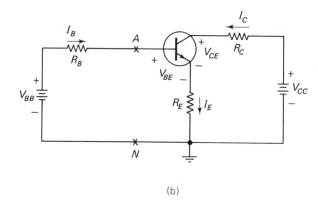

(b)

Figure 4.3 (a) The self-bias circuit of Figure 4.2 redrawn to find the Thevenin equivalent at points A and N, (b) with the Thevenin equivalent circuit replacing the voltage divider.

circuit voltage by V_{BB}, we have:

$$R_B = \frac{R_1 R_2}{R_1 + R_2} \tag{4.10}$$

and

$$V_{BB} = V_{CC} \frac{R_2}{R_1 + R_2} \tag{4.11}$$

Using these equivalent values, the circuit is redrawn as indicated in Figure 4.3(b). We can now proceed to determine the operating point by adding the voltages in the base and collector loops. For the base loop:

$$V_{BB} = I_B R_B + V_{BE} + I_E R_E \tag{4.12}$$

Before we proceed, note that in this case the base-emitter voltage V_{BE} may not always be negligible, because it is compared to V_{BB}, which is always smaller than V_{CC}. The decision whether or not to neglect V_{BE} must be made in each case and depends on the relative size of V_{BB}.

To determine the operating or quiescent base current, Equation (4.12) must be stated in terms of I_B. If we make use of the fact that the summation of the transistor terminal currents is zero, (i.e., $I_E = I_C + I_B$) and at the same time substitute βI_B for I_C (neglecting the effect of I_{CO}), then $I_E = \beta I_B + I_B$. Substituting this expression for I_E into Equation (4.12) yields:

$$V_{BB} = I_B R_B + V_{BE} + (\beta + 1)I_B R_E \tag{4.13}$$

Before we solve for the quiescent base current, note the result of the substitutions. Saying that the voltage drop from emitter to ground is caused by a current I_E flowing through R_E is equivalent to saying that it is the result of a current I_B flowing through an equivalent resistance $(\beta + 1)R_E$. If we represent the base-emitter voltage drop by a fixed voltage source, the sum of the voltages in Equation (4.13) corresponds to the equivalent loop shown in Figure 4.4, where I_B is a loop instead of branch current. This equivalent circuit may be used to simplify calculations. Furthermore, it illustrates that a resistance R_E in the emitter circuit corresponds to a resistance $(\beta + 1)R_E$ in the base circuit.

The quiescent base current is now calculated from Equation (4.13) or Figure 4.4:

$$I_{BQ} = \frac{V_{BB} - V_{BE}}{R_B + (\beta + 1)R_E} \tag{4.14}$$

As was the case in the fixed-bias circuit, we may assume that V_{BE} is essentially constant to determine I_{BQ} from Equation (4.14).

Setting the voltage rises (i.e., the supply voltage) equal to the voltage drops in the collector or output loop in Figure 4.3, we obtain the load line equation:

$$V_{CC} = I_C(R_C + R_E) + V_{CE} \tag{4.15}$$

where we have used the approximation that $I_E \cong I_C$. With the value of β known, the quiescent collector current is determined:

$$I_{CQ} \cong \beta I_{BQ} \tag{4.16}$$

This value is then used in the load line equation to solve for the quiescent collector voltage:

$$V_{CEQ} = V_{CC} - I_{CQ}(R_C + R_E) \tag{4.17}$$

The procedure outlined is illustrated in Example 4.2.

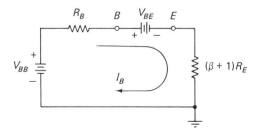

Figure 4.4 Equivalent circuit for the evaluation of base current.

Example 4.2

The circuit values in Fig. 4.2 are: $R_1 = 33$ kΩ, $R_2 = 10$ kΩ, $R_C = 2.2$ kΩ, $R_E = 1$ kΩ, and $V_{CC} = 18$ V. A silicon *NPN* transistor with $\beta = 50$ is used. We want to determine the operating point.

Solution. We first calculate the base equivalent resistance and open-circuit voltage using Eqs. (4.10) and (4.11):

$$R_B = \frac{(33)(10)}{33 + 10} \text{ k}\Omega = 7.7 \text{ k}\Omega$$

$$V_{BB} = (18)\frac{10}{33 + 10} \text{ V} = 4.2 \text{ V}$$

The quiescent base current is now calculated from Eq. (4.14) assume $V_{BE} = 0.6$ V:

$$I_{BQ} = \frac{4.2 - 0.6}{7.7 + (51)1} \text{ mA} = 61 \text{ }\mu\text{A}$$

With $\beta = 50$, the collector current is determined from Eq. (4.16) as:

$$I_{CQ} = 50(0.061 \text{ mA}) = 3 \text{ mA}$$

To complete the Q-point calculations, determine the collector voltage using Eq. (4.17):

$$V_{CEQ} = 18 - (3)(2.2 + 1) = 8.4 \text{ V}$$

Thermal stability. The thermal stability factor for the self-bias circuit is given by:

$$S = \frac{\beta + 1}{1 + \dfrac{\beta}{K}} \qquad (4.18)$$

where K is a constant that is a function of the circuit values:

$$K = \frac{R_B + R_E}{R_E} \qquad (4.19)$$

If in the limit $K = 1$, note that S is also 1. This is the absolute minimum value, because for any real resistor values, the minimum value of K is 1.

Thermal stability is achieved in the self-bias circuit in the following manner. Should the collector current tend to increase because of an increase in I_{CO}, the voltage drop across R_E would also increase. Because V_{BB} is a constant for given resistor values, it is evident from Equation (4.12) that I_B would decrease if the voltage drop $I_E R_E$ increased. Thus, a tendency for I_C to increase (because of increasing I_{CO}) is offset by the corresponding tendency for I_B to decrease. So I_C remains essentially fixed and relatively insensitive to I_{CO}. This method of achieving better thermal stability is based on sensing the output current (I_C) and causing a response in the input current (I_B). It is also an example of *negative feedback*, which will be discussed in detail in Chapter 13. Example 4.3 will let you see the improvement in thermal stability offered by the self-bias circuit. S will be determined using the values in Example 4.2.

Example 4.3

For the circuit in Ex. 4.2, determine the thermal stability factor S.

Solution. First determine K:

$$K = \frac{7.7 + 1}{1} = 8.7$$

S is now calculated from Eq. (4.18):

$$S = \frac{50 + 1}{1 + \frac{50}{8.7}} = 7.5$$

Note that this value is significantly lower than the corresponding value in the fixed-bias circuit. Moreover, by changing the values of the circuit resistors, you can obtain a different value of S. Remember in the fixed-bias circuit of Figure 4.1 that the circuit resistor values did not have any effect on S.

4.1.3 Designing a Self-Bias Circuit

In the simplest sense, designing a bias circuit involves specifying (1) resistor values to be used and (2) a certain dc supply voltage. Calculating the resistors to be used is quite similar to the analysis discussed in the previous section. The equations are just rearranged. In the design problem, the operating point quantities, voltages, and currents are known; it is the resistors that are to be determined.

To design the self-bias circuit, we will assume that the load resistor (R_C) and the supply voltage are given. In a case where these parameters are not specified, they would be chosen. If the operating point, the thermal stability factor, and the transistor parameters are known, then we have enough information to determine specific values for the remaining circuit components. The design equations are summarized here:

$$R_E = \frac{V_{CC} - V_{CEQ} - I_{CQ}R_C}{I_{CQ}} \qquad (4.20)$$

$$K = \frac{S\beta}{\beta + 1 - S} \qquad (4.21)$$

$$R_B = (K - 1)R_E \qquad (4.22)$$

$$V_{BB} = I_{BQ}[R_B + (\beta + 1)R_E] + V_{BE} \qquad (4.23)$$

$$R_1 = R_B \frac{V_{CC}}{V_{BB}} \qquad (4.24)$$

$$R_2 = \frac{R_1 R_B}{R_1 - R_B} \qquad (4.25)$$

The basic design procedure is illustrated in the Example 4.4.

Example 4.4

We want to design a self-bias circuit using a silicon transistor (V_{BE} = 0.6 V) with a β of 80. The load resistor (R_C) is to be 1.2 kΩ. With the supply voltage at 12 V, the operating point is to be: V_{CEQ} = 6 V and I_{CQ} = 4 mA. The circuit is to be designed with a thermal stability factor of 10.

Solution. We first calculate the operating-point base current corresponding to the given β:

$$I_{BQ} = \frac{I_{CQ}}{\beta} \cong \frac{4}{80} \text{ mA} \cong 0.05 \text{ mA } (50 \text{ μA})$$

We are now ready to proceed with the steps outlined in Eqs. (4.20) to (4.25):

$$R_E = \frac{12 - 6 - (4)(1.2)}{4} \text{ kΩ} \cong 300 \text{ Ω}$$

$$K = \frac{(10)(80)}{80 + 1 - 10} \cong 11.3$$

$$R_B = (11.3 - 1)(0.3) \text{ kΩ} \cong 3.1 \text{ kΩ}$$

$$V_{BB} = (0.05)[3.1 + (80 + 1)(0.3)] + 0.6 \cong 2 \text{ V}$$

$$R_1 = 3.1 \frac{12}{2} \text{ kΩ} \cong 18.6 \text{ kΩ}$$

$$R_2 = \frac{(18.6)(3.1)}{18.6 - 3.1} \text{ kΩ} \cong 3.7 \text{ kΩ}$$

The design is implemented by specifying standard resistor values. In this case, we may choose:

$$R_E = 330 \text{ Ω}, \quad R_1 = 18 \text{ kΩ}, \quad R_2 = 3.6 \text{ kΩ}$$

The choice of the thermal stability factor is governed by the following guidelines: For small-signal (low-power) stages, an S of 20 or lower is satisfactory; for power amplifiers, usually an S of 10 or less is desirable.

Designing for β spread. The design procedure just outlined is quite satisfactory when the β of the transistor to be used can be measured. However, when units are mass produced, the β of each individual transistor cannot be measured. The problem, therefore, is to design the bias circuit so that any transistor with a β within a specified range (the β spread) can be used in the circuit and produce an operating point within a desired range. If we label the difference between the minimum operating-point collector current (I_{Cm}) and the maximum operating-point collector current (I_{CM}) by ΔI_C, then the needed stability factor S_M is calculated from

$$S_M = \frac{\Delta I_C}{I_{Cm}} \frac{\beta_m \beta_M}{\Delta \beta} \quad (4.26)$$

where β_m is the minimum value of β; β_M, the maximum value of β; and $\Delta\beta$, the β spread, defined as $\beta_M - \beta_m$. This stability factor corresponds to the maximum value of $\beta(\beta_M)$. The remainder of the design procedure follows that given in the previous example. Example 4.5 demonstrates how β spread is accommodated in the design of the self-bias circuit.

Example 4.5

We want to design the self-bias circuit of Fig. 4.2 using a 2N5450 silicon NPN transistor (V_{BE} = 0.6 V). The constraints are: V_{CC} = 15 V, R_C = 330 Ω. The nominal operating point is to be I_{CQ} = 10 mA and V_{CEQ} = 7 V. The variation in the operating-point collector current, because of the uncertainty in β, cannot be lower than 9 mA or higher than 11 mA.

Solution. First, the β range for the 2N5450 transistor is determined from a data sheet to be from 50 to 150. Thus, β_M = 150, β_m = 50, and $\Delta\beta$ = 150 − 50 = 100. Also, I_{CM} = 11 mA, I_{Cm} = 9 mA, and ΔI_C = 11 − 9 = 2 mA. We can now proceed to determine the thermal stability factor needed:

$$S_M \cong \frac{2}{9} \frac{(50)(150)}{100} \cong 16.7$$

We next use this value in Eq. (4.21) with appropriate subscripts:

$$K = \frac{S_M \beta_M}{\beta_M + 1 - S_M} \cong \frac{(16.7)(150)}{150 + 1 - 16.7} \cong 18.6$$

Using the nominal Q-point values, we determine R_E from Eq. (4.20):

$$R_E = \frac{15 - 7 - (10)(0.33) \text{ k}\Omega}{10} \cong 470 \; \Omega$$

From this point on, we can use one of two methods. Each gives the same results. Because we have no idea of a "nominal" β value, we may calculate either I_{BM} or I_{Bm}:

$$I_{BM} = \frac{I_{CM}}{\beta_M} \quad \text{and} \quad I_{Bm} = \frac{I_{Cm}}{\beta_m}$$

These values are: $I_{BM} \cong 0.073$ mA and $I_{Bm} \cong 0.18$ mA. Note that I_{Bm} is larger than I_{BM}. The interpretation of the symbols is important. The M subscript on I_B denotes that I_B value that results when the transistor in the circuit has β equal to $β_M$. It does not stand for the maximum I_B, and in fact it will not usually be larger than I_{Bm}. Similarly, the m subscript on I_B denotes the I_B value that results from the transistor being used having a β equal to $β_m$. Only one value of I_B is needed for the actual solution.

V_{BB} is calculated from Eq. (4.23), using the proper combination of I_B and β values; that is, if we use I_{BM}, then we must use $β_M$; if we use I_{Bm}, then we must use $β_m$. R_B is obtained from Eq. (4.22):

$$R_B \cong (18.6 - 1)(0.47) \text{ k}\Omega = 8.3 \text{ k}\Omega$$

Using I_{Bm} and $β_m$ in Equation (4.23), we obtain:

$$V_{BB} = (0.18)[8.3 + (51)(0.47)] + 0.6 \cong 6.4 \text{ V}$$

A subsequent exercise will show that the same value of V_{BB} results if I_{BM} and $β_M$ are used in Eq. (4.22).

The calculation of R_1 and R_2 proceeds as before:

$$R_1 = 8.3 \frac{15}{6.4} \text{ k}\Omega \cong 19.5 \text{ k}\Omega$$

$$R_2 = \frac{(19.5)(8.3)}{19.5 - 8.3} \text{ k}\Omega \cong 14.5 \text{ k}\Omega$$

Again we round off the calculated values to standard available values to complete the design:

$$R_E = 470 \text{ }\Omega, \quad R_1 = 20 \text{ k}\Omega, \quad R_2 = 15 \text{ k}\Omega$$

4.1.4 Drawing the DC Load Line

Equation (4.6) is the dc load line equation for the fixed-bias circuit; Equation (4.15) is the dc load line equation for the self-bias circuit. For either circuit, we can evaluate the Q-point graphically by using the same basic method.

In order to plot the dc load line on the CE output characteristics, note first that the line is a straight line, so we need plot only two points. Secondly, the most convenient points to calculate are the two axis intercepts. We can determine the I_C axis intercept by making $V_{CE} = 0$ in the load line equation. For the self-bias circuit we find:

$$\text{point 1} \quad \left(V_{CE} = 0, I_C = \frac{V_{CC}}{(R_C + R_E)} \right)$$

In a similar fashion, we can determine the V_{CE} axis intercept by forcing I_C to be zero in the load line equation. Again for the self-bias circuit, we find:

$$\text{point 2} \quad (V_{CE} = V_{CC}, I_C = 0)$$

These two points are indicated on Figure 4.5. We then obtain the load line by drawing a straight line between them. If we know the device output characteristics

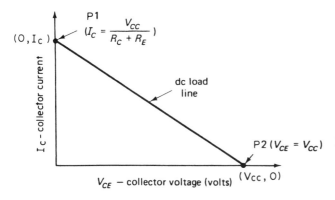

Figure 4.5 Plotting the load line on the output characteristics. (The characteristic curves have been left out for simplicity.)

(displayed on a curve tracer) and the base current, then the Q-point is the intersection of the dc load line (as plotted) and the bias curve corresponding to the base current value. This method of constructing a load line is especially useful when the curve tracer is used to determine the transistor small-signal parameters.

4.2 BJT LOW-FREQUENCY MODEL

Having described the operation and biasing of BJTs, we now examine the amplification of time-varying signals by the properly biased BJT.

To see how amplification takes place, consider the following narrative example. The fixed-bias circuit of Figure 4.1 is operated with a silicon *NPN* transistor, whose characteristics are shown in Figure 4.6. The circuit values are: $V_{CC} = 18$ V, $R_B = 450$ kΩ and $R_C = 3$ kΩ. Using the methods discussed in the previous sections, we may determine the point of operation. Assuming V_{BE} negligible as compared with the supply voltage of 18 V, the Q-point base current is found to be: $I_{BQ} = V_{CC}/R_B = 18/450$ mA $= 40$ µA. Figure 4.6(a) tells us that $V_{BEQ} = 0.6$ V (when $I_{BQ} = 40$ µA). This result justifies the earlier assumption that V_{BE} is negligible. The ac load line is obtained in the same manner as the dc load line. Because the ac resistance is exactly the same as the dc resistance in the circuit, the ac and dc load lines for this particular circuit are identical. Point 1 is simply (0 V, I_C) = (0 V, 6 mA) and point 2 is (V_{CC}, 0 mA) = (18 V, 0 mA) as noted in the load line plot of Figure 4.6(b). The Q-point is the point where the $I_{BQ} = 40$ µA curve intersects the load line. The Q-point parameters are $I_{CQ} = 4$ mA and $V_{CEQ} = 6$ V.

Now let us assume that, in some manner, the input (base-emitter) voltage is caused to vary about its quiescent value of 0.6 V. For example, let V_{BE} have a total excursion of 25 mV, that is, 12.5 mV above V_{BEQ} and 12.5 mV below V_{BEQ}. By projecting this variation on the input characteristic, we can determine the corresponding variation in the base current. We see that $\Delta I_B = 20$ µA, 10 µA above and below I_{BQ}. Using this variation in the base current on the output characteristics, we can determine the corresponding variation in the collector current and collector-emitter voltage, as shown in Figure 4.6(b). Thus, $\Delta I_C = 2$ mA and $\Delta V_{CE} = 6$ V. We conclude from this graphical analysis that voltage

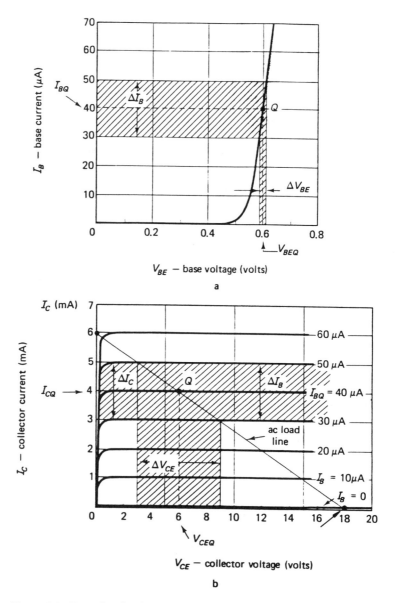

Figure 4.6 Example of voltage amplification: (a) small variation in the base-emitter voltage results in (b) a large variation in collector-emitter voltage.

amplification has taken place. A variation of 25 mV in the input (base-emitter) voltage has caused a variation of 6 V in the output (collector-emitter) voltage. The amount of voltage amplification, called *voltage gain*, is the ratio of the output voltage variation to the input voltage variation. In this case, it is 240, i.e., 6/0.025 = 240.

This graphical analysis method is an extremely useful technique in the analysis of amplifiers. We shall apply it later in our discussion of large-signal (power) amplifiers in Chapter 12. If the signals involved are small and the excursions that they cause are over an essentially linear portion of the transistor characteristics, the graphical techniques may be replaced by a systematic analysis using an approximate model of the transistor.

The common-emitter transistor configuration is shown in two-port notation in Figure 4.7, using the conventional directions for the terminal currents and polarity for the voltages. Here we see lower-case symbols being used to denote ac (time-varying) currents and voltages; these ac signals indicate small excursions around the dc Q-point. Notice that in dealing with BJT *small-signal models,* we shall consider the transistor by itself. However, remember that the transistor cannot operate without the proper bias being provided by one of the bias circuits, which we have already discussed. All these bias circuits contain resistors, which are deleted here for the sake of simplicity. At this time, we are concerned with the behavior of the transistor alone.

We shall include the effects of the bias resistors in future analyses of amplifiers and other circuits.

4.2.1 Hybrid or h Parameters

A two-port network, like the one shown in Figure 4.7, may be completely represented by a set of four parameters. The most suitable set of parameters for the BJT is the set of h parameters. The defining set of equations for the h parameters are

$$\begin{cases} v_1 = h_{11}i_1 + h_{12}v_2 \\ i_2 = h_{21}i_1 + h_{22}v_2 \end{cases} \quad (4.27)$$

Because the BJT is important and it can be operated in any one of three possible configurations (*CE*, *CB*, or *CC*), the simple numeral subscript notation is replaced by one more suitable for transistors. The two numerals are replaced by two letters, the first of which denotes the specific h parameter, and the second of which

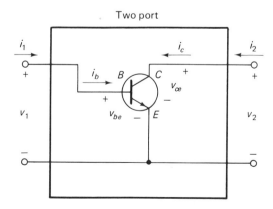

Figure 4.7 The *CE* configuration as a two-port network.

denotes the transistor terminal common to both input and output. Thus,

$$h_{11} = h_{i(\ )} \quad h_{12} = h_{r(\ )} \quad h_{21} = h_{f(\ )} \quad h_{22} = h_{o(\ )} \quad (4.28)$$

The second subscript—either *e, b,* or *c*—denotes whether the emitter, base, or collector terminal is common to the input and the output. For example, in the *CE* configuration we have

$h_{11} = h_{ie}$ —input impedance (in ohms)

$h_{12} = h_{re}$ —reverse voltage ratio (no units)

$h_{21} = h_{fe}$ —forward current ratio (no units)

$h_{22} = h_{oe}$ —output admittance (in siemens)

In a similar manner, the parameters in the *CB* configuration are: h_{ib}, h_{rb}, h_{fb}, and h_{ob}. In the *CC* configuration they are: h_{ic}, h_{rc}, h_{fc}, and h_{oc}.

CE h-parameter model. Making the appropriate substitutions in Equation (4.27) for the *CE* configuration, we obtain the defining set of equations for the CE configuration. They are:

$$\begin{cases} v_{be} = h_{ie}i_b + h_{re}v_{ce} \\ i_c = h_{fe}i_b + h_{oe}v_{ce} \end{cases} \quad (4.29)$$

Before we proceed, we must explain exactly what we mean by the voltage polarities and current directions. Because all these quantities (voltages and currents) are time-varying, their polarities and directions reverse periodically. The notation denotes the conditions at one instant of time. We would know, for example, that when the base is positive with respect to the emitter, the collector is also assumed positive with respect to the emitter.

Equation (4.29) is used to develop the small-signal model for the transistor. We identify the input and output voltages and currents from Figure 4.7 and note that the emitter is common; from this information, we have the beginning of the model, as shown in Figure 4.8(a). The first equation in Equation 4.29 is nothing more than the sum of the voltages from base to emitter. The voltage applied (v_{be}) is equal to the sum of the voltage drops. This calculation is pictured in Figure 4.8(b). The second equation is simply the sum of the currents resulting in the collector current. The current (i_c) into the output node is equal to the sum of currents leaving the same node. This result is indicated in Figure 4.8(c) where we see the complete model for transistor *CE* model. Note that both the voltage generator in the input and the current generator in the output are controlled, or dependent, generators.

The four *h* parameters in the *CE* configuration may be determined in two ways. One method involves making actual measurements in a circuit with the transistor biased and ac signals applied. The second method involves a graphical evaluation from the transistor static characteristics, which can be displayed on a curve tracer.

The conditions under which the parameters may be calculated are as follows:

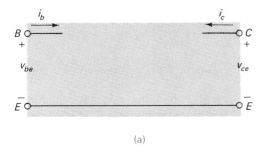

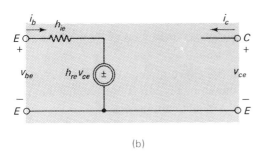

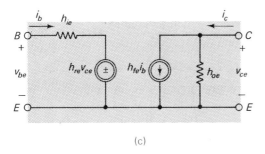

Figure 4.8 Development of the small signal CE h-parameter.

When the *ac component* of the output voltage (v_{ce}) is zero, h_{ie} and h_{fe} are calculated as

$$h_{ie} = \frac{v_{be}}{i_b}\bigg|_{V_{CEQ}} \tag{4.30}$$

$$h_{fe} = \frac{i_c}{i_b}\bigg|_{V_{CEQ}} \tag{4.31}$$

(Note: The condition $v_{ce} = 0$ means that there is no ac collector-emitter voltage; that is, the collector-emitter voltage is held constant at its (dc) Q-point value (V_{CEQ}).)

In a similar fashion, the other two parameters, h_{re} and h_{oe}, may be calculated by making the ac component of the base current zero; that is, the base current is held constant at its (dc) Q-point value (I_{BQ}).

Sec. 4.2   BJT Low-Frequency Model

$$h_{re} = \left.\frac{v_{be}}{v_{ce}}\right|_{I_{BQ}} \tag{4.32}$$

$$h_{oe} = \left.\frac{i_c}{v_{ce}}\right|_{I_{BQ}} \tag{4.33}$$

The simplest method for actually obtaining values for the four h parameters of a particular BJT is to display its input and output characteristics on a curve tracer. This technique is explored in Example 4.6.

Example 4.6

From the input and the output static characteristics (obtained on a curve tracer) of the *NPN* transistor given in Figs. 4.9 and 4.10, evaluate the *CE* h parameters.

Observation. The use of the curve tracer to display the base-emitter diode (input characteristic) is shown in Fig. 4.11. Notice that $V_{CEQ} = 6$ V is held constant by the regulated voltage supply.

Solution. The input impedance, h_{ie}, is determined from the inverse of the slope of the input characteristics at the indicated Q-point in Fig. 4.9. Thus, with a change in V_{BE} from 0.55 to 0.65 V, we have a change in I_B from 0 to 80 μA. This change corresponds to

$$h_{ie} = \frac{v_{be}}{i_b} = \frac{\Delta V_{BE}}{\Delta I_B} = \frac{0.65 - 0.55}{80 - 0} \text{ M}\Omega \cong 1.25 \text{ k}\Omega$$

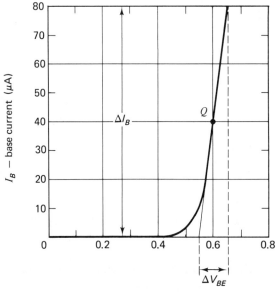

Figure 4.9 Graphical evaluation of h_{ie} from the input characteristic curve of an *NPN* silicon transistor with V_{CEQ} held constant at 6 V.

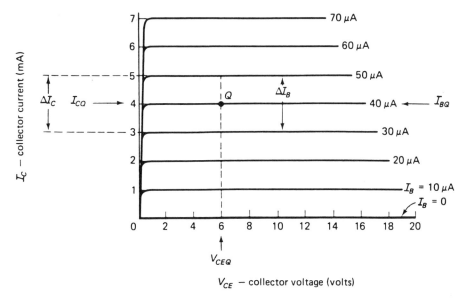

Figure 4.10 Graphical evaluation of h_{ie} from the output characteristics of an *NPN* silicon transistor.

The forward current ratio, h_{fe}, is evaluated from the output characteristics in Fig. 4.10. If the base current changes from 30 to 50 µA (which is 10 µA above and below I_{BQ} of 40 µA) for a constant V_{CE} = 6 V, the corresponding change in I_C is from about 3 to 5 mA. Thus,

$$h_{fe} = \frac{i_c}{i_b} = \frac{\Delta I_C}{\Delta I_B} = \frac{5-3}{50-30} \times 10^3 \cong 100$$

The output admittance, h_{oe}, is evaluated from the output characteristics in Fig. 4.10. It is the slope of the bias curve corresponding to $I_B = I_{BQ}$. To obtain somewhat better resolution for this calculation, the region around I_B = 40 µA is expanded along with the current scale, as shown in Fig. 4.12. We see that for a variation in V_{CE} from 2 to 10 V (which is 4 V above and below the V_{CE} Q-point value), the corresponding I_C variation taken along the I_{BQ} characteristic curve is from about 3.95 to 4.05 mA. Thus,

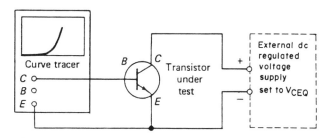

Figure 4.11 Connections for obtaining the *CE* input characteristics of a BJT.

Sec. 4.2   BJT Low-Frequency Model

$$h_{oe} = \frac{i_c}{v_{ce}} = \frac{\Delta I_C}{\Delta V_{CE}} = \frac{4.05 - 3.95 \text{ mA}}{10 - 2 \text{ V}} = 12.5 \mu S$$

This is an admittance corresponding to an impedance of 80 kΩ.

When we try to determine the reverse voltage ratio, h_{re}, we run into problems. It should be measured by noting the difference in the base-emitter voltage (at $I_B = I_{BQ}$) on the input characteristics caused by a certain variation in V_{CE}. The setup shown in Fig. 4.11 is used for this measurement, with the variation in V_{CE} accomplished by the external power supply setting. However, the change in V_{BE} thus obtained is usually imperceptible. We conclude, therefore, that h_{re} is very small, typically 1×10^{-4}. As we shall see, an actual determination for h_{re} is not usually necessary.

CE small-signal models. The complete CE h-parameter model is pictured in Figure 4.13(a). This model may be simplified by assuming $v_{ce}h_{re} = 0$ V. In most cases, this is a valid assumption since the voltage generator ($v_{ce}h_{re}$) controlled by the output voltage (v_{ce}) is negligible due to the extremely small value of v_{ce}. In this case, the transistor may be approximated by the model shown in Figure 4.13(b).

A further approximation is possible in some cases. We may neglect the effect of the output admittance (represented by h_{oe}) in some cases because it is sufficiently small when compared to the admittance of the externally connected parallel load (i.e., $h_{oe} \approx 20$ μS, $R_L \approx 200$ μS). The resulting equivalent circuit for the transistor in the CE configuration may be simplified by setting both h_{re} and h_{oe} equal to zero. This case is shown in Figure 4.13(c) and it will be discussed along with other specific cases in Chapter 9 where the validity of such approximations is discussed.

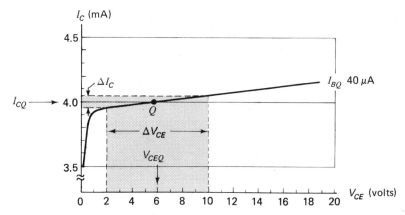

Figure 4.12 The graphical evaluation of h_{oe} from the output characteristics showing the region (shaded) of interest along with the expanded current scale in the vicinity of Q-point.

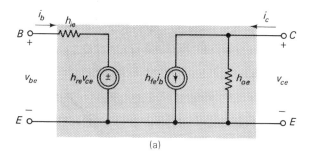

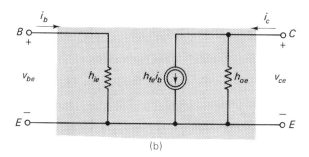

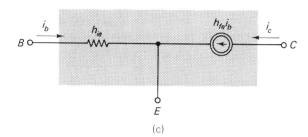

Figure 4.13 CE small-signal transistor model: (a) complete h-parameter model and successive approximate models with (b) h_{re} considered negligible and (c) both h_{re} and h_{oe} considered negligible.

4.2.2 Normalized h-Parameters

We must point out that the h parameters are not constant for any particular BJT; that is, we cannot evaluate the h parameters for the BJT at one Q-point and hope to use these same values at another Q-point. The *CE h* parameters are very much a function of the specific operating point. For this reason, the manufacturers' data sheets may include a plot of the *normalized* h-*parameter* values as a function of the Q-point collector current. Such a plot is illustrated in Figure 4.14(a). The *CE h* parameters are also sensitive to the junction operating temperature. The dependence on the junction temperature for a typical BJT is demonstrated in Figure 4.14(b) and may also be given in the manufacturers' data sheets.

The use of these normalized curves can best be illustrated through the following narrative example. Let us assume that a particular transistor is listed by the manufacturer to have $h_{ie} = 2.6 \text{ k}\Omega$, $h_{re} = 1 \times 10^{-4}$, $h_{fe} = 60$, and $h_{oe} = 10$

Sec. 4.2 BJT Low-Frequency Model 71

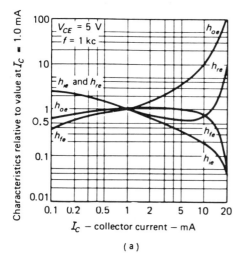
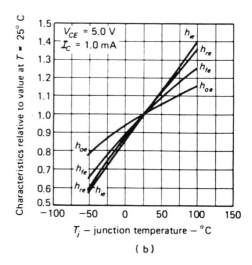

Figure 4.14 Typical variation of h parameter with (a) collector current, (b) temperature.

μS, with all values listed at 1 mA collector current and at 25° C. We can use Figure 4.14(a) to predict the values of the parameters at a different operating point. For example, the same transistor operating at a collector current of 5 mA is seen from Figure 4.14(a) to have multiplying factors of approximately 0.35 h_{ie}, 0.6 h_{re}, 1.0 h_{fe}, and 3.5 h_{oe}. Therefore, at 5 mA, we would expect $h_{ie} = (0.35)(2.6 \text{ k}\Omega) = 910 \text{ }\Omega$, $h_{re} = (0.6)(10^{-4}) = 6 \times 10^{-5}$, $h_{fe} = 60$, and finally $h_{oe} = (3.5)(10) = 35$ μS. Similarly, we can use Figure 4.14(b) to predict the parameters at temperatures other than 25° C. If the same transistor were operated at 75° C, the parameters would change by 1.28 h_{ie}, 1.24 h_{re}, 1.18 h_{fe}, and 1.11 h_{oe}.

CB h-parameter model. The BJT almost always has a set of h parameters for the *CE* configuration. For completeness, the *CB* and *CC* configurations are mentioned here. The defining set of equations for the *CB* configuration may be seen from Figure 4.15. They are

$$\begin{cases} v_{eb} = -h_{ib}i_e + h_{rb}v_{cb} \\ i_c = -h_{fb}i_e + h_{ob}v_{cb} \end{cases} \quad (4.34)$$

Notice the minus signs in the equations just given. They indicate that the definition of i_1 is in the opposite direction to the actual flow of emitter current in an *NPN* transistor. The *CB* equivalent circuit is shown in Figure 4.15(b). However, in most cases, it is more convenient to use the approximate *CB* transistor model, as shown in Figure 4.15(c).

CC h-parameter model. The *CC* configuration is illustrated in Figure 4.16. The set of defining equations in this case are

$$\begin{cases} v_{bc} = h_{ic}i_b + h_{rc}v_{ec} \\ -i_e = h_{fc}i_b + h_{oc}v_{ec} \end{cases} \quad (4.35)$$

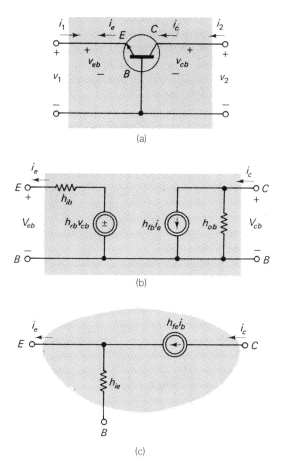

Figure 4.15 *CB* small-signal models: (a) *CB* configuration, (b) complete *CB* h-parameter model, (c) approximate *CB* model using *CE* parameters.

The BJT model in the *CC* configuration is shown in Figure 4.16(b). As was the case in the *CB* configuration, however, it is more convenient to use the *CC* approximate model, as depicted in Figure 4.16(c).

4.2.3 h-Parameter Conversions

For the few cases where the actual *CB* and *CC* (emitter follower) h-parameter equivalent circuits must be used, the needed parameter values can be determined by using the conversions listed in Tables 4.1, 4.2, and 4.3. First, calculate the *CE* h parameters, as discussed in Example 4.6. Then use these values in conjunction with the conversion tables to determine the needed parameters. The Example 4.7 demonstrates this procedure.

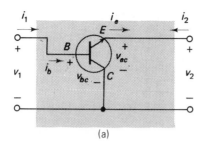

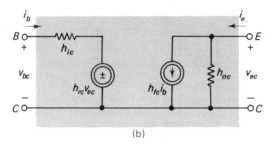

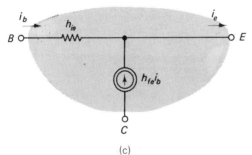

Figure 4.16 *CC* small-signal model: (a) *CC* configuration, (b) complete *CC* h-parameter model, (c) approximate *CC* model using *CE* parameters.

TABLE 4.1. CONVERSION BETWEEN *CB* AND *CE* h PARAMETERS*

$$h_{ib} = \frac{h_{ie}}{h_{fe} + 1} \qquad h_{rb} = \frac{h_{ie}h_{oe}}{h_{fe} + 1} - h_{re}$$

$$h_{fb} = -\frac{h_{fe}}{h_{fe} + 1} \qquad h_{ob} = \frac{h_{oe}}{h_{fe} + 1}$$

$$h_{ie} = \frac{h_{ib}}{h_{fb} + 1} \qquad h_{re} = \frac{h_{ib}h_{ob}}{h_{fb} + 1}$$

$$h_{fe} = -\frac{h_{fb}}{h_{fb} + 1} \qquad h_{oe} = \frac{h_{ob}}{h_{fb} + 1}$$

* The conversions contains approximations.

TABLE 4.2. CONVERSION BETWEEN CC AND CE h PARAMETERS*

$h_{ic} = h_{ie}$	$h_{rc} = 1 - h_{re}$
$h_{fc} = -(h_{fe} + 1)$	$h_{oc} = h_{oe}$
$h_{ie} = h_{ic}$	$h_{re} = 1 - h_{rc}$
$h_{fe} = -(h_{fc} + 1)$	$h_{oe} = h_{oc}$

TABLE 4.3. CONVERSION BETWEEN CB AND CC h PARAMETERS*

$h_{ic} = \dfrac{h_{ib}}{h_{fb} + 1}$	$h_{rc} = h_{rb} + 1 - \dfrac{h_{ib}h_{ob}}{h_{fb} + 1}$
$h_{fc} = -\dfrac{1}{h_{fb} + 1}$	$h_{oc} = \dfrac{h_{ob}}{h_{fb} + 1}$
$h_{ib} = -\dfrac{h_{ic}}{h_{fc}}$	$h_{rb} = h_{rc} - 1 - \dfrac{h_{ic}h_{oc}}{h_{fc}}$
$h_{fb} = -\dfrac{h_{fc} + 1}{h_{fc}}$	$h_{ob} = \dfrac{h_{oc}}{h_{fc}}$

* Certain conversions include approximations.

Example 4.7

Using the CE h parameters of the 2N2222A BJT from the data sheet in Appendix A, determine (a) the h parameters for the CC equivalent circuit and (b) the h parameters for the CB equivalent circuit.

Solution. The following h parameters were arbitrarily selected from the 2N2222A data sheet in Appendix A:

$h_{ie} = 250 \, \Omega$, $h_{fe} = 75$, $h_{oe} = 25 \, \mu S$, and $h_{re} = 4 \times 10^{-4}$.

(a) From Table 4.2:

$$h_{ic} = h_{ie} = 250 \, \Omega \quad h_{oc} = h_{oe} = 25 \, \mu S$$

$$h_{fc} = -(h_{fe} + 1) = -76 \quad h_{rc} = 1 - h_{re} = 1.0$$

(b) From Table 4-1:

$$h_{ib} = \frac{h_{ie}}{h_{fe} + 1} = \frac{250}{76} = 3.3 \, \Omega$$

$$h_{fb} = -\frac{h_{fe}}{h_{fe} + 1} = -\frac{75}{76} = -0.987$$

$$h_{rb} = \frac{h_{ie}h_{oe}}{h_{fe}+1} - h_{re} = \frac{250(25 \times 10^{-6})}{76} - 4 \times 10^{-4}$$

$$= -3.2 \times 10^{-4}$$

$$h_{ob} = \frac{h_{oe}}{h_{fe}+1} = \frac{25 \times 10^{-6}}{76} = 0.33 \; \mu S$$

4.3 BJT HIGH-FREQUENCY MODEL

In the previous section, we discussed the BJT small-signal models. We can use these models in the analysis of transistor amplifiers at low frequencies and have good agreement between the predictions of performance thus obtained and those actually observed in the laboratory. However, at high frequencies, the behavior of transistors cannot be predicted from these equivalent circuits. The actual transistor behavior at high frequencies is quite different from that predicted by the use of the *h*-parameter models. At high frequencies, both the output current and voltage are actually lower than the models might lead us to expect.

The reason for the decrease in both current and voltage gain inside a transistor at high frequencies is relatively simple to understand. As we saw in our discussion of diodes, the effect of a forward bias on a *PN* junction is to inject large numbers of carriers from one region into the other. If this process is modulated by an ac signal, a slightly larger or smaller number of carriers is injected at any given time, depending on whether the ac voltage adds to the dc bias voltage or subtracts from it. In any case, although there is an increase in the frequency at which the ac signal is changing, the injection level at the forward-biased junction cannot change instantaneously. Instead, an averaging of the peak variation in the ac signal occurs at high frequencies. We call this effect *capacitive*. Such a capacitance is termed *diffusion capacitance* (see Section 2.4). The base-emitter junction inside a transistor exhibits these effects.

In a reverse-biased junction (which is what the collector-base junction is) the capacitive effect is attributed to a transition region capacitance (again see Section 2.4). The incorporation of these two capacitive effects into the transistor *CE h*-parameter model is shown in Figure 4.17. The components are renamed

$$r_{be} = h_{ie} \tag{4.36}$$

$$r_{ce} = \frac{1}{h_{oe}} \tag{4.37}$$

Furthermore, since i_b is equal to v_{be}/r_{be} (Figure 4.17(b)), the output current generator can be replaced by

$$h_{fe}i_b = \frac{h_{fe}v_{be}}{r_{be}} = g_m v_{be}$$

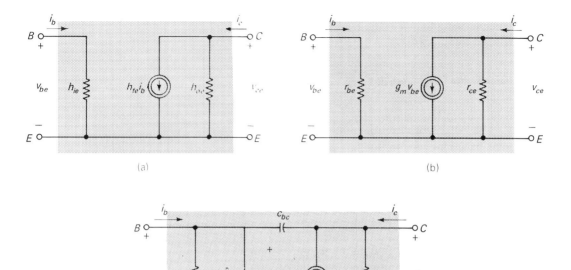

Figure 4.17 Development of the *CE* hybrid-π high-frequency model: (a) approximate low-frequency model, (b) approximate low-frequency model with relabeled parameters, (c) high-frequency hybrid-π model.

where the *transconductance* g_m is defined as

$$g_m = \frac{h_{fe}}{r_{be}} \qquad (4.38)$$

The new equivalent circuit is shown in Figure 4.17(c). It is called the *CE* hybrid-π model and it is the equivalent circuit of the BJT used for high-frequency calculations.

To verify that the CE hybrid-π model is valid at high frequencies, the short-circuit current gain is calculated. Figure 4.18(a) pictures the output shorted.

The short-circuited output current, i_o, is opposite to the current generator. Thus,

$$i_o = -g_m v_{be}$$

If the input impedance between the base and emitter terminals is labeled by z_i, then $z_i = v_{be}/i_b$ and

$$v_{be} = Z_i i_b \qquad (4.39)$$

From Figure 4.18(b) you can see that the input impedance (Z_i) is the parallel

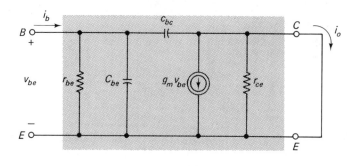

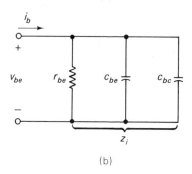

Figure 4.18 Determination of the ac short-circuit current gain: (a) the hybrid-π model with the output terminals short circuited; (b) the resulting equivalent circuit.

combination of r_{be} and the two capacitors, C_{be} and C_{bc}. Z_i may be expressed as

$$Z_i = \frac{r_{be}}{1 + j\omega r_{be}(C_{be} + C_{bc})} \quad (4.40)$$

The short-circuit current gain, $A_{isc} = i_o/i_b$, is then obtained by using Equations (4.38), (4.39), and (4.40) and it is expressed as

$$A_{isc} = \frac{-g_m r_{be}}{1 + j\omega r_{be}(C_{be} + C_{bc})} \quad (4.41)$$

In Equation (4.41), when ω is set to zero (low frequency operation), $A_{isc} = -g_m r_{be}$. Since $-h_{fe}$ is the same as $-g_m r_{be}$ (Equation (4.38)), this result provides us with the same answer at low frequencies ($\omega = 0$) as that obtained from the h-parameter circuit. Furthermore, as the frequency (ω) increases in Equation (4.41), the short-circuit current gain decreases, as is observed in the laboratory. Thus, the calculations based on the hybrid-π model agree with observed results.

β cutoff frequency. We may define a specific frequency, f_β, called the β *cutoff frequency*.

$$f_\beta = \frac{1}{2\pi r_{be}(C_{be} + C_{bc})} \quad (4.42)$$

At the β cutoff frequency (f_β), the short-circuit current gain is down 3 dB (0.707

of the gain). Although f_β is desired, the transistor manufacturer usually specifies the frequency f_T. At frequency f_T the magnitude of the short-circuit current gain is unity. The relationship between f_T and f_β is obtained by setting the magnitude of A_{isc} in Equation (4.41) equal to 1.

$$f_T \cong h_{fe} f_\beta \tag{4.43}$$

The quantity f_T is sometimes called the *gain-bandwidth product* because it is the product of the midband current gain h_{fe} and the bandwidth f_β.

Summary. With the relationships developed here, the hybrid-parameter values may be calculated from the h parameters and from the data usually provided in the manufacturers' data sheets. Such a calculation is illustrated in Example 4.8.

Example 4.8

A silicon *NPN* transistor is listed by the manufacturer as having: $C_{bc}{}^* = 12$ pF and $f_T = 50$ MHz. It has the h parameters as calculated in Example 4.6. We want to determine the hybrid-π model component values as well as f_β.

Solution. With $h_{ie} = 1.25$ kΩ, $h_{fe} = 100$, and $h_{oe} = 12.5$ μS, we have

$$r_{be} = h_{ie} = 1.25 \text{ k}\Omega$$

$$r_{ce} = \frac{1}{h_{oe}} = \frac{1}{12.5} \text{ M}\Omega \cong 80 \text{ k}\Omega$$

$$g_m = \frac{h_{fe}}{r_{be}} = \frac{100}{1.25} \text{ mS} \cong 80 \text{ mS}$$

$$f_\beta = \frac{f_T}{h_{fe}} = \frac{50}{100} \text{ MHz} \cong 500 \text{ kHz}$$

$$C_{be} + C_{bc} = \frac{1}{2\pi f_\beta r_{be}} = \frac{1}{2\pi(5 \times 10^5)(1.25 \times 10^3)} \text{ F} \cong 250 \text{ pF}$$

$$C_{be} = 250 - C_{bc} = 250 - 12 \cong 238 \text{ pF}$$

Finally, in order for a BJT to operate as an amplifier, it first must be properly biased. Once a proper and stable operating point is achieved, the transistor amplifier may be analyzed by replacing the transistor symbol in the circuit by the appropriate small-signal model. At low frequencies (say, up to a few hundred kHz) the h-parameter model is applicable; at higher frequencies, the hybrid-π model is used.

---

* C_{bc} is sometimes listed as $C_{b'c}$.

REVIEW QUESTIONS

1. What is meant by the phrase "biasing a transistor"?
2. How can neglecting one quantity with respect to another be justified?
3. How can we justify neglecting V_{BE} in the bias circuit shown in Fig. 4.1?
4. How can we justify *not* neglecting V_{BE} in the bias circuit of Fig. 4.2?
5. Why is the bias circuit shown in Fig. 4.1 called the "fixed-bias circuit"? What is "fixed" in this circuit?
6. How does the circuit shown in Fig. 4.1 bias the base-emitter junction? How does it bias the collector-base junction?
7. What is *thermal runaway*?
8. How is the thermal stability of a bias circuit measured?
9. How does the self-bias circuit achieve thermal stability? Explain.
10. On what does the thermal stability of the fixed-bias circuit depend?
11. On what does the thermal stability of the self-bias circuit depend?
12. What are the similarities and differences in designing and analyzing a bias circuit?
13. How is the dc load line plotted on the output characteristics?
14. What is meant by low-frequency model?
15. What are the factors that determine which model of the BJT (*CE*, *CC*, or *CB*) is appropriate in a specific application?
16. What factors influence the values for the *CE h* parameters of a BJT? List and explain why.
17. What are the ways in which the *CE h* parameters of a particular BJT can be determined?
18. What are the reasons why the *h*-parameter model of the BJT is not valid at high frequencies?
19. Under what conditions are the *h*-parameter and hybrid-π models used?
20. What is the gain-bandwidth product for the BJT? How is it related to the β cutoff frequency?

PROBLEMS

1. Repeat Ex. 4.1 for a silicon transistor with a minimum β of 20 and a maximum β of 100.
2. Design the bias circuit of Fig. 4.1 for an operating point of $V_{CEQ} = 7$ V and $I_{CQ} = 1$ mA if the transistor used has a β of 75 and the supply voltage is 15 V.
*3. What is the highest value of R_C that can be used in Ex. 4.1 without causing the transistor to saturate? (Note: The minimum V_{CE} is approximately 0; V_{CE} cannot be negative for an *NPN* transistor.)
4. The circuit in Fig. 4.1 is used to bias a *PNP* transistor with a $\beta = 50$. The

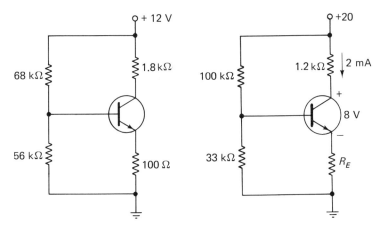

Figure 4.19 Circuit for Problems 5 and 9.

Figure 4.20 Circuit for Problems 7 and 9.

supply voltage is -20; $R_B = 200$ kΩ, $R_C = 1.8$ kΩ. Determine the operating point and compare your answers with those in Ex. 4.1.

5. Determine the operating point for the circuit in Fig. 4.19 if the transistor has a β of 40.
6. Repeat Problem 5 if the transistor β is 80.
7. Determine the value of R_E needed to set up the conditions shown in Fig. 4.20. Also determine the β of the transistor.
*8. What is the thermal stability factor for the circuit in Problem 4?
9. What is the thermal stability factor for the circuit in Fig. 4.19? In Fig. 4.20?
10. Design the bias circuit of Fig. 4.2 for the same Q-point as in Ex. 4.4 but with $S = 5$.
11. Repeat Ex. 4.5 with a β spread from 20 to 130. Compare your answers with those in Ex. 4.5 and make a conclusion.
12. Draw the load line for the circuit values in Ex. 4.2. Repeat Ex. 4.2 and draw the load line if the transistor β is 75.
*13. Determine h_{ie} for the transistor whose input characteristics are given in Fig. 4.9 for an operating-point base current of 10 μA.
14. Determine h_{fe} and h_{oe} for the transistor whose output characteristics are given in Fig. 4.10 if the Q-point is at $I_{CQ} = 1$ mA and $V_{CEQ} = 10$ V.
15. From the CE h parameters in Ex. 4.6, determine the CB h parameters.
16. From the CE h parameters in Ex. 4.6, determine the CC h parameters.
*17. A transistor is listed as having $C_{bc} = 10$ pF and $f_T = 100$ MHz. Its h parameters are $h_{ie} = 800$ Ω, $h_{fe} = 65$, $h_{oe} = 50$ μA/V (h_{re} negligible). Determine the parameters for the hybrid-π model.
18. For the transistor in Problem 17, what is the frequency at which the magnitude of the short-circuit current gain is down 3 dB from its low-frequency value?

Chapter 5

Field-Effect Transistor Operation, Biasing, and Models

This chapter introduces both the junction field-effect transistor (JFET), Figure 5.1(a), and the metal-oxide-semiconductor field-effect transistor (MOSFET), Figures 5.1(b) and 5.1(c). The construction and operation of the JFET and the MOSFET is examined so that the terminal parameters and characteristics might be understood. Different biasing schemes and methods for overcoming operating-point instabilities are covered. The small-signal model is developed using the FET terminal characteristics. Finally, the low-frequency model is modified to extend its use to high frequencies, as was done for the bipolar junction transistor (BJT).

Both the JFET and the MOSFET are voltage-controlled devices. Their operation is governed by an electric field that controls the *electron current* in an *N*-channel FET or the *hole current* in a *P*-channel FET. As indicated, the FET is a unipolar device using either holes (*P*-channel) or electrons (*N*-channel) as carriers.

Because the electric field is set up across a reverse biased *PN junction gate* in the JFET and an *insulated gate* in the MOSFET, the input terminal of the FET is isolated from the rest of the circuit. As a consequence, the FET input resistance is about 10^8 ohms for the JFET and about 10^{12} ohms for the MOSFET.

In contrast, the BJT is a current controlled device using both electrons and holes for conduction and has a very low input resistance ($\approx 10^3$ ohms). This low resistance is due to the forward biased *PN* junction of the BJT's base emitter. Our study of the FET will begin with the junction field-effect transistor (JFET).

5.1 JUNCTION FIELD-EFFECT TRANSISTOR

The structure of the two types of depletion mode (type A) JFETs is depicted in Figure 5.2. In Figure 5.2(a), an *N*-type semiconductor is formed and leads are attached to each end. One end is called the *source*; the other end is called the *drain*. A very narrow *P*-type region is diffused around the *N*-type semiconductor, forming an *N*-channel. This region of *P*-type semiconductor is called the *gate*.

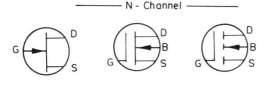

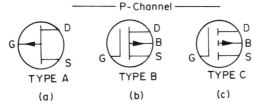

Figure 5.1 Schematic symbols for FET types: (a) JFET type A depletion mode only; (b) MOSFET type B depletion/enhancement mode; (c) MOSFET type C enhancement mode only. NOTE: G = gate; S = source; D = drain; B = bulk (substrate).

The third external connection is made to this gate region. Thus, the N-channel JFET consists of a single PN junction that is formed by the N-type channel and the P-type gate. In the P-channel JFET pictured in Figure 5.2(b), the channel is P-type semiconductor material and the gate is N-type.

5.1.1 Depletion Region

The following discussion is confined to the N-channel JFET. However, the P-channel JFET is the complement of the N-channel JFET, and the P-channel JFET is operated by reversing the polarities of all the bias supplies.

Under typical operating conditions, as noted in Figure 5.3, conventional current flows from the drain to the source due to the application of the potential V_{DD} across the drain and source. To control the drain current (I_D), the gate-source PN junction is reverse biased by the bias potential V_{GG}. As a result of the reverse bias created by V_{GG}, a depletion region is created that extends into the channel as pictured. Note that the JFET is only operated in the depletion mode.

The shape of the depletion region is governed by the amount of reverse bias on the gate-source PN junction. If the drain is positive with respect to the source, the net reverse bias on the PN junction near the drain end of the channel is larger

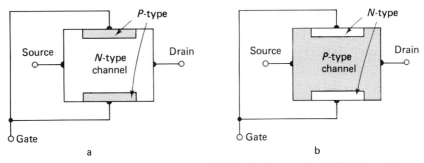

Figure 5.2 Structural representation of JFET: (a) N-channel and (b) P-channel.

Sec. 5.1   Junction Field-Effect Transistor

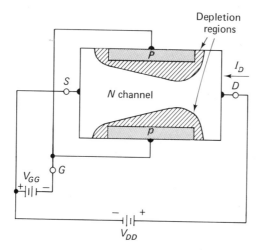

Figure 5.3 N-channel JFET with bias sources V_{DD} and V_{GG} connected. Notice that the gate-source PN junction is reverse biased.

than it is near the source end. This results because the net voltage between the drain and gate is $V_{DS} + V_{GS}$. Therefore, the depletion region extends deeper into the channel toward the drain end of the channel.

5.1.2 Channel Resistance

The effective resistance between the drain and source is called the *channel resistance*. The channel resistance depends on the resistivity of the channel and its volume. Note from Figure 5.3 that the depletion region decreases the effective volume of the channel, thus increasing the effective resistance drain to source. If we consider that the drain-to-source voltage V_{DD} is fixed at some nonzero value, then the drain current (I_D) is directly proportional to the channel resistance. Because the channel resistance depends on the size of the depletion region, the gate-source voltage may be used to control the drain current. Increasing the reverse bias (more negative) from gate-to-source causes the depletion region to occupy more of the channel, thus increasing the channel resistance, which in turn causes the drain current to decrease. Conversely, lowering the amount of reverse bias (less negative) from gate-to-source decreases the channel resistance and increases the drain current.

5.1.3 Pinchoff Voltage

As long as the gate-source junction is reverse biased, the gate conducts very little current. Thus, the gate input resistance is very high (10^8 ohms).

If a relatively low voltage is applied across the drain and source, and if the gate is made increasingly negative, then eventually the depletion region will *cut off* the channel as shown in Figure 5.4. Under these conditions, the channel resistance is extremely high, and, for all practical purposes, the drain current (I_D) is zero. The gate voltage at which this condition occurs is called the *pinchoff voltage*, V_p.

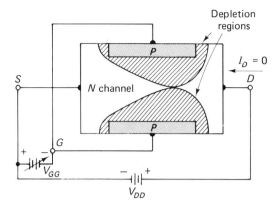

Figure 5.4 Cutoff in a JFET.

5.1.4 Constant Resistance and Constant Current Regions

If the gate-source voltage (V_{GS}) of the JFET is kept constant at some value below the pinchoff voltage while the drain-to-source voltage is increased from zero, then the resistance of the channel is essentially constant for V_{DS} equal to a few tenths of a volt. Moreover, the drain current increases linearly with increasing V_{DS}. However, at a critical value of V_{DS}, given by $V_{DS} = V_{GS} - V_p$, the separation between the depletion regions reaches a minimum as indicated in Figure 5.5(a) by the labeled w. Any further increase in the drain-to-source voltage causes more of the channel on the source end to reach this minimum width w, as noted in Figure 5.5(b). As more of the channel reaches the minimum width, the resistance

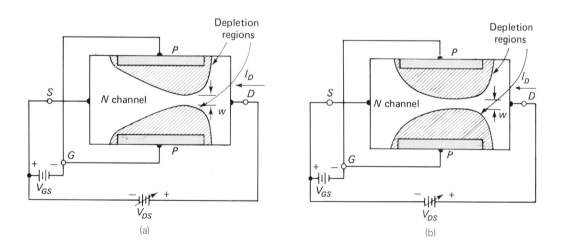

Figure 5.5 Current saturation in a JFET (a) for $V_{DS} = V_{GS} - V_p$ and (b) for $V_{DS} > V_{GS} - V_p$.

of the channel is increasing at approximately the same rate that V_{DS} is increasing. As a result, the drain current remains essentially unchanged even though V_{DS} is increased. Thus, this mode of operation provides an essentially constant current over a relatively large range of V_{DS}. The magnitude of the current is controlled only by the gate-source voltage.

5.1.5 Saturation Drain Current

If the gate-source voltage is zero, the channel is *pinched off* when the drain-to-source voltage reaches V_p. That is, with the gate tied to the source ($V_{GS} = 0$ V), the channel has reached the minimum width, w, when $V_{DS} = V_p$. Under these conditions, the drain current is called the *saturation drain current, I_{DSS}*. An approximate relationship between the saturation drain current and the drain current, I_D, for any value of V_{GS} between zero and V_p is given by Equation (5.1). Thus,

$$I_D = I_{DSS}\left(1 - \frac{V_{GS}}{V_p}\right)^2 \tag{5.1}$$

5.1.6 Transfer Curve

When Equation (5.1) is plotted, the JFET *transfer characteristic curve* of Figure 5.6 is created. This curve shows the relationship between the input voltage (V_{GS}) and the output current (I_D). The JFET transfer characteristic pictured in Figure 5.6 has a saturation drain current of 6 mA ($I_{DSS} = 6$ mA) and a pinchoff voltage of -4 V ($V_p = -4$ V). As seen from the curve of Figure 5.6, the transfer between the input voltage and output current is not linear. For example, when the input voltage changes by one volt from -3 to -2 V, the drain current changes approximately 1 mA. However, if the input voltage changes from -2 to -1 V (again a one-volt change), the drain current changes about 2 mA.

Summary. The operation of the JFET is summarized with the aid of Figure 5.7. From the *drain characteristic* of Figure 5.7 we learn that the JFET has two

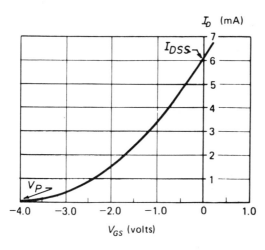

Figure 5.6 Typical *N*-channel JFET transfer characteristic.

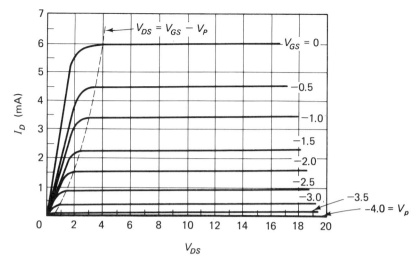

Figure 5.7 Typical N-channel JFET drain characteristic.

distinct operational regions. The region to the left of the dashed line, called the *constant resistance* region, is for low drain-to-source voltages. With the JFET in this mode of operation, the channel resistance is controlled by the amount of reverse bias at the gate. The region to the right of the dashed line, called the *constant current* region, has a V_{DS} larger than a few tenths of a volt. This mode of operation is characterized by drain currents that are *constant currents* and are only a function of the reverse bias at the gate. The drain currents are essentially independent of V_{DS} for values of $V_{DS} > V_{GS} - V_p$. The voltage-controlled resistance of the JFET has definite applications. However, the most common use of the JFET is in the constant current region of the drain characteristic—the "normal operating region."

The drain characteristic of the JFET (Figure 5.7) is a plot of drain current, I_D, as a function of drain voltage, V_{DS}, for different values of gate voltage, V_{GS}. This curve may be generated with a curve tracer. Note the similarity between the *CE* output characteristic of the BJT and that of the JFET. The most apparent difference is the control of the output current, I_D, which is controlled by the input voltage V_{GS}. In the BJT the output current, I_C, is controlled by the input current, I_B.

The JFET may be operated with the gate-source junction slightly forward biased as long as the gate does not draw appreciable current. The forward bias voltage is low, typically below 0.5 V.

5.2 BIASING THE JFET

5.2.1 Self-Bias

The analysis and design of JFET bias circuits are very similar to their BJT counterparts. The bias circuit must establish the proper voltages for operating in the

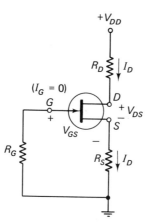

Figure 5.8 N-channel JFET self-bias circuit.

normal active region. A JFET self-bias circuit is illustrated in Figure 5.8. For this N-channel JFET, the supply voltage V_{DD} is positive in order to set up a positive voltage between drain and source. Gate bias is accomplished by the voltage drop across R_S caused by the drain current. Equating the voltage supplied (V_{DD}) to the sum of the voltage drops in the output circuit, we obtain the load line equation:

$$V_{DD} = I_D(R_D + R_S) + V_{DS} \qquad (5.2)$$

Because of the reverse bias between the P-type gate and the N-type channel, essentially no gate current flows. Thus, in the gate circuit, there is no dc voltage drop across R_G. Adding the voltages in the gate circuit gives us the *bias curve* equation for the circuit in Figure 5.8:

$$V_{GS} = -I_D R_S \qquad (5.3)$$

Equations (5.1), (5.2), and (5.3) may be used to determine the operating point for a JFET in Figure 5.8 if we assume that the JFET characteristics as well as circuit component values are known. This procedure is illustrated in Example 5.1.

Example 5.1

An N-channel JFET with $I_{DSS} = 5$ mA and $V_p = -4$ V is used in the self-bias circuit of Fig. 5.8. The circuit values are: $V_{DD} = 12$ V, $R_D = 2.2$ kΩ, and $R_S = 470$ Ω. Determine the operating point values of V_{DS}, I_D, and V_{GS}.

Solution. Using the given values of I_{DSS} and V_p and Eq. (5.1), we make a plot of the transfer characteristics. Next choose values of V_{GS} and calculate I_D in Eq. (5.1). Note that the plot of the transfer characteristics is shown in Fig. 5.9.

Finally, plot the bias curve, Eq. (5.3), on the transfer characteristics by plotting two points. Point 1: For $V_{GS} = 0$, I_D is also zero. Point 2: For $V_{GS} = -2$ V, $I_D \cong 2/0.47$ mA $\cong 4.25$ mA. The bias curve is now plotted as shown in Fig. 5.9. The Q-point is obtained at the intersection of the

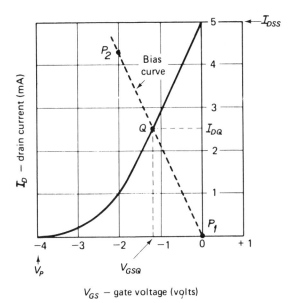

Figure 5.9 Graphical determination of the Q-point of Example 5.1.

transfer curve and the bias curve. Thus, $I_{DQ} \cong 2.5$ mA and $V_{GSQ} \cong -1.2$ V.

The drain-source voltage at the Q-point is obtained from Eq. (5.2), the load line equation:

$$V_{DSQ} = V_{DD} - I_{DQ}(R_D + R_S) \cong 12 - (2.5)(2.2 + 0.47) \cong 5.3 \text{ V}$$

The operating point values are

$$V_{DSQ} \cong 5.3 \text{ V}, \qquad I_{DQ} \cong 2.5 \text{ mA}, \qquad V_{GSQ} \cong -1.2 \text{ V}$$

5.2.2 Voltage-Divider Bias

If we examine the problem of biasing a JFET, we notice that, for a given JFET, manufacturers usually specify a range of values for both I_{DSS} and V_p. The self-bias circuit provides little flexibility for accommodating such a spread in the characteristics. However, we may use the circuit shown in Figure 5.10 to minimize the uncertainty in the Q-point because of the uncertainty in the FET parameters.

In the bias circuit of Figure 5.10, the voltage divider set up by resistors R_{G1} and R_{G2} makes the voltage from gate to ground positive. In order to reverse bias the gate-source junction, the voltage drop across R_S must be larger than the open-circuit voltage (Thevenin's voltage) across R_{G2}. The open-circuit voltage across R_{G2} is defined as V_{GG} and is determined by Equation (5.4).

$$V_{GG} = V_{DD} \frac{R_{G2}}{R_{G1} + R_{G2}} \tag{5.4}$$

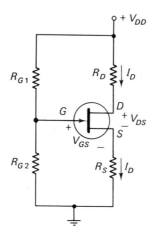

Figure 5.10 Voltage-divider bias circuit for the JFET.

To complete the Thevenization process, the equivalent resistance to the left of the gate, labeled R_G, is determined with Equation (5.5).

$$R_G = \frac{R_{G1} R_{G2}}{R_{G1} + R_{G2}} \tag{5.5}$$

The resulting equivalent circuit with the voltage divider replaced by its Thevenin equivalent is illustrated in Figure 5.11. From the output circuit, we obtain the load line equation, Equation (5.6).

$$V_{DD} = I_D(R_D + R_s) + V_{DS} \tag{5.6}$$

Notice that Equation (5.6) is the same as Equation (5.2), which is used with the circuit in Figure 5.8. Also, notice the similarities between the analysis of this bias circuit for the JFET and the analysis used for BJT circuit of Figure 4.2.

Once again we assume the gate-source junction to be reverse biased, so that no gate current flows. Since there is no gate current, there is no dc voltage drop across R_G of Figure 5.11. The summation of voltages in the gate circuit of Figure 5.11 once again yields the bias curve equation, Equation (5.7).

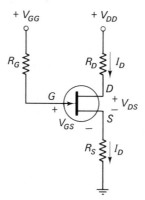

Figure 5.11 Equivalent circuit for the voltage-divider circuit shown in Figure 5.10.

$$V_{GG} = V_{GS} + I_D R_S \tag{5.7}$$

The additional flexibility in the stabilization of the Q-point offered by the voltage divider bias circuit of Figure 5.10 is illustrated in Example 5.2.

Example 5.2

The bias circuit shown in Fig. 5.10 has the following circuit values: $V_{DD} = 25$ V, $R_{G1} = 470$ kΩ, $R_{G2} = 150$ kΩ, $R_D = 3.3$ kΩ, and $R_S = 3.9$ kΩ. The manufacturer lists the N-channel JFET as having I_{DSS} between 2 and 5 mA and $V_{GS(OFF)}$ (another way of labeling V_p) between -4 V and -2 V. We want to determine the worst-case limits on the operating-point voltages and current.

Solution. The first step is to plot the worst-case transfer curves that we might encounter. This is done by pairing the two maximum values of I_{DSS} and V_p to obtain the maximum transfer curve and by pairing the minimum values of I_{DSS} and V_p to obtain the minimum transfer curve. The two plots are accomplished in the same manner as in Ex. 5.1: They are depicted in Fig. 5.12. They are significant because they provide the limits for the transfer curve that we might have for any JFET of the type specified. For example, a given JFET may have an I_{DDS} of, say, 4.5 mA and a pinchoff voltage of -3 V. The transfer curve for this particular FET would lie between the two worst-case transfer curves shown in Fig. 5.12 and would therefore yield a Q-point within the limits to be determined below for the worst-case transfer curves.

The next step is to plot the bias curve of Eq. (5.7). We find one point

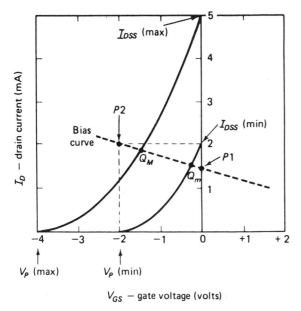

Figure 5.12 Graphical evaluation of worst-case Q-point for the JFET voltage divider bias circuit.

by setting $V_{GS} = 0$ to obtain $I_D \cong V_{GG}/R_S$. From Eq. (5.4) we determine

$$V_{GG} \cong 25 \, \frac{150}{470 + 150} \, \text{V} \cong 6 \text{ V}$$

Thus, when $V_{GS} = 0$, $V_{GG} = 6$ V, and $R_S = 3.9$ kΩ, I_D in Eq. (5.7) is

$$I_D \approx \frac{6}{3.9} \, \text{mA} \cong 1.5 \text{ mA}$$

and point 1 in Fig. 5.12 is (0, 1.5 mA).

For the second point needed to plot the bias curve, a V_{GS} value of -2 V is selected. The drain current is then computed, using Eq. (5.7), for $V_{GG} = 6$ V, $V_{GS} = -2$ V, and $R_S = 3.9$ kΩ.

$$I_D \cong \frac{6 - (-2)}{3.9} \, \text{mA} \cong 2 \text{ mA}$$

and point 2 in Fig. 5.12 is (-2 V, 2 mA).

The bias curve is drawn between points 1 and 2 as shown in Fig. 5.12. The highest worst-case operating point Q_M is the intersection of the bias curve and the highest worst-case transfer curve, as shown. Similarly, the lowest worst-case operating point Q_m is obtained at the intersection of the bias curve and the minimum worst-case transfer curve, as shown.

Under these worst-case conditions, the operating point will be between the following limits:

for Q_M: $I_{DM} \cong 1.9$ mA and $V_{GSM} \cong -1.4$ V
for Q_m: $I_{Dm} \cong 1.6$ mA and $V_{GSm} \cong -0.3$ V

The corresponding limits on V_{DS} are determined from the load line equation, Eq. (5.6). Note that when I_D is a minimum, V_{DS} will be a maximum and vice versa.

$$V_{DSM} \cong 25 - (1.6)(3.3 + 3.9) \cong 13.5 \text{ V}$$

$$V_{DSm} \cong 25 - (1.9)(3.3 + 3.9) \cong 11.3 \text{ V}$$

To summarize, any JFET with I_{DSS} and V_p within the limits given, when used in the bias circuit of Fig. 5.10 (with the values listed) will have an operating-point drain current between 1.0 and 1.6 mA, V_{DS} between 11.3 and 13.5 V, and V_{GS} between -0.3 and -1.4 V.

We can design the bias circuit shown in Figure 5.10 so that we can have specific variation in the Q-point drain current for a specified variation in the JFET parameters. This design procedure is illustrated in Example 5.3.

Example 5.3

The N-channel JFET to be used in the bias circuit of Fig. 5.10 has the following parameters: I_{DSS} between 3 and 8 mA and V_p between -2 and

−6 V. It is to be designed in such a way that if the I_{DSS} and V_p are within the limits given, the operating-point drain current will not be lower than 2.5 mA and not higher than 3.5 mA. The supply voltage, V_{DD}, is 20 V and R_D = 1.8 kΩ. The remaining circuit components. (R_S, R_{G1}, and R_{G2}) will be specified in the course of the design.

Solution. For the worst-case values of I_{DSS} and V_p given, the two transfer curves are plotted as shown in Fig. 5.13. Using the limits on the Q-point drain current of 3.5 mA maximum and 2.5 mA minimum, we project the transfer curves to obtain the worst-case operating points Q_M and Q_m, as shown.

We next have to choose a bias curve that will satisfy the requirements. We draw a straight line passing *below* Q_M and *above* Q_m, as indicated in Fig. 5.13. If the bias curve thus obtained is extended until it intersects the V_{GS} axis, the intercept yields the needed value of V_{GG}. In this case, it is 6 V. This calculation may be verified by noting that in the bias curve equation [Eq. (5.7)], when I_D is 0, $V_{GS} = V_{GG}$.

R_S is evaluated from the intersection of the bias curve and the I_D axis. When V_{GS} is 0, we see from the bias curve equation, Eq. (5.7), that

$$R_S \cong \frac{V_{GG}}{I_D} \cong \frac{6}{2.1} \text{ k}\Omega \cong 2.8 \text{ k}\Omega$$

Next we choose R_{G1} and R_{G2} to give the desired value of $V_{GG} = 6$ V

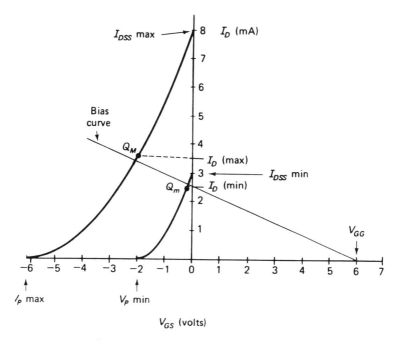

Figure 5.13 Designing for a specified maximum variation in the Q-point drain current.

when $V_{DD} = 20$ V. One possible set of values that will provide the right V_{GG} is:

$$R_{G1} \cong 680 \text{ k}\Omega \text{ and } R_{G2} \cong 330 \text{ k}\Omega$$

The answers obtained in this design example are not unique. There are many combinations of circuit values that would provide the desired performance. This is usually the case in most design problems. Finally, the operation of the JFET amplifier is considered in Sections 9.3, 10.2, and 11.1.

5.3 MOSFET

5.3.1 Depletion-Mode MOSFET

In the previous sections, you have learned that the type A FET, the JFET, is only operated in the depletion mode by reverse biasing the gate-source *PN* junction. If the gate of the JFET is operated in the enhancement-mode by forward biasing the gate-source *PN* junction, then considerable gate current will flow and the JFET parameters will be radically changed. However, if the gate is insulated from the channel between the source and drain, then no *PN* junction is formed and the FET may be operated with either a forward or reverse gate-to-source bias voltage. The insulated-gate FET (IGFET) is a type B depletion/enhancement-mode *metal-oxide-semiconductor* or, as it is commonly called, a *depletion-mode* MOSFET.

As pictured in Figure 5.14, the structure of the depletion/enhancement *N*-channel MOSFET has a lightly doped N^- type semiconductor implanted under the oxide (insulation) in the surface of the channel. The silicon dioxide insulation is an elemental glass that separates the metal-gate from the channel. The drain and source terminals are connected to the heavily doped *wells* of N^+ type semi-

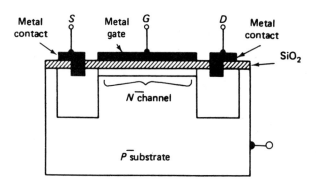

Figure 5.14 Type B *N*-channel depletion/enhancement-mode MOSFET transistor structure where N^+ indicates a region highly doped with donor atoms, N^- indicates a region lightly doped with donor atoms, and P^- indicates a region lightly doped with acceptor atoms.

conductor on either side of the channel. In the course of manufacturing the depletion/enhancement MOSFET, the lightly doped P^- type substrate is usually electrically connected to the source to produce a three-terminal device. Because some applications require added drain current control, the substrate lead is sometimes left unconnected so that a voltage may be applied between the substrate and the source.

The type B MOSFET as previously mentioned is a depletion/enhancement-mode device and it is easily recognized by its distinctive schematic symbol, transfer characteristic, and drain characteristic, as shown in Figure 5.15. To operate the device in the enhancement mode, positive voltage is applied to the gate. The N-type channel resistance decreases because additional electrons from the drain and source regions are attracted into the channel. As a result, the drain current increases. When operated in the depletion mode, a negative voltage is applied to the gate, thus causing a portion of the channel to be depleted because of the positive charge induced within the channel; as a consequence, the drain current decreases. As shown in Figure 5.15(a), with the gate-source voltage (V_{GS}) set to zero, the type B MOSFET is ON ($I_D \approx 2.2$ mA). Also notice that, like the JFET, the depletion/enhancement-mode MOSFET has a nonlinear transfer characteristic between the input voltage, V_{GS}, and the output current, I_D.

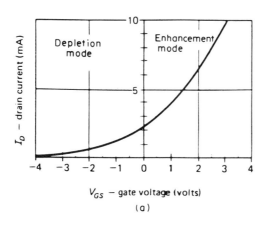

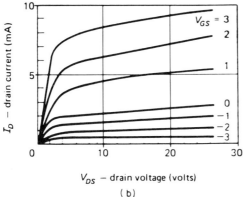

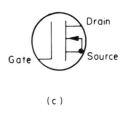

Figure 5.15 Typical N-channel depletion/enhancement-mode type B MOSFET: (a) transfer characteristic, (b) drain characteristic, (c) schematic symbol with substrate internally connected to the source.

Sec. 5.3 MOSFET

5.3.2 Enhancement-Mode MOSFET

The type C *enhancement-mode* MOSFET, shown structurally in Figure 5.16(a), contains no channel between the source and drain. However, applying a positive voltage to the gate causes free electrons from the substrate, the source, and the drain regions to be attracted into the region just below the gate, thus forming an *N*-type channel. The formation of an *N*-channel is pictured in Figure 5.16(b). We call this type of operation *enhancement*, because the number of free electrons in the region below the gate is greatly enhanced (i.e., increased) by the positive bias voltage on the gate. Applying a negative bias voltage to the gate causes no induced channel. Therefore, the enhancement-mode MOSFET can only be operated with positive gate bias.

Typical enhancement-mode MOSFET characteristics along with the schematic symbol are given in Figure 5.17. Note that a minimum positive gate voltage, called the *threshold voltage* (V_T), is necessary in order to induce enough of a

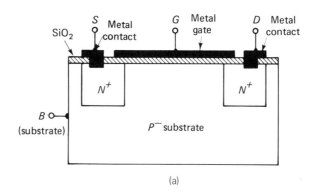

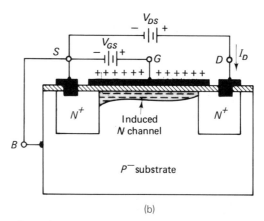

Figure 5.16 Type C *N*-channel enhancement-mode MOSFET: (a) transistor structure and (b) induced channel as a result of positive bias voltage applied to the gate.

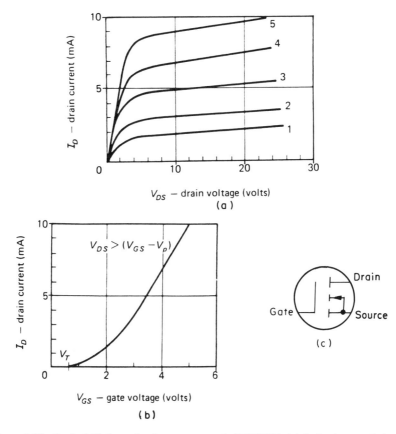

Figure 5.17 Typical N-channel enhancement-mode MOSFET: (a) drain characteristic, (b) transfer characteristic, (c) schematic symbol with substrate internally connected to the source.

channel to form a conductive path between the drain and source. If the gate voltage is less positive than the gate threshold voltage, no drain current can flow. Typically the threshold voltage is between 1 and 5 volts for the type C enhancement-mode MOSFET. Unlike the type A and B FETs that are *normally ON*, the type C FET is *normally OFF*. The schematic symbol of Figure 5.17(c) has a broken channel to indicate the absence of a permanent channel.

5.3.3 Dual-gate MOSFET

The combination of metal, insulating layer, and semiconductor form an equivalent parallel plate capacitance. The equivalent capacitance between drain and gate has the largest effect on the MOSFET high-frequency performance. To minimize this capacitive effect, a second gate region is introduced between the control gate and the drain, as shown in Figure 5.18. The resulting device is called a *dual-gate* MOSFET. Normally, the second gate is biased positively in order to reduce the

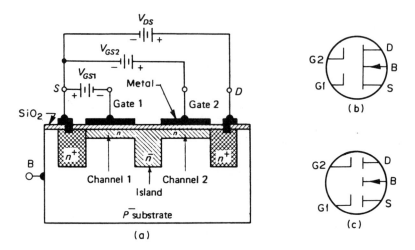

Figure 5.18 N-channel dual-gate MOSFET: (a) type B transistor structure, (b) type B circuit symbol, (c) type C circuit symbol.

resistance of the second channel and, more importantly, to break up the effective drain-to-gate capacitance. In all other respects, the operation of the dual-gate MOSFET is identical to that of the single-gate MOSFET, with gate 1 used as the input.

5.3.4 VFET

The vertical enhancement-mode MOSFET, commonly called the VFET, is a discrete power MOSFET transistor. The device acquires its name from the cross-sectional shape of the etched cavity, as pictured in Figure 5.19(a). The wide channel, which extends around the four-sided perimeter of the cavity, gives the VFET an ability to carry a large drain current. Furthermore, the short channel length (vertical dimension) provides for high switching speed.

VFETS have all the desirable properties of type C enhancement-mode MOSFETs, including good thermal stability and high input resistance. In addition, the VFET can handle substantially more power (>20 W) than the JFET or the MOSFET (<1 W). VFETs have replaced power BJT in some circuit applications. In these cases, the VFETs find application in analog systems as power amplifiers and in digital systems as high speed switches controlling power into heavy loads. A typical VFET transfer characteristic is shown in Figure 5.19(b). When compared to the MOSFET transfer characteristic of Figure 5.17(b), we see a much higher VFET drain current as well as a more linear transfer characteristic.

5.3.5 HEXFET

HEXFET® is the registered trademark for International Rectifier Corporation's power MOSFET. It is an advanced discrete enhancement mode power MOSFET that uses a planar (non-V) structure which conducts current vertically. From Figure 5.20(a), we learn that conventional drain current flows vertically up through the *N*-type drain region, then horizontally through the induced channel in the *P*-type material, then once again vertically out through the *N*-type source material.

The *reverse body-drain diode,* shown in Figure 5.20(b), may or may not be used in a particular application. However, it is turned ON when the applied drain-source voltage changes polarity. The diode then conducts with a current-handling capacity equal to that of the HEXFET transistor.

The full line of HEXFETs are available in both N-channel and P-channel devices with power ratings up to 150 W. The HEXFET has made real inroads in replacing power BJTs in many applications, including switching power supplies, motor controllers, choppers, audio amplifiers, and pulse amplifiers.

Like other MOSFET designs, the HEXFET has advantages over the BJT. Some of these advantages are: (1) no secondary breakdown; (2) very low input

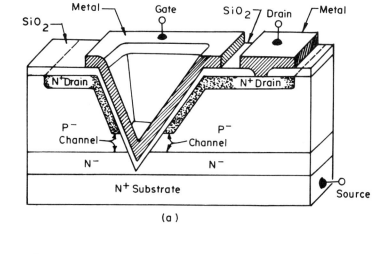

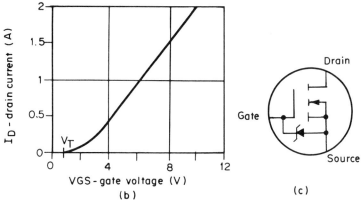

Figure 5.19 *N*-channel VFET: (a) cross-section of the transistor structure, (b) transfer characteristic, (c) schematic symbol.

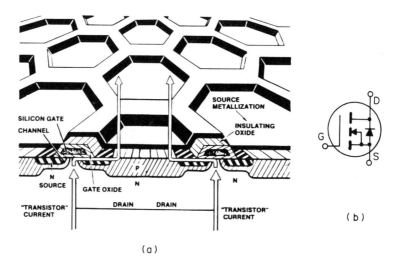

Figure 5.20 *N*-channel HEXFET: (a) cross-section of the transistor structure and (b) the schematic symbol. (*Courtesy of International Rectifier*)

currents from the signal source; (3) fast switching times; (4) ease of paralleling for higher power; and (5) non-temperature-dependent gain and response time parameters.

5.3.6 Electrostatic Discharge

Electrostatic discharge (ESD) is the discharge of static electricity that takes place when a static charge is present in an object. When an electronic component is part of the discharge path or is in the vicinity of a static charge, its parameters may be permanently changed or it may be destroyed.

MOSFETs of all types are very ESD-sensitive and must be handled carefully to prevent their destruction. To avoid possible damage to MOS devices you should: keep leads shorted together with the metal ring supplied by the manufacturer; pick up the device by the case instead of the leads; avoid inserting or removing devices from a circuit with the power ON; transport or store MOSFETs only in closed conductive containers.

Even though some MOSFETs contain protective diodes inside the device, as pictured in Figure 5.19(c), it is important that all personnel who handle MOSFETs be grounded using an OSHA-approved ground strap. Furthermore, grounded static dissipative table mats or table tops must be used when circuits containing MOSFETs are assembled and/or repaired. Finally, all soldering must be done with grounded soldering irons.

Remember that the insulating oxide layer (SiO_2) between the gate and the channel is extremely thin and even small static charges are sufficient to permanently change device parameters. Large static charges can destroy MOSFETs by fusing metal, rupturing oxide layers, and melting silicon.

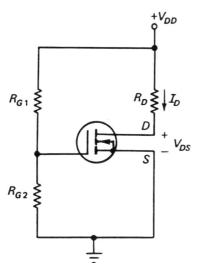

Figure 5.21 An N-channel enhancement-mode MOSFET bias circuit.

5.4 BIASING THE MOSFET

The JFET bias circuit shown in Figure 5.10 may also be used for biasing a MOSFET. As we discussed in the previous section, the enhancement mode MOSFET must have a positive gate-source voltage. Therefore, in the bias circuit of Figure 5.10, the voltage developed across R_S must be somewhat smaller than that developed across R_{G2}. An alternate bias circuit for the enhancement mode MOSFET is indicated in Figure 5.21. This circuit has R_S omitted; therefore, the gate-source voltage is equal to the open-circuit voltage across R_{G2}, which is always positive.

The type B depletion mode MOSFET may be operated without any gate bias or with either a negative or positive gate bias. Both of the JFET bias circuits, Figures 5.8 and 5.10, as well as the enhancement mode MOSFET bias circuit, Figure 5.21, may be used for biasing a depletion mode MOSFET.

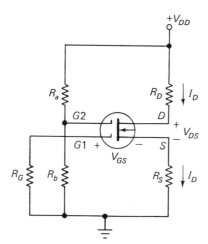

Figure 5.22 Dual-gate N-channel depletion-mode MOSFET bias circuit.

Sec. 5.4   Biasing the MOSFET   101

The analysis as well as design of MOSFET biasing circuits is almost identical to that for the JFET. Consequently, we can use the same methods and procedures for the MOSFET circuits.

The dual-gate MOSFET may be biased as indicated in Figure 5.22. The second gate can have a positive gate voltage with respect to the source if we choose the voltage divider resistors R_a and R_b so that the open-circuit voltage across R_b is larger than the voltage drop across R_S. The signal gate (gate 1) can be negatively biased with respect to the source by the voltage drop across R_S. The basic bias circuit of Figure 5.8 is used for the bias of the signal gate.

5.5 SMALL-SIGNAL FET MODEL

The FET (JFET or MOSFET) may be operated as an amplifier by applying a small time-varying signal to the gate and taking the amplified signal at the drain. The common terminal for both the input and output is the source.

Note that the basic similarity in the operation of the different FETs discussed is indicated by the similarity in their terminal characteristics. As a result, we would expect that the ac small-signal models for the different FETs would be similar. In fact, the only difference between the ac performance of different FETs is in the slightly different magnitudes of some of the parameters.

The low-frequency small-signal model for the FET is shown in Figure 5.23. At the drain terminal, the sum of the currents yields

$$i_d = g_m v_{gs} + \frac{1}{r_{ds}} v_{ds} \tag{5.8}$$

where the FET small-signal parameters g_m and r_{ds} are defined and evaluated from

$$g_m \equiv \text{transconductance} \equiv \frac{\text{change in } I_D}{\text{change in } V_{GS}} \text{ evaluated at } V_{DSQ} \tag{5.9}$$

$$r_{ds} \equiv \text{output resistance} \equiv \frac{\text{change in } V_{DS}}{\text{change in } I_D} \text{ evaluated at } V_{GSQ} \tag{5.10}$$

It is also useful to define the amplification factor of the FET:

$$\mu \equiv g_m r_{ds} \tag{5.11}$$

The quantity r_{gs} is the input resistance, which is very hard to measure,

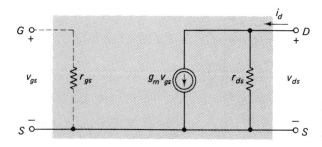

Figure 5.23 FET low-frequency small-signal model. Note: r_{gs} is usually infinite—see text for details.

especially for a MOSFET where the value may be over 1 tera ohm ($>10^{12}\ \Omega$). This quantity may be replaced by an open-circuit in most applications.

We determine the small-signal FET parameters from the FET terminal characteristics in a manner similar to that used for the BJT in Chapter 4. The graphical procedure is illustrated in Example 5.4.

Example 5.4

An N-channel JFET, whose characteristics are shown in Figs. 5.6 and 5.7, is operated at the Q-point of: $V_{GSQ} = -1$ V, $I_{DQ} = 3.5$ mA, and $V_{DSQ} = 10$ V. We want to determine the small-signal parameters for the FET model.

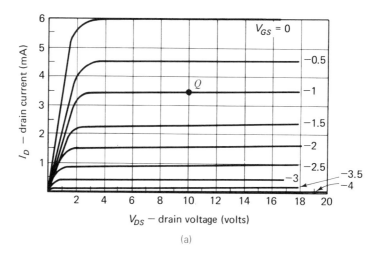

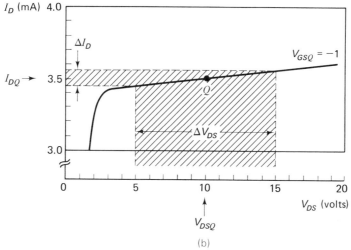

Figure 5.24 Graphical determination of r_{ds} for a JFET from: (a) the drain characteristic and (b) the characteristic with expanded current scale in the vicinity of the Q-point.

Sec. 5.5   Small-Signal FET Model

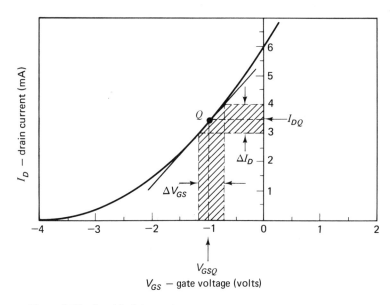

Figure 5.25 Graphical determination of g_m for the FET of Example 5.4.

Solution. The output characteristics are redrawn with the operating point (Q-point) specifically indicated on Fig. 5.24(a). In order to evaluate r_{ds}, we first expand the current scale in the vicinity of the operating point, as shown in Fig. 5.24(b). The inverse of the slope of the characteristic curve corresponding to V_{GSQ} is r_{ds}. We see that for a change in V_{DS} from 5 to 15 V, the drain current change is from about 3.45 to 3.55 mA. From Eq. (5.10):

$$r_{ds} = \frac{\Delta V_{DS}}{\Delta I_D}\bigg|_{V_{GSQ}} \quad r_{ds} \cong \frac{15-5}{3.55-3.45} \text{ V/mA} \cong 100 \text{ k}\Omega$$

From the transfer characteristics, redrawn in Fig. 5.25, we obtain g_m as the slope of the curve at the Q-point. For a change in I_D from 3 to 4 mA, the corresponding change in V_{GS} is from -1.25 to -0.75 V. From Eq. (5.9):

$$g_m = \frac{\Delta I_D}{\Delta V_{GS}}\bigg|_{V_{DSQ}} \quad g_m \cong \frac{4-3}{-0.75-(-1.25)} \cong 2 \text{ mS}$$

For completeness μ can also be determined: $\mu \cong (100)(2) \cong 200$.

Typically, JFET small-signal parameters values are g_m between 1 and 10 mS, r_{ds} between 10 kΩ and 1 MΩ, and r_{gs} between 10 and 100 MΩ.

MOSFET small-signal parameters are determined in exactly the same manner as the JFET parameters. Typically, MOSFETs have a somewhat larger range on the transconductance with the value for g_m between 1 and 30 mS. The value for r_{ds} is in the range of 1 to 100 kΩ. In a MOSFET, r_{gs} ranges from 10^{10} to 10^{14} Ω.

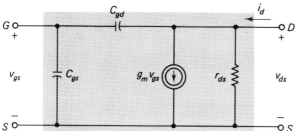

Figure 5.26 FET high-frequency small-signal model.

The FET small-signal model is adequate at low frequencies. However, at high frequencies the FET and MOSFET both exhibit capacitive effects that must be accounted for in the model.

5.6 FET HIGH-FREQUENCY MODEL

The equivalent capacitors inside a JFET appear as capacitors in the model. They are valid at high frequencies and are illustrated in Figure 5.26. Note the similarity between the FET high-frequency model and the BJT hybrid-π model. The effective capacitance between drain and source has been left out because its effects are only minor. Moreover, it would tend to complicate the use of the model.

The capacitive effects inside the JFET are caused by the reverse-biased PN junction and the charge storage associated with it. In the MOSFET, the capacitive effects result from the two conductors (the metal gate and the channel) separated by the oxide insulator, which together form an effective parallel-plate capacitor.

Typical values of C_{gs} and C_{gd} for a JFET are in the range of from 1 to 20 pF. C_{gd} is usually the smaller of the two. In a MOSFET, C_{gs} may range from 2 to 20 pF, while C_{gd} ranges between 1 and 10 pF.

The FET high-frequency model may also be used for the dual-gate MOSFET. If the second gate potential is fixed, then C_{gs} in the high-frequency model is replaced by $C_i = C_{gs} + C_{gg2}$ (where C_{gg2} is the effective capacitance between the two gates). Typical values for a dual-gate MOSFET are C_{gs} somewhat larger than for the JFET. There is also a greatly reduced drain-to-gate capacitance: in the range of from 0.005 to 0.02 pF.

Some manufacturers use different symbols for the FET parameters in their data sheet; these include the transconductance, g_m, also noted as y_{fs}; the output impedance, r_{ds}, also noted as y_{os}; the gate-to-drain capacitance, C_{gd}, also noted as C_{rss}; and the effective input capacitance, C_{iss}, which is equal to $C_{gs} + C_{gd}$. Appendix A has several FET data sheets.

REVIEW QUESTIONS

1. What are the three terminals in a JFET?
2. What is the construction of a JFET?
3. What is the difference between an N-channel and a P-channel JFET?

4. How many *PN* junctions are there in a JFET? What are they?
5. What is the mechanism for the constant resistance operation of a JFET?
6. What is the mechanism for the constant current operation of a JFET?
7. What are the conditions under which a JFET may be used as a voltage controlled resistor? Explain.
8. What is the normal operating region for the JFET? Explain in terms of the bias on it.
9. In what manner is the JFET biased for operation as an amplifier?
10. What are the conditions in the JFET described by pinchoff?
11. What is cutoff? How does it differ from pinchoff?
12. What are the conditions under which the drain saturation current I_{DSS} is measured?
13. Under what conditions is the pinchoff voltage V_p measured?
14. How is the negative gate-source voltage developed in the JFET self-bias circuit?
15. What is an IGFET? A MOSFET?
16. Describe the construction that gives the MOSFET its name.
17. What is the operation of an enhancement mode MOSFET?
18. What is the operation of a depletion mode MOSFET?
19. Discuss the operation of the JFET in terms of its capacitive behavior.
20. Discuss the operation of a MOSFET in terms of the capacitive effects inside it.
21. What is the difference between a regular MOSFET and a dual-gate MOSFET?
22. What is the purpose of the second gate in a dual-gate MOSFET?
23. How is the channel formed in an enhancement mode MOSFET?
24. How is the channel depleted in a depletion mode MOSFET?
25. What are the similarities and differences in the behavior of JFETs, MOSFETs, and dual-gate MOSFETs?

PROBLEMS

1. An *N*-channel JFET whose worst-case parameters are given in Ex. 5.2 is operated in the self-bias circuit of Fig. 5.8, with V_{DD} = 12 V, R_D = 2.2 kΩ, and R_S = 470 Ω. Determine the worst-case Q-point.
*2. Determine the bias circuit values for the self-bias circuit of Fig. 5.8 if the JFET to be used has the following parameters: I_{DSS} = 10 mA and V_p = −3 V. The desired Q-point is I_D = 7 mA and V_{DS} = V_{DD}/3. (Choose V_{DD} = 25 V.)
3. Repeat Ex. 5.2, using an *N*-channel JFET with I_{DDS} between 2 and 4 mA and $V_{GS(OFF)}$ between −1 and −6 V.
4. Repeat Ex. 5.3 for the Q-point between 3 and 3.5 mA (drain current).

5. For the JFET whose transfer characteristics are shown in Fig. 5.6, determine the transconductance at the following operating points: $I_D = 0, 1, 2, 3, 4, 5,$ and 6 mA.

6. Make a plot of the variation of the FET g_m in Problem 5 as a function of the operating-point drain current. Make a conclusion about the desired location of the Q-point for maximum g_m.

*7. Determine the FET transconductance for the transfer characteristics and Q-point shown in Fig. 5.9. Which way would the Q-point have to be moved in order to increase g_m? Why?

8. Determine the transconductance, minimum and maximum, for the JFET whose transfer characteristics are shown in Fig. 5.12. (Do this procedure for the two operating points, Q_M and Q_m, shown.)

9. Determine the small-signal parameters for the depletion mode MOSFET whose characteristics are shown in Fig. 5.15 for the following operating point: $I_D = 5$ mA and $V_{DS} = 15$ V.

10. Repeat Problem 9 for the following operating point: $I_D = 2$ mA and $V_{DS} = 10$ V.

11. Design the MOSFET bias circuit of Fig. 5.21, using the MOSFET whose characteristics are shown in Fig. 5.15 to provide an operating point of $I_D = 4$ mA and $V_{DS} = 10$ V from a supply voltage of 20 V.

*12. Repeat Problem 11 using the enhancement mode MOSFET whose characteristics are shown in Fig. 5.17.

13. Repeat Problem 9 using the enhancement mode MOSFET whose characteristics are given in Fig. 5.17.

14. Determine the minimum and maximum g_m for the enhancement mode MOSFET whose characteristics are shown in Fig. 5.17 for a gate-source voltage between the threshold value and 4 V.

15. What is the value of I_{DSS} for the JFET whose characteristics are shown in Fig. 5.7?

*16. What is the pinchoff voltage for the JFET whose output characteristics are shown in Fig. 5.7?

17. Repeat Problems 15 and 16 for the depletion mode MOSFET whose characteristics are shown in Fig. 5.15.

18. What is the threshold voltage for the enhancement mode MOSFET whose characteristics are shown in Fig. 5.17? Explain.

Chapter 6

Thyristors and Related Devices

This chapter describes the operation and terminal characteristics of several semiconductor devices. All of them have one or more *PN* junctions and some are control devices called *thyristors*. In addition to thyristors, we shall examine a group of devices whose properties closely relate to those of thyristors and whose applications are usually in conjunction with thyristors.

6.1 UJT AND PUT TRANSISTORS

6.1.1 UJT (Unijunction Transistor (UJT))

The construction of a *unijunction transistor* (UJT) consists of an *N*-type bar of silicon with ohmic (i.e., nonrectifying) contacts at either end. A single *PN* junction is formed by a small *P*-type insert near the middle of the bar, as shown in Figure 6.1. The *N*-type region, called the *base,* is of high-resistivity silicon. It has two terminals connected to it: base 1 (*B*1) and base 2 (*B*2). There is some similarity between the UJT and the JFET (discussed previously). However, the gate region of the JFET surrounds most of the channel; whereas, in the UJT, the corresponding *P* region, called the *emitter,* is much smaller. All other characteristics of the two devices are decidedly dissimilar.

We can best explain the operation of the UJT by examining the circuit of Figure 6.2. Base 2 is usually made positive with respect to base 1. The *interbase resistance* R_{BB} is the effective resistance between bases 1 and 2. It is usually high because of the light doping in the base region. In the equivalent circuit shown in Figure 6.3, this interbase resistance is represented by series resistances R_{B1} and R_{B2}. The *PN* junction between the emitter and base regions is shown by a diode in the equivalent circuit. The interbase voltage V_{BB} divides between R_{B2} and R_{B1}. The *intrinsic standoff ratio,* η, is a measure of how much voltage appears across R_{B1}. Thus,

$$\eta = \frac{R_{B1}}{R_{B1} + R_{B2}} \tag{6.1}$$

Let us now examine the consequences of applying a positive voltage to the emitter. So long as the emitter-base 1 voltage is less than the voltage across R_{B1}, the emitter diode is reverse biased. So, essentially, no emitter current flows (except for a very low leakage current). Under the conditions of zero emitter current, the interbase resistance is typically between 5- and 10-kΩ. If V_{EB1} is made larger than ηV_{BB} (the voltage across R_{B1}), the diode becomes forward biased and conducts. Emitter current flows because of the motion of (1) holes from the emitter to base 1 and (2) electrons from the base region into the emitter. The increased carrier density in the region between base 1 and emitter decreases the effective value of R_{B1}. In turn, the emitter voltage necessary for maintaining the forward bias on the emitter-base junction is reduced.

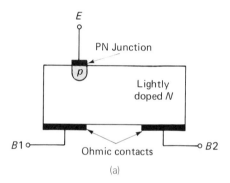
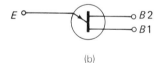

Figure 6.1 Unijunction transistor (UJT): (a) construction and (b) circuit symbol.

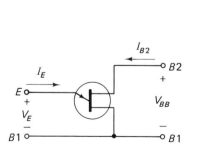

Figure 6.2 UJT voltages and currents.

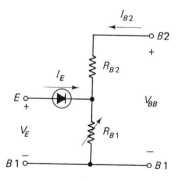

Figure 6.3 UJT equivalent circuit.

Sec. 6.1 UJT and PUT Transistors

Once the diode forward voltage reaches the cut-in voltage V_a, a large emitter current flows and the effective value of the interbase resistance is drastically decreased. At this point, the emitter voltage, called the *peak voltage* (V_p), is

$$V_p = \eta V_{BB} + V_a \tag{6.2}$$

Typical input characteristics are displayed in Figure 6.4. The decrease in voltage resulting from an increase in current corresponds to a negative resistance. We can see that the UJT has a negative resistance region in its characteristics. From Equation (6.2) you may note that the peak point depends on the interbase bias voltage V_{BB}. Typically, UJTs have V_a of between 0.5 and 0.7 V, η between 0.5 and 0.8, and a peak point current of between 5 and 25 μA.

The output or interbase characteristics of a typical UJT are shown in Figure 6.5. The curve for zero emitter current corresponds to a fixed high interbase

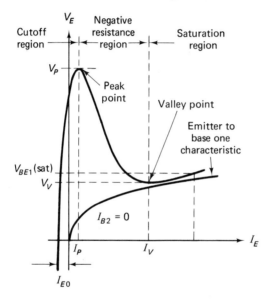

Figure 6.4 Typical UJT emitter characteristics.

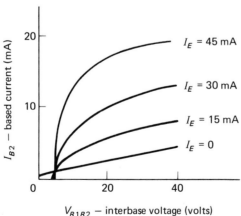

Figure 6.5 UJT output (interbase) characteristics.

resistance. The successive curves for higher I_E show a region of low interbase resistance (where the curves are almost perpendicular to the voltage axis). As we pointed out, this low value of interbase resistance is caused by the large number of carriers injected into the base region. At higher interbase voltages, the characteristic curves level off and some transistor action is evident. However, the UJT α (defined as the ratio of the change in I_{B2} to the change in I_E at a fixed V_{BB}) is quite small, in the range of 0.1 to 0.5. Therefore, the UJT is not normally used as an amplifier. However, the two distinct regions of the emitter characteristic—one of extremely high resistance (V_E below V_p) and the other of very low resistance (V_E above V_p)—make the UJT very useful in many applications.

Complementary UJT (CUJT) devices, with P-type base and N-type emitter regions, are also available. Their circuit symbol is shown in Figure 6.6.

6.1.2 Programmable Unijunction Transistor (PUT)

The structure of a programmable unijunction transistor (PUT) is totally different from that of the unijunction transistor (UJT). The PUT is a four-layer *PNPN* device with a gate (G) terminal connected to the N_1 layer, as indicated in Figure 6.7(a).

The PUT and UJT are different in both their structure and the method of operation. However, they share a similar characteristic curve. It is this similarity in operating characteristics that causes them to have similar names. Figure 6.8 pictures the characteristic of the programmable unijunction transistor.

The operation of the PUT can be explained by looking at the circuit of Figure 6.9. The resistors R_2 and R_1 are used to program the intrinsic standoff ratio, η, (Equation (6.1)) which controls the gate voltage, V_G. Thus when $I_G = 0$,

$$V_G = \eta V_{BB} \qquad (6.3)$$

Since R_2 and R_1 may assume any value; the value of η is a variable. This, of course, is not the case with the η of the UJT of Equation (6.2) where R_{B2} and R_{B1} are internal device parameters and are not adjustable. The PUT, therefore, has a programmable turnon point. The value of voltage V_p (V_{AK} in Figure 6.9)

Figure 6.6 Complementary UJT (CUJT) circuit symbol.

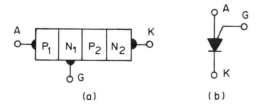

Figure 6.7 Programmable unijunction transistor (PUT): (a) construction and (b) circuit symbol.

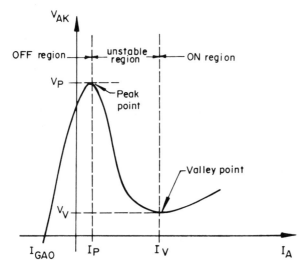

Figure 6.8 Typical PUT characteristic.

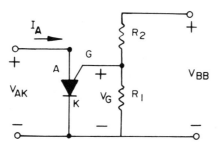

Figure 6.9 Bias circuit for the PUT.

needed to turn on the PUT is given by Equation (6.4) where $V_a = 0.7$ V. Thus,

$$V_p = V_{AK} = \eta V_{BB} + V_a \qquad (6.4)$$

An example of a programmable unijunction transistor data sheet is located in Appendix A. The application of the UJT and the PUT as trigger circuits for thyristors may be found in Chapter 18.

6.2 MULTILAYERED DIODES

In Chapter 2 we discussed the simple *PN*-junction diode as well as some special types of diodes. All of those diodes have the same basic construction: a single junction between their *P*- and *N*-type regions. In this section, we shall discuss devices with two or more *PN* junctions that still qualify for the term *diodes* because they have only two external terminals.

6.2.1 AC Bilateral Trigger Diode

The ac bilateral trigger diode, or bilateral trigger diac as it is sometimes called, is a three-layer, two-junction device quite similar in construction to a BJT. But the region that would correspond to the base has no external connections. Furthermore, the ac bilateral trigger diode has identical N regions, thus distinguishing it from a BJT.

An *NPN* bilateral trigger diode is illustrated schematically in Figure 6.10. As its name implies, it is a bilateral (or bidirectional) device; that is, it exhibits identical (typically within 10%) characteristics in both directions. Its two terminals are indistinguishable.

When a voltage of either polarity is applied to the bilateral trigger diode, one junction is forward biased, the other reverse biased. The current through the diode is limited to a small leakage current by the reverse biased junction. However, when this junction is reverse biased to the extent that breakdown occurs, a large current through the diode results. Because the reverse biased junction breaks down, a smaller total voltage across the diac is necessary to maintain the current. This corresponds to a negative resistance region.

Typical bilateral trigger diac characteristics are shown in Figure 6.11. Notice

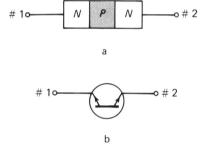

Figure 6.10 An ac bilateral trigger diode: (a) construction and (b) circuit symbol.

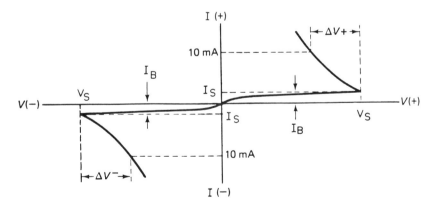

Figure 6.11 Typical characteristics for an ac bilateral trigger diode.

the two distinct regions: one of low current (and high resistance), the other of high current (and low resistance). At the transition point between the two regions, the trigger diode is said to *fire* and begin to conduct. This action is that of a switch. The bilateral trigger diode is an effective open circuit until the voltage across its terminals reaches the breakdown voltage, at which point the diode resistance becomes low and the switch is effectively closed. Because this action occurs in both directions, the device is sometimes called an *ac switch*. The data sheet for an MPT20 bilateral trigger diode may be found in Appendix A.

6.2.2 Four-Layer (Shockley) Diode

The Shockley diode, shown in Figure 6.12, is a four-layer, three-junction device. Using conventional diode terminology, we call the connection to the P_1 layer the *anode* and the lead at the N_2 layer the *cathode*. But here the similarity between a conventional *PN* diode and the *PNPN* diode stops.

The *PNPN* structure of the four-layer diode can be considered the equivalent of two BJTs: one a *PNP*, the other an *NPN*, as shown in Figure 6.13(a). With a

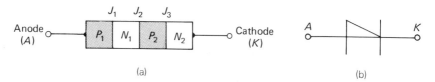

Figure 6.12 Four-layer (Shockley) diode: (a) construction and (b) circuit symbol.

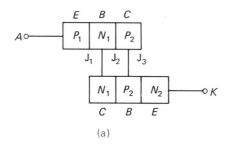

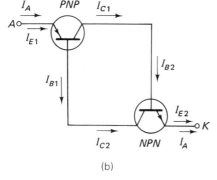

Figure 6.13 (a) Two-transistor representation for the four-layer (Shockley) diode and (b) the resulting circuit.

forward bias voltage across the diode (anode positive with respect to the cathode), the $P_1N_1(J_1)$ and $P_2N_2(J_2)$ junctions are forward biased, while the $N_1P_2(J_2)$ junction is reverse biased. From the equivalent circuit in Figure 6.13(b), we see that $I_A = I_{E1} = I_{E2}$ and also $I_{C1} = I_{B2}$. We can recall from Chapter 3 the relationship of transistor currents as

$$I_{C1} = \alpha_1 I_{E1} + I_{CO1} \tag{6.5}$$

and for transistor 2 as

$$I_{C2} = \alpha_2 I_{E2} + I_{CO2} \tag{6.6}$$

Using the relationships from the equivalent circuit in Equation (6.1), we obtain

$$I_{B2} = \alpha_1 I_A + I_{CO1} \tag{6.7}$$

Remembering that $I_{B2} = I_{E2} - I_{C2}$ from Equation (6.5), we get

$$I_A - I_{C2} = \alpha_1 I_A + I_{CO1} \tag{6.8}$$

We now substitute Equation (6.2) for I_{C2} in Equation (6.6):

$$I_A - \alpha_2 I_A - I_{CO2} = \alpha_1 I_A + I_{CO1} \tag{6.9}$$

Rearranging and solving this equation for I_A yields

$$I_A = \frac{I_{CO1} + I_{CO2}}{1 - (\alpha_1 + \alpha_2)} \tag{6.10}$$

Because of the reverse bias on J_2, the anode current is low. Remember that the transistor α is dependent on the current (emitter current specifically). For low currents, α is very small, and so the anode current may be approximated by

$$I_A \cong I_{CO1} + I_{CO2} \tag{6.11}$$

Thus, the anode current is quite small, being the sum of two reverse-leakage currents.

Let us now consider the results of further increasing the voltage in the forward direction. Almost all of the increase in voltage appears across the reverse-biased junction J_2. At a critical voltage, called the *breakover voltage*, this junction breaks down by the avalanche mechanism. The result is an abrupt increase in current and both α_1 and α_2 increase. Equation (6.10) tells us that obviously when $(\alpha_1 + \alpha_2)$ becomes equal to 1, the anode current can increase without limits. Consequently, we must use a series resistor to limit the anode current. Under the breakdown conditions described, the voltage required to maintain the increasing anode current rapidly falls off, and the diode resistance becomes drastically reduced.

Typical four-layer diode characteristics are depicted in Figure 6.14. In the forward direction, the diode behaves like a solid-state switch with a high-resistance region (switch OFF) and a low-resistance region (switch ON). In the reverse direction, it effectively blocks current for reverse voltages less than the reverse breakdown voltage, V_{RB}.

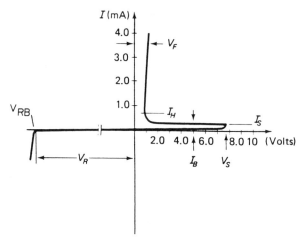

Figure 6.14 Typical characteristics of a four-layer (Shockley) diode.

Once the four-layer diode is in its forward conducting state, we can turn it off either by interrupting or reducing the anode current (below the holding current I_H) or by removing or reducing the anode voltage (below the holding value V_H). Any one of these actions causes the four-layer diode to open and return to its high-resistance state.

6.2.3 Bilateral Diode Switch (Diac)

The construction and circuit symbol for the bilateral or bidirectional diode are shown in Figure 6.15. Although the term *diac* may be applied to all two-terminal devices having identical (or nearly identical) characteristics in both directions, the bilateral diode switch is commonly called a diac.

To understand the operation of the diac, consult the diagram in Figure 6.16. We can represent the bilateral diode as two *PNPN* diodes connected in parallel. (The two are $P_2N_2P_1N_1$ and $P_1N_2P_2N_3$.) When terminal 1 is made positive with respect to terminal 2, junctions J_1 and J_3 are reverse biased, whereas junctions J_2 and J_4 are forward biased. The *PNPN* diode between K_1 and A_2 is in the off state, because K_1 is positive with respect to A_2. However, the other *PNPN* diode between A_1 and K_2 is forward biased and will conduct heavily when the voltage is high enough to cause avalanche breakdown at junction J_3. (The ohmic drop, caused by a current flow in the P_1 region, forces J_1 to be reverse biased.) Thus,

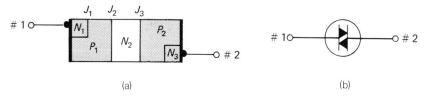

Figure 6.15 Diac (bilateral diode): (a) construction and (b) circuit symbol.

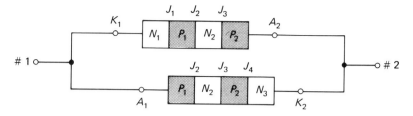

Figure 6.16 Two-PNPN-diode representation of a bilateral diode (diac).

with terminal 1 positive with respect to terminal 2, the characteristics are governed by the four-layer diode $P_1N_2P_2N_3$.

When terminal 2 is made positive with respect to terminal 1, the other four-layer diode (composed of $P_2N_2P_1N_1$) controls the characteristics; that is, it turns on when the voltage applied is high enough to cause avalanche breakdown at J_2. The diac thus exhibits turn-on characteristics in either direction, accompanied by a low resistance, in much the same way as the ac trigger diode. Figure 6.17 gives typical terminal characteristics for a diac.

Once the diac has turned on in either direction, we can turn it off in the same manner as for the simple four-layer (Shockley) diode: (1) by interrupting the current, (2) by reducing the current below I_H, or (3) by reducing the voltage below V_H. Commercially available diacs have forward and reverse characteristics that are closely matched.

6.2.4 Silicon Unilateral (SUS) and Bilateral (SBS) Switches

The silicon unilateral switch (SUS) and the silicon bilateral switch (SBS) are used in designing low-cost regenerative triggering circuits for thyristors. Like the three-layer trigger diode, the unijunction transistor (UJT), and the programmable uni-

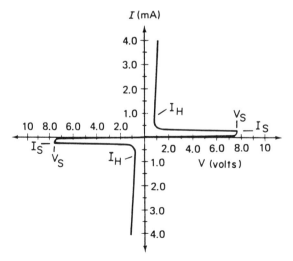

Figure 6.17 Typical characteristics for a diac (bilateral diode).

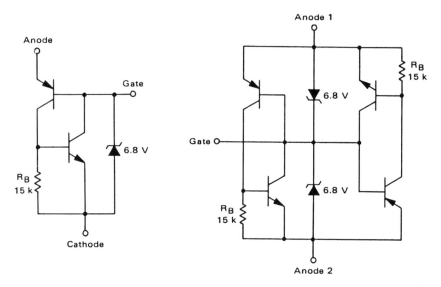

Figure 6.18 Integrated switching circuits: (a) silicon unilateral switch (SUS) and (b) silicon bilateral switch (SBS). (*Courtesy of Motorola, Inc.*)

junction transistor (PUT), the silicon unilateral and bilateral switches provide the high-current and fast rise-time gate pulses needed to properly turn on thyristors (SCRs (silicon-controlled rectifiers) and triacs). The SUS and SBS actually have better performance properties than any of the *NPN* or *PNPN* multilayered triggering diodes, and at a lower cost. Unlike the *stacked structure* of the multilayered diodes, the SUS and SBS are *integrated circuits* made up of resistors, diodes, and transistors, as shown in Figure 6.18.

From the schematic symbols of Figure 6.19, we learn that the SUS and SBS have three leads, anode (*A*), cathode (*K*), and gate (*G*). The addition of the gate allows for a downward adjustment of the nominal 8 V switching voltage. Also, by taking a small current out of the gate, the SUS or SBS may be *gated* ON when the applied anode to cathode voltage is less than the switching voltage.

The electrical characteristics pictured in Figure 6.20 indicate the unilateral characteristic of the SUS and the bilateral characteristic of the SBS. Here we see

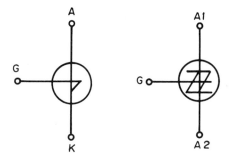

Figure 6.19 Schematic symbols: (a) SUS, and (b) SBS.

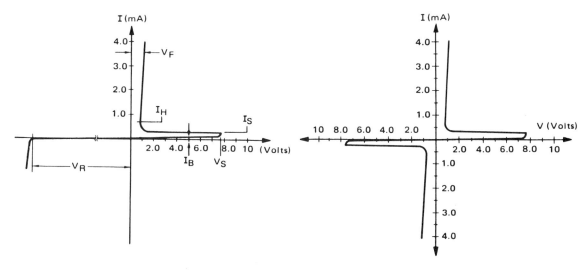

Figure 6.20 Forward and reverse characteristics: (a) SUS and (b) SBS. (*Courtesy of Motorola, Inc.*)

that the device switches from an OFF state to an ON state at a switching voltage (V_S) of approximately 8 V. Prior to switching states, only a small leakage current (I_B) flows. However, once V_S is reached, a much larger switching current (I_S) is present. In the conducting state, the current rises dramatically above the holding current (I_H) and the voltage drops to the forward voltage (V_F). Because of the low ON resistance ($\approx 3\ \Omega$), the ON current must be limited by adding a resistance in series with the voltage supply.

Chapter 18 has several examples of triggering circuits used to trigger thyristors. Also, a specification sheet for a silicon bilateral switch may be found in Appendix A.

6.3 THYRISTORS

The term *thyristor* is usually applied to a family of solid-state devices having turnon characteristics that can be externally controlled by either current or voltage.

6.3.1 Silicon-Controlled Rectifier (SCR)

The silicon-controlled rectifier (SCR) is also called the *reverse blocking thyristor*. It is a four-layer device similar in construction to the programmable unijunction transistor. The difference is that the third terminal is connected to the P_2 layer, as indicated in Figure 6.21(a). The circuit symbol is shown in Figure 6.21(b).

The operation of the SCR is quite similar to that of the Shockley (four-layer) diode. With a forward voltage applied between anode and cathode, the SCR is in the normally off or high-resistance state. As the forward voltage is increased

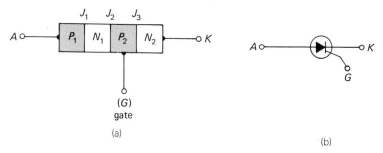

Figure 6.21 Silicon-controlled rectifier (SCR): (a) construction and (b) circuit symbol.

and reaches a critical value (the *breakover voltage*), J_2, which is reverse biased, avalanches. Then the anode current becomes high; the SCR has turned on. The third terminal is called the *gate*. It regulates the voltage at which breakover occurs—or the switching between the off and on states.

To understand how this control is achieved, let us examine the two-transistor analogy of the *PNPN* device. This analogy is shown in Figure 6.22, where the gate lead is attached to the base of the *NPN* transistor. You might also refer to a similar discussion in Section 6.22 concerning the Shockley diode. In this case, however,

$$I_{B2} = I_{C1} + I_G \qquad (6.12)$$

It can be shown that the anode current for the SCR is given by

$$I_A = \frac{I_{CO1} + I_{CO2}}{1 - (a_1 + a_2)} + I_G \frac{\alpha_2}{1 - (\alpha_1 + \alpha_2)} \qquad (6.13)$$

Obviously, where $I_G = 0$, the anode current for the SCR is the same as that for the Shockley diode [Equation (6.10)]. In the normal forward-blocking (off) state, the anode current is low and the αs are small. The term containing I_G signifies that the anode current for a given forward voltage is higher than in the Shockley diode. Moreover, the quantity $(\alpha_1 + \alpha_2)$ reaches unity for a lower value of forward voltage because of the more rapid increase in current. Thus, the gate current effectively controls the voltage at which the SCR switches from its nonconducting

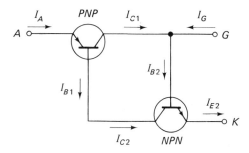

Figure 6.22 Two-transistor representation of an SCR.

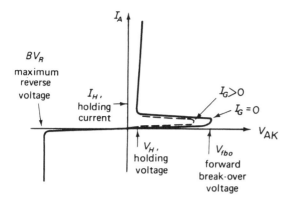

Figure 6.23 Typical anode characteristics of an SCR.

off state to its forward-conducting on state. We determine the breakover voltage by the magnitude of the gate current: The higher the gate current, the lower the breakover voltage. Figure 6.23 supplies these typical SCR characteristics.

The reverse characteristics of the SCR are essentially unaffected by the gate signal and resemble those of the Shockley diode.

Once the SCR is in its forward-conducting state, the gate loses control. In fact, to turn on an SCR, gate current need flow for only a few microseconds to a few milliseconds. So long as $(\alpha_1 + \alpha_2)$ is equal to one (or J_2 is under avalanche breakdown), the anode current will be high and the SCR will remain on, regardless of the gate potential and current. Even a negative gate potential cannot turn off the SCR once it has fired. To turn off the SCR, the anode current must be interrupted or reduced below a certain minimum value. But this minimum value, called the *holding current* (I_H), is dependent on the gate bias, as shown in Figure 6.23. With zero gate current, the minimum anode current to keep the SCR on is called the *latching current* (I_L).

In addition to the anode characteristics, the manufacturer may also specify the gate characteristics for an SCR. Figure 6.24 provides an example. Because

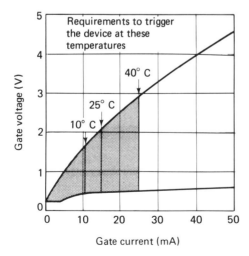

Figure 6.24 Typical gate characteristics for an SCR family.

of the variation between devices of even the same family, slightly different combinations of gate (to cathode) voltage and gate current are necessary to fire a given SCR. In normal applications, to ensure firing of the SCR, the gate is usually overdriven; that is, we apply a gate current in excess of the minimum specified.

Commercially available SCRs range from very low-power units to some capable of controlling anode currents in excess of 1000 Amperes.

6.3.2 Gate Turnoff Switch

As we stated earlier, the SCR cannot normally be turned off at the gate. A device very much similar to the SCR, called the *gate turnoff switch* (GTO) can be turned off by a negative bias on the gate. In construction, the GTO is identical to the SCR as noted in Figure 6.25. However, in the terminology of the two-transistor representation of an SCR in Figure 6.22, the α of the *PNP* transistor is made small. The GTO is a low-current device when compared to an SCR.

The turnon characteristics of a GTO are identical to those of a comparable SCR. The difference between the two lies in the turnoff. To see how turnoff can be accomplished in a GCS, we use the two-transistor representation in Figure 6.22. When the GTO is in its conducting state, anode current flows into the anode and essentially the same current flows out of the cathode. Turnoff can be accomplished by causing a negative gate current, equal to I_{C1}, to flow. So we apply a negative bias to the gate. (A negative gate current flows in the direction opposite to that shown in Figure 6.22.) Equation (6.12) shows that this procedure in effect eliminates the base current, I_{B2}, from the *NPN* transistor. The *NPN* transistor emitter current is sufficiently decreased to cause turnoff. If the α of the *PNP* transistor is small, than I_{C1} will be relatively small and the negative gate current required for turnoff will also be small. The ratio of the anode current in the on state to the negative gate current required to turn off the GTO is called the *turnoff current gain*. Typical values for the turnoff current gain of commercially available GTOs are between 5 and 10. The GTO circuit symbol is shown in Figure 6.25(a).

As a sidelight to the operation of a GTO, note that a high-current SCR cannot be turned off by a negative gate bias, but the application of a negative gate bias can significantly reduce the turnoff time of an SCR (that is being turned off at the anode).

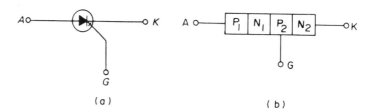

Figure 6.25 Gate turnoff switch (GTO): (a) circuit symbol and (b) construction.

6.3.3 Silicon-Controlled Switch (SCS)

The silicon-controlled switch (SCS) is sometimes called the *tetrode thyristor*. It is a four-layer device and is very similar to the SCR. As the tetrode name would suggest, the SCS has four terminals, one on each of its four layers, as shown in Figure 6.26(a). The SCS circuit symbol is shown in Figure 6.26(b). In addition to the cathode gate, G_k, the SCS has an anode gate, G_a, connected to the N_1 region.

The SCS is a versatile device. It can be operated in numerous different ways. One obvious mode of operation is as an SCR with the anode gate not connected. It may also be operated as a complementary SCR. In this case, a negative anode gate voltage causes the SCS to fire; the cathode gate is not connected, however. (Complementary SCRs, with only an anode gate, are manufactured, but the conventional SCR is by far the more common of the two.) Neither of the preceding applications warrants the existence of two gates. The full potential of the SCS is realized only when both gates are used. If we operate the SCS like an SCR, the anode gate can turn off the SCS once it has fired. We would apply a sufficiently positive potential to the anode gate junction J_1 to cause reverse biasing, thus turning off the SCS.

Perhaps the most important use of the anode gate is to make the SCS turn on quicker. In the off state, J_2 is reverse biased. As such, it has a capacitive effect that slows down the rate at which the SCS turns on when the anode becomes positive enough. If we apply a positive voltage to the anode gate, we charge this junction capacitance, so that when the anode signal exceeds V_{BO} and initiates the firing of the SCS, the turnon rate is drastically reduced.

The SCS is a low-power device. As such, it may be turned off at the cathode gate by a negative voltage, in a way much like the GTO.

6.3.4 Bilateral Triode (Triac)

The triac is a bilateral (or bidirectional) three-terminal device. To fully understand the operation of a triac, we first need to examine the principles involved in the operation of thyristors having a *remote gate* or a *junction gate*.

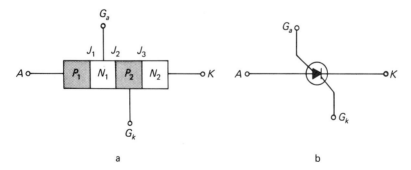

Figure 6.26 Silicon-controlled switch (SCS): (a) construction and (b) circuit symbol.

A remote gate thyristor is illustrated schematically in Figure 6.27. It is a four-layer device comprised of $P_1N_1P_2N_2$, with the gate not at either N_1 or P_2 as in a conventional SCR, but at N_3, as indicated. In the forward-blocking state (where A is positive with respect to K) and a negative voltage applied to the gate, the N_3P_1 junction is forward biased. In addition, J_1 and J_3 are forward biased and J_2 is reverse biased. Because of the forward bias on J_4, electrons from N_3 are injected into P_1, diffuse through P_1, and are collected at J_1. The effect is to increase the forward bias on J_1, so additional holes are injected from P_1 into N_1. These holes diffuse across N_1 and are collected at P_2. As a result, the forward bias across J_3 is increased. Additional electrons are injected from N_2 into P_2. They are collected at J_2, thus further increasing the forward bias across J_1. We have then a cycle started by a negative bias on the gate, causing a larger forward bias across J_1, which, by the action described, is self-perpetuating; it increases the anode current until the thyristor fires in the same manner as a conventional SCR. The remote gate thyristor fires in the forward direction as a result of negative gate bias.

A schematic diagram of a junction gate thyristor is shown in Figure 6.28. Let us assume that the device is off, with a forward bias on the anode and a negative bias on the gate. Because of the gate bias V_G, the gate is more negative with respect to the anode than the cathode. The four-layer device comprising $P_1N_1P_2N_3$ begins to turn on. As a result, gate current starts to flow in the direction shown. As the gate current increases, so does the voltage drop across the gate bias resistor R. When the drop across R exceeds V_G, the gate becomes positive and the large flow of holes from P_2 into N_3 is diverted to N_2. Thus, the forward bias of J_3 is significantly increased, and the four-layer $P_1N_1P_2N_2$ turns on. In effect, the negative gate bias initiates turnon of $P_1N_1P_2N_3$. As voltage across R builds up, the large current from the anode is diverted to the cathode, accomplishing turnon of $P_1N_1P_2N_2$.

We can now proceed with the discussion of the bilateral triode thyristor or triac, shown schematically in Figure 6.29. The triac is truly a bidirectional device.

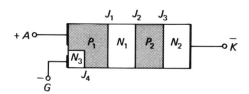

Figure 6.27 Schematic representation of a remote gate thyristor.

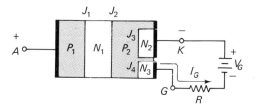

Figure 6.28 Schematic representation of a junction gate thyristor.

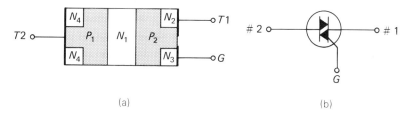

Figure 6.29 Triac (bilateral triode): (a) construction and (b) circuit symbol.

It can fire with either positive or negative gate bias. Let us examine the four possible configurations:

Terminal 2 positive, gate positive (both with respect to $T1$): The active regions are $P_1N_1P_2N_2$, forming a conventional SCR with $T2$ as the anode, $T1$ as the cathode, and G connected to P_2 as the cathode gate.

Terminal 2 negative, gate negative: The operation is that of a remote gate thyristor with active regions $N_4P_1N_1P_2$. $T2$ acts as the cathode, $T1$ as the anode, and the remote gate function is accomplished by the N_3P_2 junction.

Terminal 2 positive, gate negative: Operates as the junction gate thyristor, with active regions $P_1N_1P_2N_2$. $T2$ acts as the anode, $T1$ as the cathode, and the junction gate is formed by the junction of N_3 and P_2.

Terminal 2 negative, gate positive: The operation is that of a remote gate thyristor, shown in Figure 6.30. The active regions are $P_2N_1P_1N_4$, with $T1$ acting as the anode, $T2$ as the cathode, and the remote gate function accomplished by the gate contact at P_2. The forward bias on the P_2N_2 junction initiates the firing.

The terminal characteristics of the triac are identical in both directions. They resemble the forward characteristics of an SCR. Typical characteristics are given in Figure 6.31.

We can turn off a triac operating in any of the possible configurations by either interrupting or lowering the current below the holding value of the $T1$-$T2$ current.

Chapter 18 has several application circuits using thyristors. Also, Appendix A has several thyristor data sheets (SCR and triac) that contain valuable information.

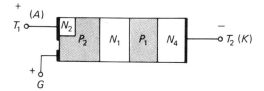

Figure 6.30 Remote gate operation of a triac with T_2 negative and G positive.

Sec. 6.3   Thyristors

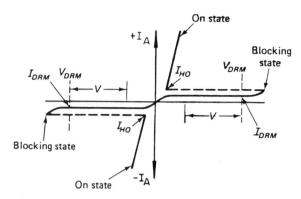

Figure 6.31 Typical triac characteristics.

REVIEW QUESTIONS

1. In terms of its construction, justify the naming of a UJT.
2. What are the three terminals of a UJT?
3. What is the intrinsic standoff ratio? How is it defined?
4. What is the relationship of the interbase resistance of a UJT to its intrinsic standoff ratio?
5. How are the peak, cut-in voltage, and the intrinsic standoff ratio for a UJT related?
6. Briefly explain the "firing" of a UJT.
7. What is the structural difference between an ac trigger diode and a Shockley diode?
8. Briefly describe the firing of an ac trigger diode. Is it a unilateral or a bilateral device?
9. How does the operation of a Shockley diode differ from that of the ac trigger diode? How are they similar?
10. What is the construction of a bilateral diode? How do its characteristics differ from an ac trigger diode? How are they similar?
11. Once in conduction, how can we turn off the ac trigger diode, the Shockley diode, and the bilateral (diac) diode?
12. In what way does an SCR differ from the diodes (ac trigger, Shockley, and diac)? How is it similar?
13. Explain the conditions leading to triggering (firing) in an SCR?
14. Describe the two-transistor analogy for an SCR.
15. What determines the firing voltage (anode-to-cathode) in an SCR?
16. Once an SCR has been fired, what are the ways in which it can be turned off?
17. Is the SCR a unilateral or a bilateral device? Why?
18. How does a gate turnoff switch GTO differ from an SCR? In what ways are they similar?

19. What controls the firing characteristics for a GTO? How can it be turned off once it is conducting?
20. What is the turnoff current gain in a GTO? What is it numerically?
21. How are an SCR and an SCS similar? In what ways are they different?
22. What are the ways in which an SCS can be turned on? What control is provided by the two gates?
23. What are the advantages of using an SCS with the additional anode gate over using an SCR?
24. What are the firing characteristics of a triac?
25. In what ways are the characteristics and operation of an SCR and a triac similar? In what ways are they different?
26. What are the ways in which a triac may be turned off once it is conducting?

Chapter 7

Photoelectric Devices

In this chapter we shall discuss semiconductor devices that are activated by light (visible and infrared) and that emit light. These devices are used for industrial control, information transmission, visual displays, and even power generation.

7.1 PHOTOCONDUCTIVE CELLS

Photoconductive cells are sometimes called *photoresistors*. They operate on the principle of *photoresistivity*. Certain semiconductors, when illuminated by light, show a decrease in resistance. Typical examples of photoresistive semiconductors are made of cadmium sulphide (CdS) and cadmium selenide (CdSe). The cells are fabricated by depositing a thin layer of semiconductor on a ceramic base, with two interleaved electrodes, as shown in Figure 7.1(a). The enclosure contains a glass window or lens that allows light to reach the semiconductor.

As we noted in Chapter 1, the resistivity of a semiconductor is a function of the number of free charge carriers that are available for conduction. Before the semiconductor is illuminated, the number of free charge carriers is very small, so resistivity is high. When light, in the form of photons,* is allowed to strike the semiconductor, each photon delivers energy to the semiconductor. If the energy exceeds the energy gap of the semiconductor, free mobile charge carriers are liberated; and, as a result, the resistivity of the semiconductor is decreased.

The response of a given photoconductive cell is determined by its energy gap. For a specific energy gap (E_G), only photons with energies larger than E_G can liberate additional carriers, so the cell reacts only to those photons. The frequency (f) of light is related to the wavelength, λ, as follows:

---

* For the purposes of our discussion, visualize the light striking the semiconductor as being made up of a stream of particles, called *photons*. Each photon carries with it an amount of energy hf, where h is Plank's constant (about 4×10^{-15} eV-sec) and f is the frequency of the light.

$$f = \frac{c}{\lambda} \tag{7.1}$$

where c is the speed of light (3×10^{10} cm/s). The energy of a photon can then be expressed in terms of the wavelength:

$$E = \frac{hc}{\lambda} \tag{7.2}$$

We calculate the minimum wavelength needed to decrease the resistivity as

$$\lambda = \frac{hc}{E_G} = \frac{12}{E_G} \times 10^{-5} \, cm = \frac{12{,}000}{E_G} \, \text{Å} \tag{7.3}$$

where E_G is in electron volts and 1 cm is equal to 10^8 Angstron units (Å). Figure 7.2 shows the spectrum visible to the human eye. Typical photoconductive semiconductors have energy gaps on the order of 2eV The spectral response of CdS and CdSe is illustrated in Figure 7.3, together with the response of the human eye. Cadmium sulphide has a response similar to the human eye, so CdS cells are often used to simulate the human eye. For example, they are used in light-metering circuits in photographic cameras.

The terminal characteristics of a photoconductive cell are a plot of cell resistance as a function of light intensity. Typically, cells have a *dark resistance* of 1 MΩ or higher. The dark resistance, as the name implies, is the cell resistance when there is no light striking it. Under illumination, the cell resistance drops to a value between 1 and 100 kΩ, depending on the surface illumination. Units of surface illumination are lumens per square meter or *lux* (1 lx = 1 lm/m²). Typical characteristics, showing the nonlinear decrease in resistance with increasing surface illumination, are given in Figure 7.4.

We must point out that the resistance value under a specific light intensity depends on the previous light intensity to which the cell was subjected. Moreover,

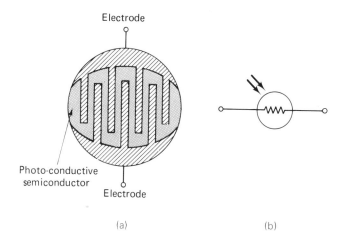

Figure 7.1 Photoconductive cell: (a) typical construction and (b) circuit symbol.

Sec. 7.1   Photoconductive Cells

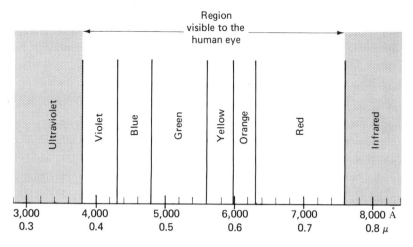

Figure 7.2 Visible frequency spectrum with the wavelength specified in both Angstroms (1A = 10^{-10} m) and microns (1 micron = 10^{-6} m).

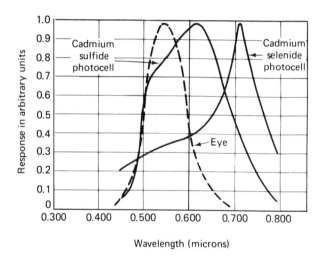

Figure 7.3 Photoconductive cells (photocell) spectral response curves.

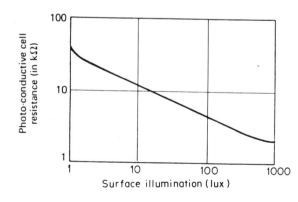

Figure 7.4 Typical characteristics for a photoconductive cell.

there is a finite time required for the cell to react to a change in light intensity. The *rise time* is the time required for a cell starting in the dark condition to reach the lower resistance value (actually 62% of this value) at a higher illumination level. Typical cells have rise times of less than a millisecond.

7.2 PHOTOVOLTAIC CELLS

Photovoltaic, or *solar*, cells are devices characterized by the generation of a small voltage upon illumination. The solar cell contains a *PN* junction. When the junction is illuminated, holes and electrons are liberated in much the same way as in photoconductive cells. However, these newly liberated charge carriers are acted upon by the junction contact potential. Electrons are forced toward the *N* side, holes toward the *P* side. If external wiring is connected, we can observe that a current results. The displacement of holes toward the *P* side and electrons toward the *N* side lowers the junction contact potential. This effect is manifested externally by the appearance of a voltage across the terminals of the cell. The electrode connected to the *P* side becomes positive with respect to the electrode connected to the *N* side. Thus, a terminal voltage is generated as a result of light striking the *PN* junction of the cell. The voltage is directly proportional to the intensity of the light; it may be as high as 0.6 V, depending on the load placed across the cell.

Silicon is a commonly used material in the fabrication of photovoltaic cells. Another cell construction features *P*-type selenium, with a layer of *N*-type cadmium oxide over it to form the *PN* junction. The construction of these two cells is depicted in Figure 7.5. Typical characteristics of photovoltaic cells are shown

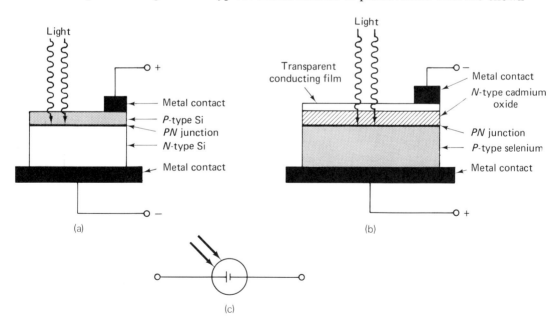

Figure 7.5 Construction of photovoltaic cells: (a) silicon, (b) selenium, (c) circuit symbol.

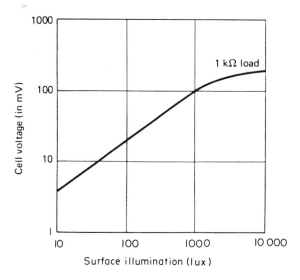

Figure 7.6 Typical photovoltaic cell characteristics.

in Figure 7.6. Note that the rise in voltage is not linearly related to increases in illumination.

As the name "solar cell" might suggest, photovoltaic cells can be used to convert solar energy into electrical energy. In such applications, literally thousands of individual cells can be connected together to generate kilowatts of power.

7.3 PHOTODETECTORS

Silicon photodetectors are responsive to a wide range of electromagnetic radiation in the optical spectrum, as pictured in Figure 7.7. Besides responding to wavelengths of the entire visible spectrum, silicon photodetectors also respond to the *near* infrared wavelengths of the tungsten lamps and the gallium arsenide (GaAs) infrared emitting diode.

Since tungsten lamps are often used as radiation emitters for photodetectors, it is important that the tungsten lamp be operated at above 2600 kelvin (visible white light) to ensure sufficient radiation for the photodetector.

Figure 7.7 shows the narrow band spectral characteristic (curve B and D) of the solid state emitting diodes. Of the two emitting diodes, the gallium arsenide infrared emitting diode is often used with the silicon photodetectors to form an efficient *coupled pair*. As seen in Figure 7.7, the peak emission of the GaAs diode is close to the peak sensitivity of the silicon photodetector.

A variety of silicon photodetectors are available from various manufacturers in the form of photodiodes, phototransistors, photodarlingtons, optically triggered triac drives, and light-actuated SCRs. The applications for photodetectors include card and tape readers, pattern recognition, shaft encoders, counters, position sensors, sorters, and industrial inspection and control.

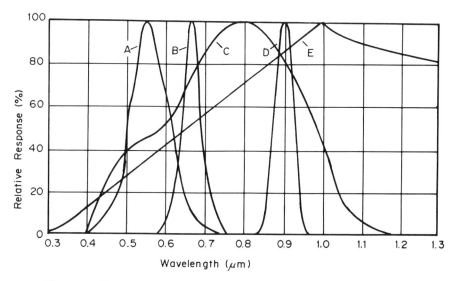

Figure 7.7 Spectral characteristics: (A) eye of a standard observer, (B) GaAsP light-emitting diode (red light), (C) silicon photodetector, (D) GaAs infrared-emitting diode, (E) 2800 K tungsten lamp.

7.3.1 Photodiodes

Photodiodes are almost identical in construction to photovoltaic cells. However, photodiodes are operated in the reverse-biased condition. Under these conditions they behave similarly to photoconductive cells.

To understand the operation of photodiodes, recall that a conventional diode under reverse bias sustains a small (reverse-saturation) current that is the result of the flow of minority carriers. The same process occurs in a photodiode when it is not illuminated. So the dark current of a photodiode is low and essentially the same as the normal reverse-saturation current. When light is allowed to fall on the *PN* junction of a photodiode, those photons whose energy exceeds the energy gap of the semiconductor produce additional free charge carriers: holes and electrons. These newly created free carriers have no appreciable effect on the majority carrier concentrations in the *N* and *P* regions, but they greatly increase the minority carrier concentrations. The action of the reverse bias forces the newly created holes toward the cathode and the electrons toward the anode. The net effect is a large increase in the reverse current.

In the photodiode, the increase in the reverse current (referred to as light current, I_L) is almost directly related to the total incident radiation energy called *irradiance* (radiation flux density), which is measured in units of power per unit area*, mW/cm². Figure 7.8 pictures the irradiated $V - I$ characteristic for a photodiode.

---

* Visible light energy is called luminous flux and is measured in lumens. Radiant light energy is called radiant flux and is measured in watts.

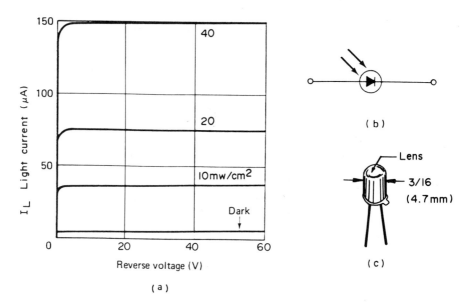

Figure 7.8 Photodiode: (a) typical characteristics, (b) circuit symbol, (c) metal hermetic package with convex lens.

7.3.2 Phototransistors

Phototransistors are made in much the same way as their conventional BJT counterparts, but they usually have no external connection to the base. Their operation is based on the photodiode that exists at the collector-base junction.

Silicon *NPN*s are the most common phototransistor configurations. Usually, the phototransistor is made by the diffusion process, with a transparent window or lens in the case, like the one pictured in Figure 7.8(c). The light incident on the collector-base junction creates additional charge carriers. The phototransistor is operated with the *N*-type emitter negative with respect to the *N*-type collector. The base-emitter junction is then forward biased, whereas the collector-base junction is reverse biased. If the incident light falls on the collector-base junction and the photons have enough energy to create additional free charge carriers, the reverse bias at the junction causes electrons to go into the collector region. The remaining holes in the base region, being positive, raise the potential of the base with respect to the emitter. As a result, electrons from the emitter are injected into the base. Some of these electrons are lost through recombination in the base, but most of them do reach the collector. Electrons reaching the collector flow out through the collector lead and make up the output current.

The operation of a phototransistor is similar to that of a conventional BJT. However, in a phototransistor, light-liberating charge carriers constitute the input (base) current.

The dark current in a phototransistor is larger than in a photodiode. It is essentially equal to I_{CEO}, typically between a nanoAmpere and a microAmpere.

However, the collector current in a phototransistor under illumination is essentially β times the photocurrent of a similar photodiode. Figure 7.9 displays typical output characteristics of a phototransistor, with radiation flux density (irradiance) instead of base current as the parameter. The same figure also shows the circuit symbol.

7.3.3 Photodarlington

The photodarlington is used in applications where maximum sensitivity to incident radiant energy is required. Although the typical switching speed (rise and fall time) of 50 μs is much longer than the phototransistor (2 μs) or the photodiode (1 ns), the photodarlington configuration (Figure 7.10) results in a photodetector

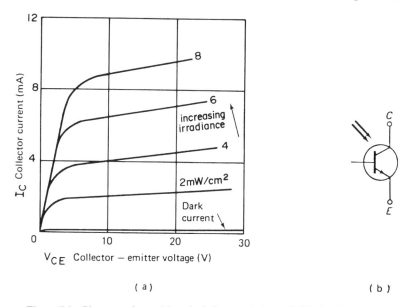

Figure 7.9 Phototransistor: (a) typical characteristics and (b) circuit symbol.

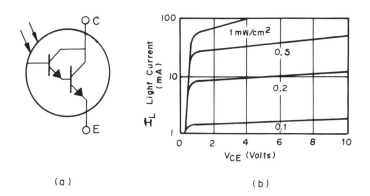

Figure 7.10 Photodarlington: (a) circuit symbol and (b) typical characteristics.

Sec. 7.3   Photodetectors

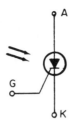

Figure 7.11 Circuit symbols for light-activated SCR (LASCR).

with an order of magnitude improvement in sensitivity. The base-collector junction of the photodarlington drive transistor (left side of Figure 7.10(a)) is controlled by incident radiation (mW/cm^2). The drive transistor, in turn, controls the following transistor. Because of the high β of the two transistor darlington configuration, the darlington phototransistor is very sensitive to radiant flux.

7.3.4 Light-Actuated SCR

As we saw in Chapter 6, switching in an SCR takes place when we increase the anode-to-cathode current. In a conventional SCR, this switching is controlled by current supplied to the gate. In a light-actuated SCR (Figure 7.11), the anode-to-cathode current is increased when light strikes near the middle junction. Photons with enough energy create additional free carriers near the reverse-biased middle junction (J_2 in Figure 6.21). These additional carriers are acted upon by the reverse bias at J_2. Holes are forced toward the cathode; electrons, toward the anode. As a result, the SCR current increases, in turn increasing α_1 and α_2 in Equation (6.13). Switching is initiated as if the increase in current were being provided by the gate.

The light current at the middle junction is directly proportional to the radiation flux density. Therefore, the rise in incident radiation energy on a light-actuated SCR (LASCR) makes the switching occur at an ever-decreasing voltage. As is the case with a conventional SCR, once the LASCR is in its conducting state, it will continue to conduct, whether light is present or not. It can be turned off only by interrupting the anode current or by any of the other ways of turning off the SCR.

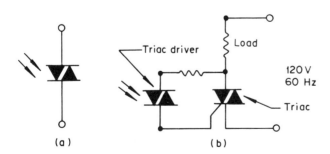

Figure 7.12 Optically triggered triac driver: (a) circuit symbol and (b) connection to a triac.

7.3.5 Optically-Triggered Triac Driver

The optically-triggered triac driver, pictured in Figure 7.12, is a light-sensitive integrated circuit (IC) that acts as a trigger for a triac (Figure 7.12(b)).

7.4 PHOTOEMITTERS

Figure 7.7 pictured the spectral characteristics of two photoemitters (curves B and D). One is the gallium arsenide phosphide (GaAsP) diode (curve B) which emits visible light and is called a light-emitting diode (LED); the other (curve D) is the *PN* gallium arsenide (GaAs) diode, which emits radiant light in the near infrared (0.9 μm) and is called an infrared-emitting diode (IRED).

7.4.1 Infrared-Emitting Diode

The gallium arsenide IREDs are spectrally matched for use with silicon photodetectors. Figure 7.13 pictures the circuit symbol and the spectral response.

7.4.2 Light-Emitting Diode (LED)

One possible construction of an LED is depicted in Figure 7.14. An alloy of *N*-type gallium arsenide phosphide (GaAsP) is grown epitaxially on a gallium arsenide substrate. The *P* region is then diffused and topped off with a metal electrode. The electrode is shaped in such a way that it gives an even current distribution and allows most of the light to escape.

When the LED is forward biased, the junction potential is reduced and current flows. The current is made up of electrons flowing into the *P* region and holes into the *N* region. These carriers become the minority carriers in the respective regions. When electrons cross the junction, they may be lost through recombination. At the same time that they recombine they give off energy, which is the difference between their "free" or conduction-band state and their bound valence-band state. This energy is given off in the form of radiation. In some cases, the energy difference corresponds to radiation in the visible spectrum, that is, light. Depending on the LED material, the light may be in the green, yellow, or (more commonly) red part of the visible spectrum.

(a)

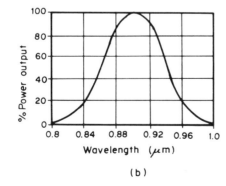

(b)

Figure 7.13 Infrared-emitting diode (0.9 μm) *PN* gallium arsenide: (a) circuit symbol and (b) relative spectral output.

The brightness of the light given off by the LED is contingent on the number of photons released by the recombination of carriers inside the LED: The higher the forward bias voltage, the larger the current and the larger the number of carriers that recombine. Therefore, we can increase the brightness by increasing the forward current. Typically, LEDs have a forward voltage rating of between 1 and 2 V and a current rating of about 50 mA. Figure 7.15 gives typical LED characteristics, showing the light intensity as a function of forward current, as well as the LED circuit symbol.

Light-emitting diodes may be used as circuit status indicators or, when used in arrays, they are suitable for visual alphanumeric readout displays.

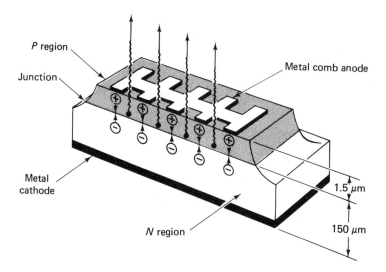

Figure 7.14 LED construction. The comb-type anode gives even current distribution while blocking less than 25 percent of the available light. (*Courtesy of Hewlett-Packard*)

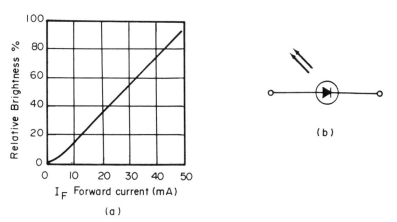

Figure 7.15 LED: (a) brightness versus forward current and (b) circuit symbol.

TABLE 7.1 OPTICAL COUPLERS/ISOLATORS

Device	Ckt construct emitter/detector	Applications
Optotransistor coupler	IRED/Si transistor	Electrical isolation in switching circuits, coupling circuits, and interface circuits.
Optodarlington transistor coupler	IRED/Darlington transistor	Electrical isolation and high current transfer ratio in interfacing systems, solid-state relays, and switching circuits.
Optotriac driver	IRED/SBS	Electrical isolated triac trigger.
Opto-SCR coupler	IRED/SCR	Electrical isolation between IC logic circuits and ac power line.
Optodigital logic coupler	IRED/IC	Electrical isolation in interfacing computer peripheral equipment, digital control of motors and servos.
Opto-ac linear coupler	IRED/ac linear amp	Electrical isolation in line coupling, medical equipment, and audio sytems.

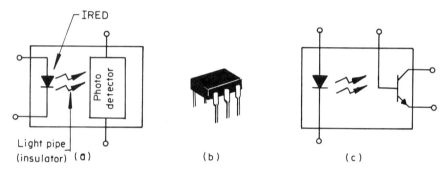

Figure 7.16 Optical coupler/isolator: (a) block diagram, (b) typical six-pin dual-in-line (DIP) package, (c) schematic of the optotransistor coupler.

7.5 OPTICAL COUPLERS/ISOLATORS

Optical couplers encompass a wide array of light-coupled devices, including transistor couplers, triac drivers, and logic couplers. Table 7.1 lists the various optical couplers/isolators along with typical applications. Figure 7.16(a) pictures the general structure of the optocoupler/isolator. The optocoupler is made up of an IR-emitting diode (IRED) connected (and electrically isolated) by a *light pipe* to the silicon photodetector. The optocoupler is usually packaged in a six-pin dual-in-line (DIP) package. For your reference, a data sheet for an optodarlington coupler is reproduced in Appendix A.

7.6 FIBER OPTIC EMITTERS AND DETECTORS

Because of their outstanding performance in transmitting data, fiber optic components are finding application in microprocessor systems, medical electronics, automobiles, and industrial control circuits. The glass optical cable used in the fiber optic system is lightweight and small. It is immune to electromagnetic and radio frequency interference. It can also transmit a great deal of data on each fiber since light frequencies (10^{13}-10^{14} Hz) are used.

In its simpler form, a fiber optics system consists of a single emitter and detector connected by a single optical fiber. Figure 7.17 pictures this system. The optical fiber is a thin, flexible thread-like fiber of clear plastic or glass.

The fiber optic active component (FOAC), which may be either an emitter or a detector, is manufactured with a sleeve or *ferrule* around its perimeter to facilitate mating with the fiber optic connector. Figure 7.18 pictures the construction of the fiber optic active component.

7.6.1 Emitters

The fiber optic IR-emitter diodes, which are housed in an FOAC case, transmit at one of two wavelengths (0.820 μm or 0.900 μm) and have a 200 μm (0.008 in.) diameter *optical port*.

7.6.2 Detectors

The fiber optic IR-detectors, which are housed in an FOAC case like the fiber optic IR-emitter, respond to one of two frequencies (0.820 μm or 0.900 μm) and

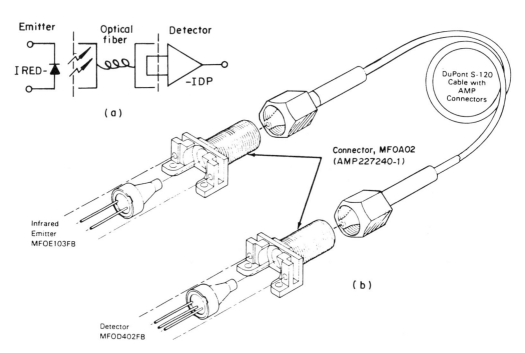

Figure 7.17 Simple fiber optic system: (a) system schematic and (b) component assembly. (*Courtesy of Motorola, Inc.*)

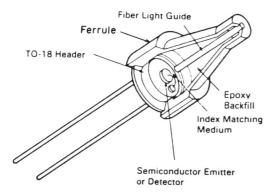

Figure 7.18 The construction of the fiber optic active component (FOAC). (*Courtesy of Motorola, Inc.*)

have a 200 μm (0.008 in.) diameter optical port. The detector may be a PIN* photodiode, a phototransistor, a photodarlington transistor, or an integrated detector preamplifier (IDP). A fiber optic system data sheet for an integrated detector/preamplifier is located in Appendix A.

REVIEW QUESTIONS

1. What is the principle of photoresistivity? Name some substances exhibiting it.
2. How does a photoconductive cell behave when illuminated?
3. What is meant by the term *dark resistance* as applied to photoconductive cells?
4. What are the characteristics of photovoltaic cells?
5. In what ways are photoconductive cells different from photovoltaic cells?
6. What are some commonly used materials in the production of photovoltaic cells?
7. What is a solar cell? Name some applications for solar cells.
8. How does a photodiode differ from a photoconductive cell? In what ways are they similar?
9. What is the meaning of the term *dark current* as applied to a photodiode?
10. When a photodiode is illuminated, which one of its properties changes and how?
11. What is a phototransistor and how is it different from a photodiode? In what ways are they similar?
12. What similarity and differences with a BJT does a phototransistor have?

---

* The *PIN* diode differs from the typical *PN* diode in that a region of intrinsic material separates the *P*- and *N*-type materials. In cross-section, the structure of the diode is *P*-Intrinsic-*N* materials, thus the acronym PIN.

13. What advantages does a phototransistor have over a photodiode?
14. What is the advantage of the photodarlington over the phototransistor?
15. What is the operation of an LASCR? How is it similar to the operation of an SCR? How is it dissimilar?
16. In what part of the spectrum do IREDS transmit?
17. From what materials are the photoemitters constructed?
18. What is the mechanism that causes the LED to emit light?
19. What controls the brightness of the LED light output?
20. List several applications for optocouplers.
21. Using the information in Appendix A, describe the circuit of an integrated detector preamplifier (IDP).
22. What do the letters *PIN* mean when used with IR-emitters?

Part II

Circuits

This section of the book covers assorted electronic circuits. We shall discuss basic applications of the many devices introduced in Part I, together with some of the most common integrated circuits.

In order to understand these electronic circuits, you will study such basic concepts as rectifiers and filters, amplifier fundamentals, oscillators, differential amplifiers, and operational amplifiers. Besides providing applications for the devices covered in Part I, these circuits serve as the building blocks for the more complex systems treated in Part III.

Chapter 8

Rectifiers and Filters

In prior chapters, we have seen that most devices need a dc voltage supply to set the appropriate operating point for operation on time-varying signals. In this chapter, we shall examine some of the basic building blocks of a system that converts 60-Hz ac line voltage into dc voltage. Such a system is called a *power supply*, and it varies in complexity depending on the application. The complete power supply will be taken up in Chapter 17.

The first and most important component of a power supply is called the *rectifier*. It converts or rectifies the alternating current into a unidirectional current that has a time-varying component as well as a dc (or average) value. The time-varying component of the rectifier output, called the *ac ripple*, is undesirable; its effects are minimized by the use of appropriate *filters*. An ideal filter, in this context, is a two-port network that passes direct current but effectively blocks any alternating current. Because an ideal filter does not exist, we shall apply the term *filter* to any two-port (four-terminal) network that has a smaller ac ripple component at its output terminals than at its input terminals.

8.1 HALF-WAVE RECTIFIER

The half-wave, single-phase rectifier circuit is shown in Figure 8.1. The ac input is typically 115 V-rms at 60 cycles. It is usually coupled into the rectifier by a transformer, as shown. Depending on the magnitude of the dc voltage desired, the transformer either steps up or steps down the ac voltage. The *dot convention* on the transformer indicates that if the current at the input is entering the side with the dot, the current at the output is leaving the side with the dot, and vice versa. Furthermore, if the input voltage is such that the terminal with the dot is positive, then the voltage at the output dotted terminal will also be positive.

8.1.1 Half-Wave Operation

When a sinusoidal ac voltage, such as shown in Figure 8.2(a), is impressed on the input, a similar ac voltage—either larger or smaller in magnitude—appears

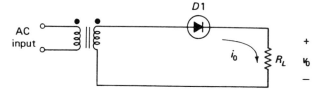

Figure 8.1 Half-wave rectifier.

at the output terminals of the transformer. During the first half-cycle of the input voltage, a current will flow in the circuit, because the anode of the diode is driven positive with respect to the cathode. Although there is a small voltage drop across the forward-biased diode (≈ 0.6 V), in most cases it can be neglected. During the positive half-cycle, all of the applied voltage appears across the load resistor R_L and almost none across the forward-biased diode. The relationship between voltage and current in R_L is linear. During this time, therefore, the voltage and current waveforms are identical to the input waveform, as illustrated in Figures 8.2(b) and 8.2(c).

During the second half-cycle of the input voltage, the polarity is reversed;

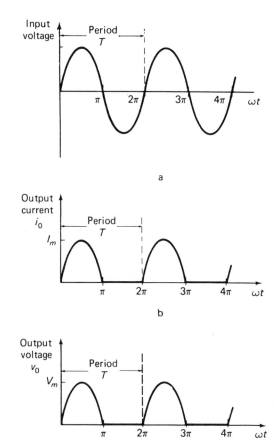

Figure 8.2 Waveforms in the half-wave rectifier circuit.

the anode of the diode is made negative with respect to the cathode. The diode is then reverse biased and acts as an effective open circuit. An extremely small reverse-saturation current does flow. But it is many orders of magnitude smaller than the forward current, so it can be neglected. All the voltage appears across the diode during the negative half-cycle of the input. There is no current through nor any voltage across the load resistor R_L.

As the cycle of the input voltage is repeated, the diode alternately conducts current (during the positive half-cycle) and blocks current (during the negative half-cycle). The current and voltage waveshapes at the load, as shown in Figures 8.2(b) and 8.2(c), result. The output of a half-wave rectifier is a unidirectional time-varying current, also called *pulsating direct current*. It is also periodic or repetitive. We can express the output current over one period as

$$i_o = \begin{cases} I_m \sin(\omega t) & \text{for } 0 \leq \omega t \leq \pi \\ 0 & \text{for } \pi < \omega t < 2\pi \end{cases} \quad (8.1)$$

where I_m is the peak value of the current, $\omega = 2\pi f$, and usually $f = 60$ Hz. The period, T, of the waveshape appears at $\omega T = 2\pi$, so $T = 1/f$.

8.1.2 Average Current

The dc or average value of the current is given by the area under the waveshape in one period divided by the period. The area under the waveshape for one period is determined by taking the integral of the current over one period. The average value for half-wave operation is

$$I_{dc} = \frac{I_m}{\pi} = 0.318 I_m \quad (8.2)$$

8.1.3 Ripple Factor

A measure of the alternating or fluctuating component in the rectified waveshape, called the *ripple factor* (labeled r), is defined as

$$r = \frac{\text{effective (rms) value of alternating component of the wave}}{\text{average (dc) value of the wave}} \quad (8.3)$$

The effective value of a wave is the rms (root-mean-square) value. It is determined by taking the square root of the average value of the square of the wave. It is not always easy to determine the rms value of the alternating component of the wave.

Equation (8.3) is rewritten for the ripple factor using symbols. Thus,

$$r = \frac{\sqrt{I_{rms}^2 - I_{dc}^2}}{I_{dc}} = \sqrt{\left(\frac{I_{rms}}{I_{dc}}\right)^2 - 1} \quad (8.4)$$

where the rms value of the current, for the half-wave rectifier, is

$$I_{rms} = \frac{I_m}{2} = 0.5\, I_m \quad (8.5)$$

Using the rms value given in Equation (8.5) and the dc value given in Equation (8.2), we can calculate the ripple factor for the half-wave rectifier as

$$r = \sqrt{\left(\frac{I_m/2}{I_m/\pi}\right)^2 - 1} = \sqrt{\left(\frac{\pi}{2}\right)^2 - 1} \cong 1.21 \qquad (8.6)$$

Thus, the ripple factor for a half-wave rectifier is very high. Ideally, we want the alternating component to be as small as possible or the ripple factor to be as close to zero as possible.

8.2 FULL-WAVE RECTIFIER

In applications requiring low ripple and a high dc component, the half-wave rectifier is not used because of its high ripple factor and its low dc component. Instead, a somewhat improved performance is obtained from a full-wave rectifier, like the one shown in Figure 8.3.

8.2.1 Full-Wave Center-Tapped Rectifier

In the full-wave center-tapped rectifier circuit (FWCT), the ac input is applied to a transformer having a center-tapped secondary. The two-diode circuit functions like a tandem connection of two half-wave rectifier circuits; that is, during the positive half-cycle of the input, one half-wave rectifier is conducting; the other rectifier is off. The operation is reversed during the negative half-cycle of the input.

If we apply a sinusoidal input of the type illustrated in Figure 8.4 during the positive half-cycle (between 0 and π), diode $D1$ is forward biased and diode $D2$ is reverse biased. Thus, $D1$ passes current; whereas, $D2$ is effectively an open circuit. When the input voltage goes negative (between π and 2π), $D1$ is reverse biased and effectively an open; whereas, $D2$ is forward biased and conducts. The

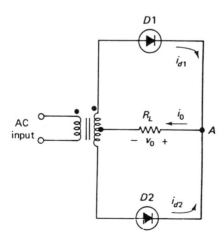

Figure 8.3 Full-wave center tapped rectifier (FWCT).

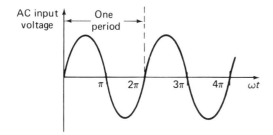
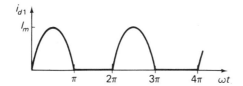
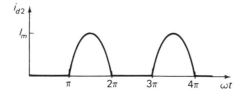
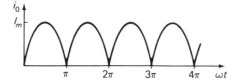

Figure 8.4 Waveforms in the full-wave rectifier circuit of Figure 8.3.

output current, i_o, can be determined by adding the currents at node A in Figure 8.3:

$$i_o = i_{d1} + i_{d2} \tag{8.7}$$

Because i_{d1} flows during the first half-cycle and i_{d2} flows during the second half-cycle, the output [Figure 8.4(d)] flows for the full cycle—thus, the name full-wave rectifier.

8.2.2 Full-Wave Bridge Rectifier

Another form of the full-wave rectifier, called the *full-wave bridge rectifier* (FWB), is depicted in Figure 8.5. The bridge rectifier offers some advantages over the two-diode full-wave rectifier. It has a lower reverse voltage per diode, and it eliminates a center-tapped transformer; however, it does need four diodes.

The operation of the bridge rectifier is similar to that of the conventional full-wave rectifier. Diodes operate in pairs. During the first half-cycle of the input

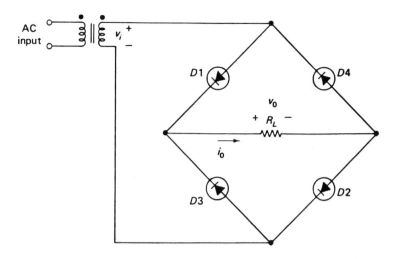

Figure 8.5 Full-wave bridge rectifier (FWB).

while v_i is positive, $D1$ and $D2$ are forward biased, and $D3$ and $D4$ are reverse biased and effectively open. $D1$ and $D2$ provide a current path through the load resistor R_L, as shown in Figure 8.6. When the input voltage reverses polarity during the second half-cycle, the situation is reversed. $D1$ and $D2$ are reverse biased and effectively open; $D3$ and $D4$ are forward biased and provide a current path through R_L, as shown in Figure 8.7. The current through R_L, labeled i_2, is in the same direction (in this case, from left to right) during both the negative and positive half-cycles. The pertinent waveshapes for the bridge rectifier are shown in Figure 8.8.

The load current, i_o, for either of the full-wave rectifier circuits is given by

$$i_o = \begin{cases} I_m \sin x & \text{for } 0 \leq x \leq \pi \\ -I_m \sin x & \text{for } \pi < x < 2\pi \end{cases} \quad (8.8)$$

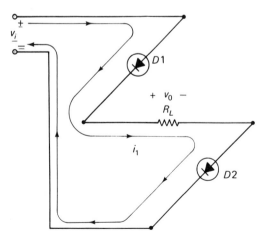

Figure 8.6 Full-wave bridge rectifier circuit during the positive half-cycle.

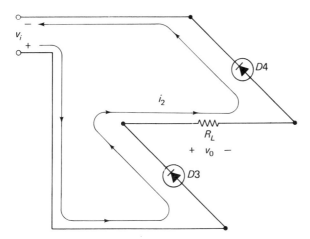

Figure 8.7 Full-wave bridge rectifier circuit during the negative half-cycle.

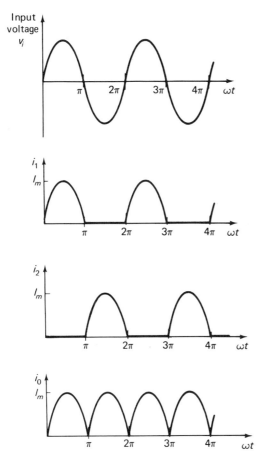

Figure 8.8 Waveshapes for the full-wave bridge rectifier circuit.

The dc and rms currents are given by:

$$I_{dc} = \frac{2I_m}{\pi} = 0.637\, I_m \tag{8.9}$$

$$I_{rms} = \frac{I_m}{\sqrt{2}} = 0.707\, I_m \tag{8.10}$$

Comparing Equations (8.2) and (8.9), we can readily see that for the same peak input current, the full-wave rectifier yields a somewhat larger dc output current.

Another comparison between the two is afforded by the ripple factor. Using Equation (8.4), we can obtain the ripple factor for the full-wave rectifier:

$$r = \left[\left(\frac{I_m}{\sqrt{2}}\frac{\pi}{2I_m}\right)^2 - 1\right]^{1/2} = \sqrt{1.2337 - 1}$$
$$r = 0.483 \tag{8.11}$$

We can readily see that the ripple factor is lower for the full-wave rectifier, so the ac component in the rectified output will also be smaller for the full-wave rectifier than for the half-wave rectifier.

Even though the full-wave rectifier reduces the ac component, the ac ripple is still too high for most applications. With either a half-wave or a full-wave rectifier, we usually filter the rectified output to further reduce the ac component.

8.2.3 Center-Tapped Bridge Rectifier

Figure 8.9 pictures the *dual complementary rectifier circuit,* which is made up of two full-wave center-tapped circuits. Because this circuit is usually implemented using a full-wave bridge rectifier assembly, it is commonly called a *center-tapped bridge rectifier* (CTB).

During the positive half-cycle of the input ac sinusoidal wave, diodes D_1 and D_2 are forward biased (ON) and diodes D_4 and D_3 are reverse biased (OFF). Current flows out of the top of the transformer, down through D_1 and R_{L1}, and back to the center-tap of the transformer. At the same time, current flows into the bottom of the transformer and out through the center-tap down through R_{L2}, around and down through D_2.

During the negative half-cycle of the ac sinusoidal wave, diodes D_4 and D_3 are forward biased (ON) and diodes D_1 and D_2 are reverse biased (OFF). Current flows into the top of the transformer and out through the center-tap, down through R_{L2}, around and up through D_4. At the same time, current flows out of the bottom of the transformer, up through D_3 and down through R_{L1} and back to the center-tap of the transformer. Thus, D_1 and D_3 form a positive FWCT rectifier to feed R_{L1}, and D_4 and D_2 form a negative FWCT rectifier to feed R_{L2}.

The center-tapped bridge rectifier is a very simple way of obtaining two equal voltage sources of opposite polarity that share a common reference.

The dc and rms currents in the CTB rectifier are determined by Equations

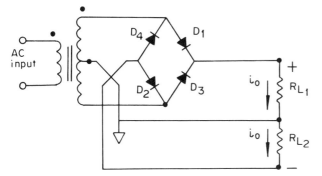

Figure 8.9 Center-tapped bridge rectifier (CTB).

8.9 and 8.10, respectively. The ripple factor for the CTB, like that of the FWB and FWCT, is 0.483.

8.3 RECTIFIER FILTERS

Many different two-port networks can be and are used to filter the output of a rectifier. No matter how complex or how simple these filters are, their operation is based on the action of certain reactive elements: capacitors, inductors, or both. Two simple single-element filters along with an *"electronic filter"* are shown in Figure 8.10.

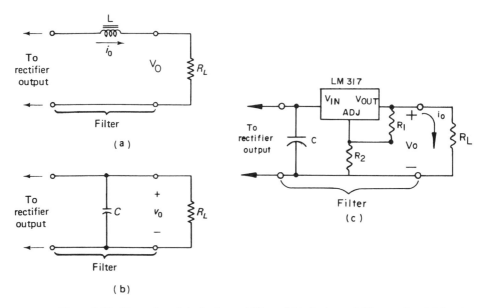

Figure 8.10 Examples of single-element filters: (a) inductor and (b) capacitor. (c) An electronic filter.

8.3.1 Inductor Filter

In the inductor filter, an inductor is inserted in series with the load. The rectifier voltage, containing both dc and ac components, is applied to the inductor-resistor combination. It divides between the two; the inductor having almost zero dc resistance has essentially negligible dc voltage across its terminals. Almost all of the dc voltage appears across the load resistance (R_L). The impedance of the inductor is extremely high to the ac component of the current so most of the ac voltage appears across the inductor. We can also view the filtering action of the inductor in another way. Recall that the inductor tends to oppose changes in current, thereby smoothing out the current waveshape. In effect, this process reduces the ac component in the output current.

8.3.2 Capacitor Filter

In the capacitor filter, a capacitor is inserted in parallel with the load. The rectifier output current, containing both direct and alternating components, divides between the capacitor and resistor. The capacitor offers almost infinite impedance to the dc current. Consequently, almost all of the dc current flows through the load resistance (R_L). The impedance of the capacitor is very low to the ac component of the current, so that a large portion of the alternating component of the current flows through the capacitor. Another way of viewing the filtering action of the capacitor is to remember that the capacitor acts to prevent the voltage across its terminals from changing. Because it is in parallel with the load resistor, the output voltage is smoothed by the action of the capacitor.

8.3.3 Electronic Filter

The simplest and most versatile of the electronic filters is the LM 117 (LM 317) 3-terminal adjustable integrated-circuit positive voltage regulator [Figure 8.10(c)] and its complement, the LM 137 (LM 337) negative voltage regulator. These regulators provide filtering equivalent to a large multiple element LC filter, like those pictured in Figure 8.11. Like a gigantic capacitor, the active regulator provides a low impedance path to ground at low frequencies for the ac component of the rectified wave while, at the same time, it passes the dc component onto the load resistance (R_L). Both discrete and integrated regulators are presented in Chapter 17.

With the availability of low cost, high quality full-wave rectifier assemblies in a single package (Figure 8.12), the use of the half-wave rectifier circuit is practically nonexistent in today's applications. Because of the lower ripple factor (0.48) of the full-wave rectifier circuits, the passive filter components (inductors and capacitors) are smaller and less expensive than those of the half-wave rectifier.

8.3.4 Multiple-Element Filters

A multiple-element filter may be used in some special applications. When this is the case, the action of the inductor (choke) and capacitor are combined in a single filter to improve the performance of the rectifier and lower the ripple content of

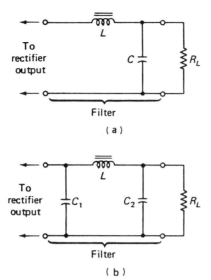

Figure 8.11 Multiple-element filters: (a) L-section and (b) π-section.

the load current and voltage. Two filters in common use are the *L-section* and *π-section* filters, illustrated in Figure 8.11. In both filters, the choke is employed in series with the load to smooth out the current. The capacitor (or capacitors) is placed in parallel to smooth out the voltage. Pertinent data for the four basic *L*, *C*, and *LC* filters are summarized in Table 8.1.

Frequently, the choke that needs to be used in the π-section filter is quite high in value (Henrys) and, therefore, very bulky and expensive. In such cases, the choke may be replaced by a resistor whose resistance is equal to the reactance of the choke. However, this is possible only when the voltage drop across the series resistor (which replaces the choke) is not large ($\leq 10\% \, R_L$) and does not represent an unwarranted loss of power. Example 8.1 develops the values for the components in the rectifier-filter circuit of Figure 8.13.

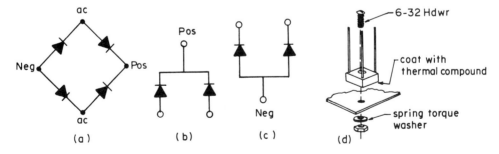

Figure 8.12 Circuit schematics of commercially available full-wave rectifier assemblies: (a) FWB, (b) FWCT for a positive dc voltage, (c) FWCT for a negative dc voltage, (d) Typical assembly of a 2-A epoxy bridge rectifier onto an aluminum heat sink.

Sec. 8.3 Rectifier Filters

TABLE 8.1 FILTER PERFORMANCE*

Filter type	Ripple factor (r)	dc voltage (V_{dc})
Choke (inductor)	$\dfrac{R_L}{1{,}600\, L}$	$0.636\, V_m$
Capacitor	$\dfrac{2{,}400}{R_L C}$	$V_m - \dfrac{4{,}200\, I_{dc}}{C}$
L-section	$\dfrac{0.83}{LC}$	$0.636\, V_m$
π-section	$\dfrac{3{,}300}{C_1 R_L C_2 L}$	$V_m - \dfrac{4{,}200\, I_{dc}}{C}$

* For a full-wave rectifier with resistance in ohms, capacitance in μF, and inductance in Henries. The line frequency is assumed to be 60 Hz.

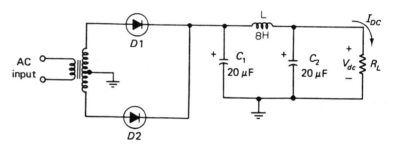

Figure 8.13 Full-wave rectifier with a π-section filter (see Example 8.1).

Example 8.1

For the capacitor input π-filter circuit shown in Fig. 8.13, we want to determine: (1) the ripple factor and the ripple voltage, (2) the diode ratings, and (3) the transformer rating. The output dc voltage is 50 V at a current of 100 mA.

Solution. (1) From Table 8.1, we see that the ripple factor for the π-section filter is

$$r = \frac{3300}{(20)(R_L)(20)(8)} = \frac{1.03}{R_L}$$

The load resistance R_L can be determined from the ratio of V_{dc} to I_{dc}: $R_L = 50/0.1 = 500\, \Omega$. Thus,

156 Rectifiers and Filters Chap. 8

$$r = \frac{1.03}{R_L} = \frac{1.03}{500} = 0.0021$$

The rms ripple voltage, V_r, may be determined from the definition of the ripple factor, Eq. (8.3), as $r = V_r/V_{dc}$. Thus,

$$V_r = rV_{dc} = (0.0021)(50) \text{ V} = 0.105 \text{ V}$$

(2) Each diode must be capable of sustaining a forward current of at least 100 mA and must have a peak-reverse voltage rating in excess of 2 V_m. From Table 8.1, we can determine V_m.

$$V_m = V_{dc} + \frac{4200 I_{dc}}{C} \cong 50 + 21 \cong 71 \text{ V}$$

The exact peak forward current that each diode experiences is V_m/R_L or 142 mA. Diodes with a current rating of 150 mA and reverse voltage rating in excess of 150 V may be used. The IN4003 diode meets these requirements (see data sheet in Appendix A).

(3) The transformer rating is determined from Table 8.2 and Eq. 8.12. From Eq. 8.12, determine the transformer rms voltage (assume that the primary is a nominal 115 V):

$$V_{trans} = \frac{0.707(V_{load} + V_{rect} + V_r + V_{ind})}{\eta_{rect}} \qquad (8.12)$$

where: V_{rect} = 1.2 V FWCT and 2.4 V FWB
V_{ind} assumed to be $0.10 \times V_{load}$
η_{rect} is the transformer efficiency—typically 0.81 for an FW rectifier

(Note: Eq. (8.12) may only be used with capacitive input filters.)

$$V_{trans} = \frac{0.707(50 + 1.2 + 0.1 + 5)}{0.81} = 49 \text{ V}$$

Thus, a 100-V CT (center-tapped) transformer (50-0-50) would be used. From Table 8.2:

$$I_{trans} = 1.2 \times I_{load} = 1.2 \times 100 \text{ mA} = 120 \text{ mA}$$

Had an FWB been used instead of the FWCT, then the transformer would

TABLE 8.2 TRANSFORMER SECONDARY CURRENT

Rectifier	Input component	Secondary current
FWCT	Inductor	0.7 × dc load current
FWCT	Capacitor	1.2 × dc load current
FWB	Inductor	1.0 × dc load current
FWB	Capacitor	1.8 × dc load current
CTB	Use FWCT factors	

be

$$V_{trans} = \frac{0.707(50 + 2.4 + 0.1 + 5)}{0.81} = 50 \text{ V}$$

and from Table 8.2:

$$I_{trans} = 1.8 \times I_{load} = 1.8 \times 100 \text{ mA} = 180 \text{ mA}$$

Summary.

    Diode: $V_R = 150$ V, $I_F = 150$ mA

    Transformer: FWCT 100 V CT at 120 mA = 12 VA

                       FWB 50 V at 180 mA = 9 VA

Observation. It is not possible to replace the choke by an appropriate resistor in the preceding example because the resistance needed to keep the same ripple factor would be $2\pi fL \approx 3000$ Ω. The dc current would fall to about $V_{Dc}/3$ kΩ, and the output dc voltage would be almost negligible.

REVIEW QUESTIONS

1. What is the role of a rectifier in a dc power supply?
2. What devices are used in rectifiers?
3. What are the basic types of rectifiers?
4. What are the characteristics of a half-wave rectifier? How many diodes are necessary?
5. What is the waveshape output from a half-wave rectifier? Sketch it.
6. Sketch a half-wave rectifier circuit that provides a positive output voltage. Sketch a circuit for a negative output voltage.
7. How does a half-wave rectifier operate? Discuss in terms of the conduction and nonconduction of the diode for a sinusoidal input.
8. What is meant by *ripple*?
9. What provides a numerical measure of the ripple?
10. What are the different types of full-wave rectifiers?
11. In what ways are half-wave and full-wave rectifiers similar? How are they different?
12. What advantages does a full-wave rectifier offer over a half-wave rectifier?
13. At what cost does the full-wave rectifier offer advantages over a half-wave rectifier?
14. What is the difference between a full-wave rectifier and a bridge rectifier? What are the similarities?
15. What are the advantages of a bridge rectifier over a full-wave rectifier? What are the disadvantages?

16. Describe the operation of a full-wave rectifier (Fig. 8.3) in terms of the individual diodes.
17. What is the role of a filter in a dc power supply?
18. What basic properties of inductors and capacitor are used in filters?
19. What is the action of a capacitor filter?
20. What is the action of an inductor filter?
21. What are some other commonly used filters?
22. What are the advantages of multiple-element filters over the simple capacitor filter? What are the disadvantages?
23. What is the effect of any filter on the ripple?
24. What are the elements of an *L*-section filter?
25. What are the elements of a π-section filter?

PROBLEMS

1. The half-wave rectifier circuit shown in Fig. 8.1 has an input that is 115 V-rms. It is used to drive a 100 Ω load. Neglecting the voltage drop across the diode, determine the dc output current and the ripple.
*2. The full-wave rectifier circuit shown in Fig. 8.3 has a center-tapped transformer delivering 50 V-rms from each end to the tap. The load resistance is 30 Ω. Determine the output dc voltage and the ripple factor.
3. Repeat Problem 2 with an input of 50 V-rms across the complete secondary of the transformer.
4. The transformer delivers 10 V-rms to the bridge rectifier circuit shown in Fig. 8.5. Use silicon diodes with a forward drop of 1 V and a 20 Ω load. Determine the dc output current and the ripple.
5. Repeat Problem 4, neglecting the diode forward drop. In terms of percentage, how much error is introduced?
6. The input to an *L*-section filter is a full-wave rectified voltage, 45 V peak. Determine the ripple factor if the inductor is 1 henry (H) and the capacitor is 100 microfarads (μF).
7. The output of the rectifier in Problem 2 is applied to a capacitor filter, with a dc load of 100 Ω. Determine the capacitor needed to produce a ripple factor of 0.01. For this value of capacitor, determine the output dc voltage.
8. The input to a π-section filter is a full-wave rectified voltage with a 50 V peak. The load is 1 kΩ. An 8 H inductor is available. Determine the capacitors needed to produce a ripple of 0.001. (Assume $C_1 = C_2$.)
*9. A π-section filter driving a 100 Ω load produces a ripple factor of 0.05. Determine the ripple factor if the load is: (a) 50 Ω and (b) 200 Ω.
10. A 5000 μF capacitor is placed at the output of a full-wave rectifier with a voltage of 8 V peak. Determine the load resistance if the output dc voltage of 6 V is measured. Under these circumstances, what is the ripple factor?

Chapter 9

Amplifier Fundamentals

The amplifier is one of the most basic and common blocks used in an electronic system. As the name implies, an amplifier is a circuit that raises the level of a signal. A small signal present at the input of an amplifier is processed by the amplifier to provide a signal at its output that is identical in all aspects to the input signal except that it is larger in magnitude.

The complexity and type of circuit used in the amplifier is dictated by the type of signal to be amplified and by the amount of amplification (called *gain*) needed. Many different classifications are possible: current amplifiers, voltage amplifiers, power amplifiers (where both current and voltage are amplified), dc amplifiers, ac amplifiers (where time-varying signals are amplified), and so on. However, all amplifiers have certain properties in common: (1) They amplify—therefore, they all use at least one *active* device; (2) they all require a dc voltage supply; and (3) their operation can be summarized by specifying the gain, input impedance, output impedance, and frequency response.

Gain. The *gain* is a measure of the amount of amplification and it is defined as the numerical ratio of the output signal to the input signal. Thus, current gain is the ratio of output current to input current, voltage gain is the ratio of output voltage to input voltage, and power gain is ratio of output power to input power.

Impedance. The input *impedance* of an amplifier is the impedance seen by the source of the signal to be amplified. It is defined as the ratio of voltage to current at the input terminals of the amplifier. The output impedance of an amplifier is the impedance seen by the load, and it is defined as the ratio of the voltage to the current at the output terminals of the amplifier. Both the input and output amplifier impedance may be *real* (i.e., purely resistive) or *complex* (i.e., partly capacitive and/or inductive as well as resistive).

Frequency response. The *frequency response* of an amplifier is closely related to the gain. Usually the gain is not constant over all possible frequencies;

that is, signals of different frequencies are not amplified by the same amount. Therefore, when specifying the gain, we usually specify the range of frequencies (called the *bandwidth*) over which the gain is essentially constant. The bandwidth may be expressed in terms of frequency. For example, an audio amplifier may have a bandwidth of 20 kHz. The bandwidth may also be specified in terms of the lower and upper limits, called the lower and upper *cutoff frequencies*. In the case of the audio amplifier, these limits may be specified as a certain gain from 30 Hz to 20 kHz.

We can summarize our introduction of amplifiers by saying that our analysis will concern the gain (either current, voltage, or power), the input and output impedance, and the frequency response.

9.1 GAIN CALCULATIONS—SYSTEMATIC ANALYSIS

In the broadest sense, we can distinguish between two basic types of amplifiers according to the method of analysis that we find most suitable. In the previous discussion, we assumed that the gain of the amplifier did not depend on the amplitude of the input signal. If this is indeed the case, the amplifier properties can be determined by applying circuit analysis to the small-signal equivalent circuit or model of the active devices used in the amplifier. This type of amplifier is said to be *linear* or to have a *linear gain*. For example, if the amplifier gain is 100 and the input voltage signal is 10 mV, the output voltage will be 100×10 mV, or 1 V. In the same amplifier, if the input signal is 20 mV, the output will still be 100 times larger, or 2 V. This type of amplifier is called a *small-signal amplifier*.

In the second type of amplifier, called *large-signal* or *power* amplifier, the input signal is so large that the device (or devices) inside the amplifier will not operate in a linear fashion. For example, the same amplifier that gives an output current of 5 A for an input current signal of 0.2 A may give an output of 6 A for an input signal of 0.5 A. (The current gains are 25 and 12, respectively.) The large-signal or power amplifier is also characterized by varying degrees of *distortion*. Distortion is said to occur when an output-signal waveshape is not a true reproduction of the input-signal waveshape. Graphical analysis is best suited to large-signal or power amplifiers. We shall treat these subjects in more detail in Chapter 12 when we study power amplifiers.

There is no clear-cut separation between small-signal and large-signal amplifiers. In general, amplifiers having an output power level up to a few hundred milliwatts are analyzed as small-signal amplifiers. Amplifiers with output power levels around a watt or more are analyzed as large-signal amplifiers using graphical analysis. You should, however, consider these power levels as only the first steps in determining the type of analysis to be used. In the long run, the only valid analysis is the one that gives verifiable answers. Therefore, you must examine the characteristics of the devices used in the amplifier and ascertain whether the input signal is large enough to cause nonlinear operation. The test is not the characteristics of the devices alone nor the power or signal level alone; it is a combination of the two.

Let us assume for the moment that we are dealing with an amplifier operating under small-signal conditions. We then proceed with a systematic analysis as follows:

Step 1. Draw the schematic of the complete amplifier with the appropriate symbols for the active devices.

Step 2. Label all device terminals with a letter and number. For example, B for base, D for drain. If there are two or more of one type of device, use numbers as well. If there are two or more BJTs, use $B1$, $E1$, and $C1$ for the first, $B2$, $E2$, and $C2$ for the second, and so on.

Step 3. Redraw the schematic, keeping the relative positions of the components as close to the original as possible, with the following changes:
  (a) Replace all device symbols with the device model (or equivalent circuit).
  (b) Replace all dc voltage supplies with their equivalent impedance. In most cases, this equivalent impedance can be considered zero. Therefore, replace all dc voltage supplies with a short circuit.

Step 4. Determine the selected amplifier parameter (i.e., gain) by applying conventional circuit analysis. At this point, the circuit consists of ac voltage and current generators, capacitors, inductors, and resistors.

The procedure outlined here is quite valid; however, its implementation requires thought. The first stumbling block you will experience is in determining which model or equivalent circuit to use for the device [Step 3(a)]. You should use the model that is valid for the calculation to be made. For example, in calculating the dc operating point, the model of the base-emitter of the BJT (Figure 4.4) is a dc battery. Because it is dc, such a model would be inappropriate for ac analysis. When selecting an ac model, determining which form of which ac model to use for the BJT in a particular circuit may not be so obvious. The first step in selecting a model is to decide whether the BJT is connected in the CE, CB, or CC configuration. The model is then chosen from among those depicted in Figures 4.13, 4.15, and 4.16. The specific form of the model to be used depends on the desired degree of accuracy and the desired amplifier parameter. If, for example, we consider the *CE* configuration, we can choose among the three models in Figure 4.13. The model in Figure 4.13(a) is the most accurate; the one in Figure 4.13(c) is the least accurate; and the model of Figure 4.13(b) is somewhere in between.

In the final application of circuit analysis specified in Step 4, it is not always clear which way or method leads to the most direct answer. Because there is no best way, you may use many different methods to obtain the solution. The preferable method is the one that is the shortest. However, you should begin by using the methods you are most familiar with and with which you feel the most comfortable. The following sections will be devoted to the application of the procedure outlined in the analysis of single-stage BJT and FET amplifiers.

9.2 SINGLE-STAGE BJT AMPLIFIER

A single-stage BJT (common-emitter) amplifier is shown in Figure 9.1. Before we illustrate how this small-signal amplifier may be analyzed, let us discuss the reasons for the circuit components in the amplifier.

Resistors R_1, R_2, R_E, and R_C are bias resistors needed to set up an appropriate operating point (Q-point) as discussed in Chapter 4. Capacitors C_1 and C_2 are called *coupling capacitors*. Their role is to block direct current, so that the dc performance of the amplifier will not be disturbed by the source or the load. They also prevent direct current from entering the source and load. The capacitor C_B is called a *bypass* capacitor because it is usually chosen large enough to effectively short out R_E at signal frequencies.

We can proceed with the steps for small-signal analysis as outlined, noting that steps one and two are already incorporated into Figure 9.1. The results of Step 3 are shown in Figure 9.2. If we replace the voltage supply V_{cc} by its equivalent impedance (a short), then R_2 is placed in parallel with R_1, and R_C is connected from collector to ground, as indicated in Figure 9.2(a). The transistor symbol is replaced by the hybrid parameter equivalent circuit of Figure 4.13(b). This model for the transistor is valid only at low and medium frequencies, not at high frequencies. Figure 9.2(a) is an equivalent low-frequency circuit for the amplifier shown in Figure 9.1 because the coupling and bypass capacitors are included. The coupling and bypass capacitors are large, usually in the microfarad range, so that at midfrequencies their impedance approaches a short and the equivalent circuit of the amplifier then becomes the one shown in Figure 9.2(b). Here we see that C_1 and C_2, as well as C_B, are replaced by short circuits. Note that this equivalent circuit contains no reactive elements, and it will yield a gain that is independent of frequency. The range of frequencies over which the equivalent circuit of Figure 9.2(b) is valid is called the *midband region*, and the gain over these frequencies is called the *midband gain*. At some higher frequencies (perhaps in the megahertz range, depending on the transistor), the h-parameter model is

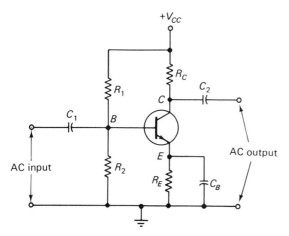

Figure 9.1 Single-stage common-emitter bipolar junction transistor (BJT) amplifier.

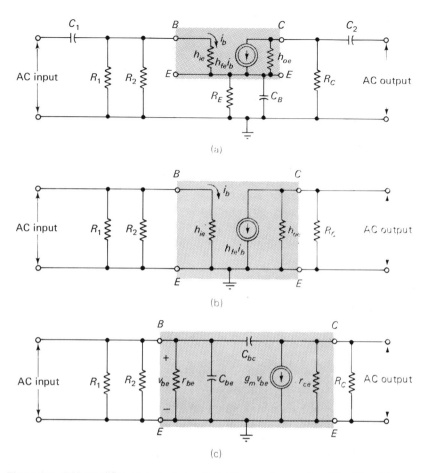

Figure 9.2 BJT amplifier equivalent circuits (a) for low frequencies, (b) midband frequencies, and (c) high frequencies.

no longer valid and the hybrid-π model must be used. The hybrid-π model yields the high-frequency equivalent circuit of the amplifier, shown in Figure 9.2(c).

We now have three possible equivalent circuits for the amplifier illustrated in Figure 9.1. Each equivalent circuit is valid and may be used to predict the properties of the amplifier. For example, if we want to determine the amplifier gain, we should use the midband equivalent circuit shown in Figure 9.2(b). We can use the equivalent circuit of Figure 9.2(a) to determine the frequency at which the gain begins to fall off, that is, the lower limit of the midband region. Similarly, we can use the equivalent circuit of Figure 9.2(c) to predict the upper limit for the midband region.

The midband equivalent circuit of Figure 9.2(b) will be used to determine the midband gain. Combining R_1 and R_2 and labeling their parallel combination as R_B, we have the equivalent circuit shown in Figure 9.3. We can now implement Step 4 of the small-signal analysis procedure by finding the current gain of the

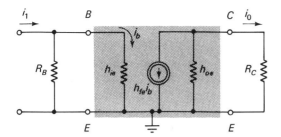

Figure 9.3 BJT amplifier simplified equivalent circuit.

amplifier. We first define the input current i_1 as shown and the output current i_o as shown. The current gain, labeled A_I, is then the ratio of i_o to i_1. The simplest way to evaluate the current gain is to express both i_i and i_o in terms of i_b. At the input we have current division of i_1 between R_B and h_{ie}. Thus,

$$i_b = i_1 \left(\frac{R_B}{R_B + h_{ie}} \right) \tag{9.1}$$

At the output we have current division between R_C and the resistance $1/h_{oe}$. Thus,

$$i_o = (-h_{fe} i_b) \left[\frac{\dfrac{1}{h_{oe}}}{\dfrac{1}{h_{oe}} + R_C} \right] \tag{9.2}$$

The minus sign is due to the direction of the $h_{fe} i_b$ current generator. Equation (9.2) can be rearranged by multiplying the numerator and denominator by h_{oe} to obtain

$$i_o = i_b \frac{-h_{fe}}{1 + h_{oe} R_C} \tag{9.3}$$

We can now substitute Equation (9.1) for i_b into Equation (9.3):

$$i_o = i_1 \left(\frac{R_B}{R_B + h_{ie}} \right) \left(\frac{-h_{fe}}{1 + h_{oe} R_C} \right) \tag{9.4}$$

Dividing both sides of the preceding equation by i_1 yields the desired result—the midband current gain:

$$A_I = \frac{i_o}{i_1} = \frac{-h_{fe} R_B}{(R_B + h_{ie})(1 + h_{oe} R_C)} \tag{9.5}$$

Equation (9.5) gives the midband current gain of the amplifier in Figure 9.1 in general form. For a specific case, to get a numerical answer for the midband gain, we simply need to insert the resistor values together with the transistor parameter values into the same equation. This procedure is illustrated in Example 9.1.

Example 9.1

A BJT amplifier like the one in Fig. 9.1 is constructed with the following resistor values: $R_1 = 150$ kΩ, $R_2 = 100$ kΩ, $R_E = 1.5$ kΩ, and $R_C = 4.7$ kΩ. It has been determined that the transistor parameters at the resulting operating point are: $h_{ie} = 1$ kΩ; h_{re} is negligibly small; $h_{fe} = 80$; $h_{oe} = 20$ μS. We want to determine the midband current gain.

Solution. Before we can use Eq. (9.5), we need to determine R_B:

$$R_b = \frac{R_1 R_2}{R_1 + R_2} = \frac{(150)(100)}{150 + 100} \text{ k}\Omega = 60 \text{ k}\Omega$$

Now, making use of Eq. (9.5), we find that

$$A_I = \frac{(-80)(60)}{(60 + 1)[1 + (0.02)(4.7)]} \cong -72$$

Observation. The minus indicates 180° phase shift between input and output.

You can gain additional insight into determining the gain if you examine how we obtained the gain expression in Equation (9.5), considering the typical values in the preceding example. When obtaining Equation (9.1), note that if R_B is very large compared to h_{ie} (which it usually is), we can get a very good approximation by neglecting the effect of R_B. So Equation (9.1) becomes

$$i_1 \cong i_b$$

In a similar manner, Equation (9.2) may be made into an approximation. That is, the resistance corresponding to $1/h_{oe}$ may be much larger (as is the usual case with a silicon BJT) than R_C and, as such, may be neglected. Equation (9.2) becomes

$$i_o \approx -h_{fe} i_b$$

The midband current gain can then be approximated by

$$A_I \approx -h_{fe} i_b / i_b \approx -h_{fe} \tag{9.6}$$

In general, when two resistors are connected in parallel, if one resistor is at least ten times larger than the other, we can neglect the effect of the larger resistor and have no more than a 10 percent error in the approximation.

9.3 SINGLE-STAGE FET AMPLIFIER

A single-stage FET amplifier is shown in Figure 9.4. The basic similarity between the FET and BJT amplifiers is obvious. The procedure in determining the properties of the FET amplifier is essentially the same as for the BJT amplifier.

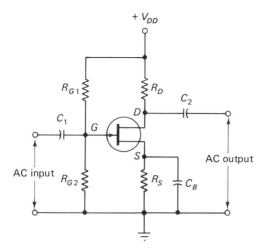

Figure 9.4 Single-stage FET amplifier.

We can redraw the circuit, replacing the FET symbol by its low-frequency equivalent circuit, as shown in Figure 9.5(a). This equivalent circuit, containing the coupling and bypass capacitors, is valid at low frequencies. As in the case of the BJT amplifier, we can replace all coupling and bypass capacitors with short circuits to obtain the midband equivalent circuit for the FET amplifier, as indicated in Figure 9.5(b). For completeness, the high-frequency equivalent circuit for the FET amplifier is illustrated in Figure 9.5(c). It is distinguished by the use of the high-frequency model for the FET.

Let us now proceed with analyzing the FET amplifier, that is, determining the midband voltage gain. The midband equivalent circuit of Figure 9.5(b) is redrawn, combining resistors R_{G1} and R_{G2} into a single resistor: R_G. In addition, the input and output voltage polarities, v_1 and v_o respectively, are defined. These steps are shown in Figure 9.6. In this figure, we see that

$$v_1 = v_{gs} \tag{9.7}$$

At the output, the total resistance, R_t, is the parallel combination of r_d and R_D, so

$$R_t = \frac{r_d R_D}{r_d + R_D} \tag{9.8}$$

The output voltage v_o is then given by the product of the current and total resistance:

$$v_o = -g_m v_{gs} R_t \tag{9.9}$$

The minus sign is due to the direction of the current generator. The voltage gain A_v is defined as the ratio of the output to input voltage. Thus, by making use of Equation (9.7) in Equation (9.9), we can obtain

$$A_v = v_0/v_1 = -g_m v_{gs} R_t/v_{gs} = -g_m R_t{}^* \tag{9.10}$$

---

* g_m is often specified on data sheets as y_{fs} or y_{os}.

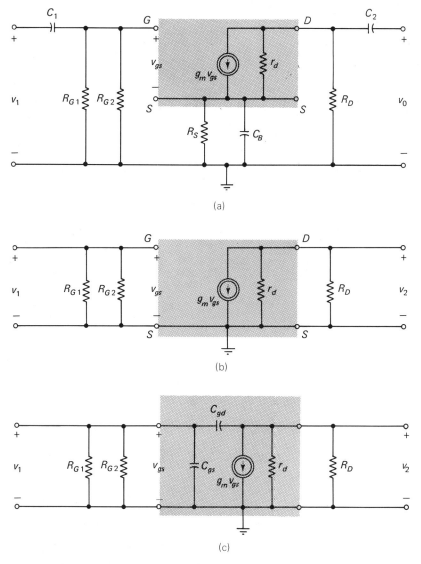

Figure 9.5 FET amplifier equivalent circuits for (a) low frequencies, (b) midband frequencies, and (c) high frequencies.

where R_t is defined in Equation (9.8). As was the case with the BJT amplifier, the gain expression obtained here is general. In a specific case, to obtain a numerical solution for the midband gain, we need only insert proper values into Equation (9.10). This procedure is illustrated in Example 9.2.

Example 9.2

A FET amplifier, as illustrated in Fig. 9.4, is constructed with resistor values:

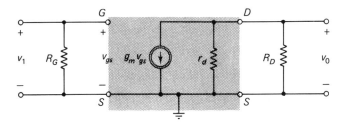

Figure 9.6 Simplified FET amplifier equivalent circuit.

$R_{G1} = R_{G2} = 1 \text{ M}\Omega$, $R_S = 5.6 \text{ k}\Omega$, and $R_D = 6.8 \text{ k}\Omega$. At the resulting operating point, the FET parameters have been determined as $g_m = 2 \text{ mS}$ and $r_d = 160 \text{ k}\Omega$. We want to calculate the midband voltage gain of the amplifier.

Solution. We first find R_t. (Note that r_d is much larger than R_D, so we could neglect r_d and approximate the total resistance as essentially R_D.)

$$R_t = \frac{(160)(6.8)}{160 + 6.8} \text{ k}\Omega = 6.52 \text{ k}\Omega$$

Making use of Eq. (9.10), we obtain the midband voltage gain:

$$A_V = -(2)(6.52) = -13$$

Had we neglected r_d, we would have come out with a slightly greater voltage gain of -13.6. Again the minus indicates 180° phase shift between input and output.

We can use the procedures outlined and illustrated here to evaluate the current or voltage gain of any small-signal amplifier stage, no matter how many stages there are or how complete the circuit is. The same general approach can also be followed to evaluate the current and voltage gain at either high or low frequencies; just use the appropriate model of the device (BJT or FET) under consideration. The difference between midfrequencies and high or low frequencies is the complexity of the analysis.

Be careful not to apply formulas that may appear elsewhere until you have gained experience. Although there is nothing wrong with using available formulas, you may not interpret the symbols correctly. For the time being, it is much safer to follow the procedures outlined in the previous sections.

9.4 FREQUENCY RESPONSE

Any amplifier provides gain over a certain range of frequencies. For a particular amplifier, the specific range of frequencies over which it can provide gain is a function of two quantities: (1) the response of the device (or devices) used in the amplifier and (2) the response of the reactive elements selected to be used in conjunction with the amplifier.

The simple single-stage amplifier illustrated in Figure 9.1 has three distinct frequency ranges: the midband (already discussed), the high, and the low. These ranges have definite boundaries: The gain begins to decrease from its midband value as the frequency is either increased (high-frequency end) or decreased (low-frequency end).

9.4.1 High Frequency

The high-frequency response is determined from the high-frequency equivalent circuit, which is obtained by following the rules just outlined and using the high-frequency model for the transistor (see Figure 4.17). The high-frequency equivalent circuit is that shown in Figure 9.2(c). Analysis of this circuit gives us the following results:

$$A_{i(hf)} \cong \frac{A_{i(mid)}}{1 + j\frac{f}{f_2}} \qquad (9.11)$$

where $A_{i(hf)}$ stands for the high-frequency current gain; $A_{i(mid)}$ is the midband current gain: f is the frequency and can take on any value; f_2 is the upper cutoff frequency determined from the following relationship:

$$f_2 = \frac{1}{2\pi R_t C_t} \qquad (9.12)$$

and R_t and C_t are the circuit equivalent resistance and capacitance calculated from the high-frequency equivalent circuit. In Figure 9.2(c), the effect of C_{bc} may be included in the input circuit, as shown in Figure 9.7, by modifying the capacitance to a value C_{bc} times the voltage gain from base to collector without C_{bc} in the circuit. Thus, this equivalent capacitance, called the *Miller capacitance,* is given by

$$C_M \cong g_m R_L C_{bc} \qquad (9.13)$$

where R_L is the total resistance at the output (r_{ce} in parallel with R_C). Observe, therefore, that the total equivalent capacitance at the input is the parallel com-

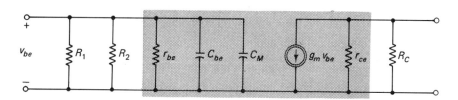

Figure 9.7 High-frequency equivalent circuit for BJT amplifier of Figure 9.1.

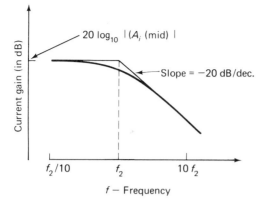

Figure 9.8 Magnitude response for the current gain.

bination of C_{be} and C_M:

$$C_t \cong C_{be} + g_m R_L C_{bc} = C_{be} + C_M \quad (9.14)$$

Similarly, the total equivalent resistance at the input is the parallel combination of R_1, R_2, and r_{be}:

$$R_t = \cfrac{1}{\cfrac{1}{R_1} + \cfrac{1}{R_2} + \cfrac{1}{r_{be}}} \quad (9.15)$$

The midband gain can be written in terms of the hybrid-π parameters:

$$A_{i(\text{mid})} = -g_m \frac{R_L R_t}{R_C} \quad (9.16)$$

This expression is exactly equivalent to the current gain expression obtained in Equation (9.5). The equivalence may be shown by making the appropriate substitutions from the hybrid-π parameters to h-parameters. In the case where r_{ce} is at least 10 times larger than R_C, r_{ce} may be neglected and $R_L = R_C$. We can then simplify the midband current gain expression to

$$A_{i(\text{mid})} = -g_m R_t \quad (9.17)$$

A plot of the high-frequency (hf) current gain given by Equation (9.11) can be made if we allow the variable f to take on different values. This plot usually has two parts: (1) the magnitude of the gain and (2) the phase shift of the gain expression. Figure 9.8 shows the magnitude response as a function of frequency. The gain is seen to be essentially constant until the upper cutoff frequency f_2. Then it falls off at a rate of 20 dB per decade (6 dB/*octave).

Let us explain these two units of measure for the magnitude of the response and its plots. The gain is usually measured in decibels (dB). Decibels measure

---

* An octave is the interval between frequencies with a 2:1 ratio. i.e., an octave above 2kHz is 4 kHz, while an octave below 2 kHz is 1 kHz.

the power gain, but the term may be applied to either a voltage or current gain. The number of decibels is determined from the following equation:

$$dB = 20 \log_{10} \left(\frac{I_o}{I_{in}} \right) \tag{9.18}$$

Instead of currents here, we could use voltages. The magnitude plot is made on semilogarithmic graph paper. This paper has one linear scale, on which the number of dB is plotted. The other scale is logarithmic; that is, the distance from 1 to 10 is the same as the distance from 10 to 100. We use this scale to plot the frequency. Such a graph is shown in Figure 9.8. Because the frequency scale is logarithmic, the basic unit is a tenfold change in frequency, called a *decade*.

At the upper cutoff frequency ($f = f_2$), the magnitude of the hf current gain is equal to 0.707 $A_{i(\text{mid})}$. In terms of dB (decibels), this figure corresponds to a value which is 3 dB below the midband dB gain. Thus, the upper cutoff frequency is also referred to as the upper 3 dB frequency. This procedure is illustrated in Example 9.3.

Example 9.3

Assume that the amplifier shown in Fig. 9.1 has the following circuit values: $R_1 = R_2 = 47$ kΩ and $R_C = 2.2$ kΩ. Transistor parameters are: $r_{be} = 1$ kΩ, $g_m = 50$ mS, $r_{ce} = 60$ kΩ, $C_{be} = 200$ pF, and $C_{bc} = 10$ pF. We want to determine the upper cutoff frequency and sketch the current gain magnitude response.

Solution. First, we must decide whether or not any approximations can be made. Because R_1 and R_2 constitute a parallel resistance of over 20 kΩ, they may be neglected as compared to r_{be}. Similarly, because r_{ce} is more than 10 times larger than R_C, we may neglect r_{ce}. The midband current gain is now calculated:

$$A_{i(\text{mid})} = -g_m R_t = -(50)(1) = -50$$

where we have approximated R_t by r_{be}. The dB value of $A_{i(\text{mid})}$ is obtained next as

$$20 \log |(-50)| = 20 \log(50) \cong 20(1.7) \cong 34 \text{ dB}$$

The total equivalent capacitance at the input is calculated from Equation (9.13):

$$C_t = 200 + (50)(2.2)(10) \cong 1{,}300 \text{ pF}$$

The upper cutoff frequency is determined from Eq. (9.12):

$$f_2 = \frac{1}{2\pi(10^3)(1.3 \times 10^{-9})} \text{ Hz} \cong 122 \text{ kHz}$$

The sketch of the magnitude response is shown in Fig. 9.9. Note that if the gain falls at 20 dB/decade after 122 kHz, it will be 14 dB at 1.22 MHz

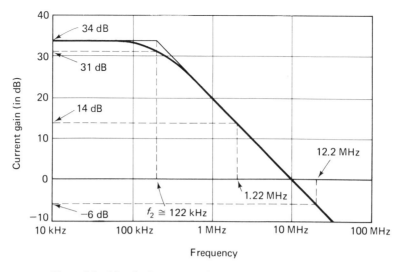

Figure 9.9 Magnitude response for the amplifier in Example 9.3.

and −6 dB at 12.2 MHz. A negative dB value means that the gain is less than 1. Moreover, at 122 kHz, the gain is 3 dB down from its midband value of 34 dB, that is, the gain is at 31 dB.

We can find the high-frequency response for an FET amplifier in a manner similar to that used with the BJT by using the FET high-frequency model.

9.4.2 Low Frequency

As was stated earlier, the low-frequency response of the amplifier is governed by the coupling and bypass capacitors. For typical resistor and transistor parameter values, even for a relatively small coupling capacitor value (say, 1 μF), the lower cutoff frequency is determined primarily by the bypass capacitor. If we assume that the coupling capacitors are shorted, the low-frequency equivalent circuit appears as shown in Figure 9.10. The low-frequency current gain is then given by

$$A_{i(lf)} \cong \frac{A_{i(mid)} \dfrac{jf}{f_1}}{1 + j\dfrac{f}{f_1}} \qquad (9.19)$$

where the lower cutoff frequency f_1 is approximately given by

$$f_1 = \frac{h_{fe}}{2\pi(R_B + h_{ie})C_B} \qquad (9.20)$$

with the stipulation that $h_{fe}R_E$ be at least 10 times greater than R_B and $R_B > h_{ie}$

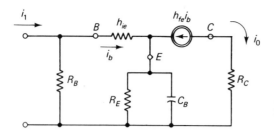

Figure 9.10 Approximate low-frequency equivalent circuit for the amplifier in Figure 9.1.

(which is usually the case). The midband current gain is calculated as before. The plot of the low-frequency response is illustrated in Example 9.4.

Example 9.4

The amplifier shown in Fig. 9.1 has the following circuit and parameter values: $R_B = 10$ kΩ, $R_c = 3.3$ kΩ, $h_{ie} = 1$ kΩ, $h_{fe} = 100$, and $h_{oe} = 1/50$ kΩ. The bypass capacitor is 5 μF. We want to determine the lower cutoff frequency and sketch the current gain magnitude response at low frequencies.

Solution. The midband current gain is calculated from Eq. (9.5) by noting that the $h_{oe}R_c$ term (0.066) is negligible with respect to 1. Thus,

$$A_{i(\text{mid})} \cong -\frac{100(10)}{10 + 1} \cong -91$$

This value in dB is:

$$20 \log(91) \cong 20(1.96) \cong 39.2 \text{ dB}$$

The lower cutoff frequency is now calculated:

$$f_1 = \frac{100}{2\pi(10 + 1) \times 10^3(5 \times 10^{-6})} \text{ Hz} \cong 300 \text{ Hz}$$

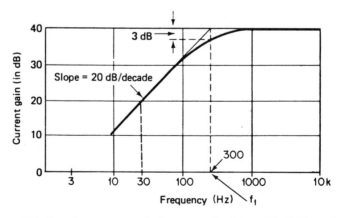

Figure 9.11 Low-frequency magnitude response for the amplifier in Example 9.4.

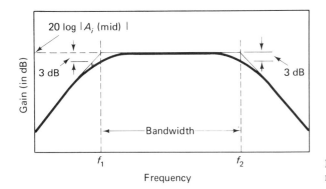

Figure 9.12 Complete frequency response.

The magnitude response for the low-frequency current gain is shown in Fig. 9.11.

The complete frequency response of an amplifier is obtained by combining the three individual responses: the midband response and the upper and the lower frequencies. This combination is shown in Figure 9.12. The gain is constant over a range of frequencies called midband. It is 3 dB down from this value at both the upper and lower cutoff frequencies, and it falls at an approximate rate of 20 dB/decade (6 dB/octave) on either side of these frequencies. The bandwidth is the difference between the upper and lower 3-dB (cutoff) frequencies.

REVIEW QUESTIONS

1. What is the role of an amplifier?
2. Name some of the important amplifier parameters.
3. What is a current amplifier? A voltage amplifier? A power amplifier?
4. What is the difference between a linear and a nonlinear amplifier?
5. How are small-signal amplifiers analyzed? How are nonlinear amplifiers analyzed?
6. Of what is the gain in a linear small-signal amplifier a function?
7. Of what is the gain in a nonlinear amplifier a function?
8. In what way can we distinguish between linear and nonlinear amplifiers? Give an example to illustrate.
9. What are the steps to be followed in the systematic analysis of small-signal amplifiers?
10. What are the steps to be followed in order to obtain the midband equivalent circuit for an amplifier? How is this different from the steps to obtain the low-frequency response? To obtain the high-frequency response?
11. In what ways is the systematic analysis of BJT amplifiers different from the analysis of FET amplifiers? In what ways are the two similar?
12. How is the midband region defined? Illustrate.

13. How is the high-frequency region defined?
14. How is the low-frequency region defined?
15. What transistor parameters determine the BJT amplifier midband response?
16. What FET parameters determine the midband response of an FET amplifier?
17. What circuit values determine the high-frequency response of a BJT amplifier?
18. What circuit values determine the low-frequency response of a BJT amplifier?
19. What is the amplifier bandwidth? What determines it?
20. What determines the complete response of an amplifier?

PROBLEMS

1. Identify the amplifier configuration in Fig. 9.13. Draw the midband equivalent circuit.
*2. Determine the midband voltage gain in Problem 1 if $R_E = R_C = 10$ kΩ, $h_{ie} = 1$ kΩ, $h_{fe} = 80$, with h_{oe} and h_{re} neglected.
3. Draw the low-frequency equivalent circuit for the amplifier shown in Fig. 9.14.
4. Draw the high-frequency equivalent circuit for the amplifier in Fig. 9.14. De-

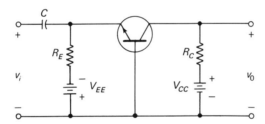

Figure 9.13 Circuit for Problem 1.

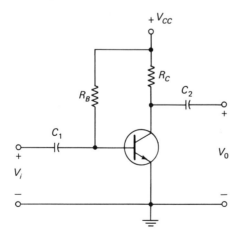

Figure 9.14 Circuit for Problem 3.

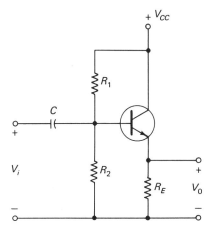

Figure 9.15 Circuit for Problem 7.

termine the upper cutoff frequency if the transistor parameters are the same as listed in Ex. 9.3 and $R_B = 220$ kΩ and $R_C = 1$ kΩ.

5. Repeat Problem 4 with $R_C = 10$ k.
6. Determine the midband current gain in Problem 4.
7. Identify the amplifier configuration in Fig. 9.15. Draw the midband equivalent circuit.
*8. If the circuit parameters in Problem 7 are $R_1 = R_2 = 33$ kΩ, $R_E = 1$ kΩ, and the transistor has the parameters listed in Ex. 9.4, determine the voltage gain v_o/v_i at midband.
9. Draw the low-frequency equivalent circuit for the amplifier shown in Fig. 9.15. Show that the low-frequency current gain expression has the same form as given in Eq. (9.19).
10. Determine the lower cutoff frequency in Problem 8 if $C = 0.1$ µF.

Chapter 10

Practical Amplifier Considerations

In this chapter we shall focus on some of the practical aspects of amplifiers, which include determining input and output impedance, real and apparent gain, frequency response, and the effects of cascading amplifier stages.

10.1 INPUT AND OUTPUT IMPEDANCE

A practical transistor amplifier contains the active device, bias resistors, and sometimes coupling and bypass capacitors. The active device itself has an input and output impedance. These impedances are determined by the physical shape and makeup of the internal device and by the bias applied externally. However, as indicated in Figure 9.3, external bias resistors appear in parallel with both the input and output terminals of the device. Thus, the actual input and output impedance of the amplifier contains both the effect of the device itself and the bias resistors. The midband equivalent circuit of the amplifier of Figure 9.1 is repeated here as Figure 10.1, with the device shaded and the amplifier outlined with dotted lines. At midfrequencies, the input and output impedance is considered resistive.

We can determine the input impedance of the amplifier by finding the ratio of v_1 to i_1. In this case, the input impedance is just the parallel combination of h_{ie} and the bias resistor R_B. Therefore, looking into the input terminals of the amplifier, we see an effective impedance R_i, where R_i is defined as the input impedance. Here it is given by

$$R_i = \frac{h_{ie} R_B}{h_{ie} + R_B} \tag{10.1}$$

Looking into the output terminals, the effective output impedance is the parallel combination of the bias resistor R_C and the impedance corresponding to the transistor output admittance h_{oe}. The output impedance, R_o, is

$$R_o = \frac{R_C \dfrac{1}{h_{oe}}}{\dfrac{1}{h_{oe}} + R_C}$$

Multiply the numerator and denominator by h_{oe} to obtain

$$R_o = \frac{R_C}{1 + h_{oe}R_C} \qquad (10.2)$$

To complete the characterization of the amplifier in terms of its input and output impedances, we need to recalculate the gain.

10.2 REAL AND APPARENT GAIN

Let us imagine that the amplifier of Figure 10.1 has a short circuit across the output terminals (i.e., $v_o = 0$). Using the techniques discussed in Chapter 9, we can determine the current gain under these conditions. Note that the analysis of Section 9.2 applied for the input but not the output, because no current will flow in either R_C or h_{oe} with the output shorted. The current gain with a short circuit at the output is determined as follows: Where A_{ISC} is the short-circuit current gain, i_{osc} is the output short-circuit current, and i_1 is the input current. Thus,

$$A_{ISC} = \frac{i_{OSC}}{i_1} = -\frac{h_{fe}i_b}{i_1}$$

However,

$$i_b = \frac{i_1 R_B}{h_{ie} + R_B}$$

Substitute:

$$A_{ISC} = -\frac{h_{fe}\dfrac{i_1 R_B}{h_{ie} + R_B}}{i_1}$$

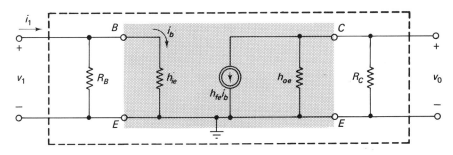

Figure 10.1 BJT amplifier equivalent circuit.

Multiply the numerator and denominator by $1/i_1$ to obtain

$$A_{ISC} = -\frac{h_{fe}R_B}{h_{ie} + R_B} \qquad (10.3)$$

We can now completely characterize the amplifier according to its input and output impedances and the short-circuit current gain, as indicated in Figure 10.2. The current generator in this figure must be the short-circuit current gain multiplying the input current. In effect, we have found the Norton's equivalent of the amplifier's output.

There are several advantages to representing an amplifier by the equivalent circuit shown in Figure 10.2. First, *the short-circuit current gain is the largest possible current gain that any amplifier can provide*. It is important to know the magnitude of the short-circuit current gain for a particular amplifier. From such an understanding, we can determine if a single amplifier is capable of providing sufficient gain or if additional stages of amplification will be needed. It is also desirable to represent an amplifier as shown in Figure 10.2 because it is a standard form into which any current amplifier can be placed.

Consider next the case where we want to know the voltage gain of the original amplifier of Figure 9.1. From Figure 10.2 we can calculate the open-circuit voltage gain; that is, there is no resistance in parallel with the output. Note that the output voltage is given by

$$v_o = A_{ISC} i_1 R_o \qquad (10.4)$$

and the input voltage is given by

$$v_1 = i_1 R_i$$

The open-circuit voltage gain (A_{VOC}) is then

$$A_{VOC} = \frac{v_0}{v_1} = A_{ISC}\frac{R_o}{R_i} \qquad (10.5)$$

Under the open-circuit voltage gain, we can transform Figure 10.2 into a voltage generator in series with a resistor at the output, as shown in Figure 10.3. This procedure corresponds to finding the Thevenin's equivalent circuit for the amplifier output terminals. The circuit of Figure 10.3 is the voltage *dual* of the circuit in Figure 10.2. Its importance is similar to that stated for Figure 10.2. Because it is a general form it can represent any amplifier. Moreover, *the open-circuit*

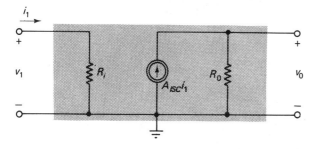

Figure 10.2 Current amplifier.

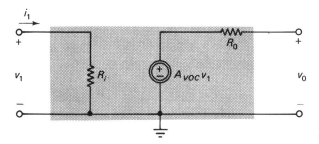

Figure 10.3 Voltage amplifier.

voltage gain is the largest possible voltage gain for any amplifier. Like the short-circuit current gain, the open-circuit voltage gain can be used in predicting the number of amplification stages needed in a particular application.

We can think of the short-circuit current gain and the open-circuit voltage gain as only apparent gains. In a real application, the actual gain will be lower; that is, the actual current gain will always be lower than A_{ISC}, and the actual voltage gain will always be lower than A_{VOC}.

This procedure is illustrated in Example 10.1.

Example 10.1

Use a BJT amplifier shown in Fig. 9.1. The circuit values as listed in Ex. 9.1 are $R_1 = 150$ kΩ, $R_2 = 100$ kΩ, $R_B = 60$ kΩ, $R_E = 1.5$ kΩ, and $R_C = 4.7$ kΩ. The BJT h-parameters are $h_{ie} = 1$ kΩ, h_{re} negligibly small, $h_{fe} = 80$, and $h_{oe} = 20$ μS. We want to determine the input and output impedance, the short-circuit current gain, and the open-circuit voltage gain of the BJT amplifier.

Solution. Use Eq. (10.1), so $R_i = (60)(1)/(60 + 1)$ k$\Omega \cong 1$ kΩ. Eq. (10.2) gives

$$R_o = \frac{4.7 \text{ k}\Omega}{1 + (0.02)(4.7)} \cong 4.3 \text{ k}\Omega$$

Calculating A_{ISC} from Eq. (10.3), we get

$$A_{ISC} = -\frac{(80)(60)}{1 + 60} \cong -79$$

Calculating A_{VOC} from Eq. (10.5), we get

$$A_{VOC} = (-79)\frac{(4.3)}{(1)} \cong -340$$

For comparison purposes, note that the actual current gain (considered R_C as the load) found in Example 9.1 was -72, whereas the short-circuit current gain of Example 10.1 was -79.

The methods outlined in this section can be used to determine the input impedance, the output impedance, the short-circuit current gain, and the open-circuit voltage gain for any amplifier, whether it is a BJT in any of the three configurations (*CE*, *CB*, or *CC*), a JFET, or any MOSFET. The representations of an amplifier, as shown in Figures 10-2 and 10-3, are universal and apply equally well to any amplifier.

10.3 AMPLIFIER LOADING

We next consider the effect of source impedance on the input current and the effect of load impedance on the output current. In any real application, the input signal is derived from a transducer or another amplifier. However, the source has a definite impedance (resistance) that must be accounted for in the analysis. This fact is illustrated in Example 10.2.

Example 10.2

The amplifier of Ex. 10.1 is connected to a source of 100 µA, which has an impedance of 20 kΩ. The amplified signal is to drive a load (R_L) of 1 kΩ. We want to determine the output current, i_o, and the overall current gain.

Solution. The equivalent circuit is shown in Fig. 10.4. At the input, the source current i_s divides between R_s and R_i. Only the component of i_s that flows into the amplifier (i_1) is amplified:

$$i_1 = i_s \frac{R_s}{R_i + R_s} \tag{10.6}$$

For the values listed, i_1 = (100 µA) 20/(20 + 1), or approximately 95 µA. Thus, of the 100 µA available from the source, the amplifier receives 95 µA. With a short-circuit current gain of -79, the magnitude of the current generator is $(-79)(95 \text{ µA}) = -7.5$ mA. This available current divides at the output between R_o and R_L. The output current is given by

$$i_o = A_{ISC} i_1 \frac{R_o}{R_o + R_L} \tag{10.7}$$

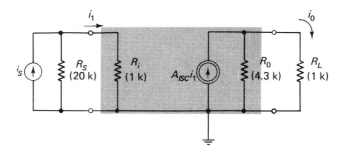

Figure 10.4 Loading in a current amplifier (see Example 10.2).

For the values given $i_o = (-7.5 \text{ mA}) 4.3/(4.3 + 1)$, or approximately -6 mA.

To review, with 100 μA available at the input, the output current is 6 mA. The overall current gain (real current gain) of the amplifier from source to load is 6 mA/100 μA = 60.

If we examine the previous example, we can make some generalizations:

1. *The lower the input impedance of an amplifier, the higher the real current gain.*
2. *The higher the output impedance of an amplifier, the higher the real current gain.*

Thus, for given source and load impedances, we can improve the performance of a current amplifier (i.e., make the actual current gain larger) by either *decreasing* the input impedance or *increasing* the output impedance, or both.

We can also specify the points at which we call an amplifier a *current amplifier*: (1) when the input resistance of the amplifier is low in relation to the source impedance, (2) when the output impedance is high in relation to the load, and (3) when the amplifier itself has a significant short-circuit current gain available (A_{ISC}).

The loading effect in a voltage amplifier is illustrated in Example 10.3.

Example 10.3

An amplifier is characterized by an input impedance of 50 kΩ, an output impedance of 2 kΩ, and an open-circuit voltage gain of -100. It is driven from a source of 10 mV at an impedance of 5 kΩ. It is to drive a 10 kΩ load. We want to determine the output voltage and the overall voltage gain.

Solution. The equivalent circuit is depicted in Fig. 10.5. Only the part of the source voltage that appears across R_i is amplified. Thus,

$$v_1 = v_S \frac{R_i}{R_i + R_S} \tag{10.8}$$

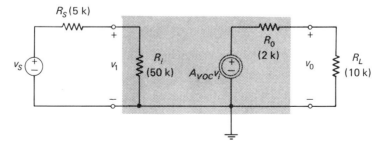

Figure 10.5 Loading in a voltage amplifier (see Example 10.3).

Sec. 10.3   Amplifier Loading

With the values given, $v_1 = 9.1$ mV. The voltage available at the output is $A_{VOC}v_1$, which is -0.91 V. The output voltage is only the part of the available voltage that appears across R_L.

$$v_o = A_{VOC} v_1 \frac{R_L}{R_L + R_o} \tag{10.9}$$

With the values given in our example, we see that v_o will be approximately -0.76 V. The overall voltage gain (actual voltage gain) from source to load is then $v_o/v_s \approx -760 \text{ mV}/10 \text{ mV} \approx -76$.

As was the case with the current amplifier, we can make some generalizations for the voltage amplifier:

1. *The higher the input impedance of an amplifier, the higher the actual voltage gain.*
2. *The lower the output impedance of an amplifier, the higher the actual voltage gain.*

Thus, for given source and load impedances, we can increase the voltage gain of an amplifier by either *increasing* the input impedance or *decreasing* the output impedance, or both.

We can also recognize the specific characteristics of what we term a *voltage amplifier*. It has a high input impedance in comparison to the source impedance and an output impedance that is low in comparison to the load impedance. It also has an amplifier with a significant open-circuit voltage gain (A_{VOC}).

10.4 IMPEDANCE MATCHING

We saw in the previous section that for high current gain, R_i should be less than R_S and R_o should be greater than R_L. For high voltage gain, the reverse should be true, i.e., $R_i > R_S$ and $R_o < R_L$. In many applications, we are interested in both a high voltage as well as high current gain. Consequently, we would like to maximize the power gain.

Consider the circuit of Figure 10.6. Let us assume for the time being that

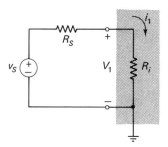

Figure 10.6 Maximum power transfer (voltage amplifier).

both v_S and R_S are fixed and specified. We would like to determine a value of R_i that will cause the largest power to be transferred to the input of our amplifier represented by R_i. The power supplied to the amplifier is the product of v_1 and i_1. It is this product that we need to maximize. Let us first examine the extreme values of R_i. If we first consider R_i to be zero, we have obviously caused the largest current i_1 to flow [i_1 is given by $v_S/(R_S + R_i)$], while v_1 would be zero. Although the current is a maximum, the voltage is still zero, so no power is delivered to the amplifier.

We would have R_i at the other extreme when it is infinite. Under these conditions, i_1 would be zero, while v_1 would be a maximum and equal to v_S. Again, there is no power delivered to the amplifier, in this case because the current is zero. Obviously, the maximum power to the input does not occur when either the voltage or the current is a maximum.

If we decide to make R_i equal to R_s or "match impedances," it turns out that the power delivered to the input is maximized. Under these conditions,

$$i_1 = \frac{v_s}{2R_s} \quad \text{and} \quad v_1 = \frac{v_s}{2} \tag{10.10}$$

The power delivered to the input P_1 is

$$P_1 = i_1 v_i = \frac{v_s^2}{4R_s} \tag{10.11}$$

The total power supplied by the v_s generator is twice the power delivered to the amplifier. (The proof of this statement is left as an exercise.) It may seem that one-half of the available power is being uselessly lost. This is not the case, because we cannot discount the power lost in R_s, which is an inseparable part of the source. The power in R_s is exactly the same as that in R_i, so that the power supplied by the source is exactly the same as that dissipated in R_i.

One of the exercises will show that maximum power is transferred to the amplifier in Figure 10.7 when the impedances are matched, i.e., R_i made equal to R_s.

We can summarize by saying that if we want maximum power transfer, we must match impedances.

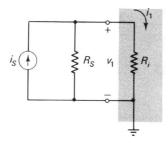

Figure 10.7 Maximum power transfer (current amplifier).

10.5 CASCADING OF AMPLIFIERS

10.5.1 Current Gain

In most applications, the gain provided by a single amplifier state is insufficient. Consequently, two or more amplifier stages are cascaded (or connected serially, from output 1 to input 2, etc.), as shown in Figure 10.8. Although we shall confine our discussion to two stages, it should be apparent that the same procedure may be extended to handle any number of cascaded stages. This procedure is illustrated in Example 10.4.

Example 10.4

Two amplifiers are cascaded as indicated in Fig. 10.8. The amplifier parameters are: $R_{i1} = 1$ kΩ, $R_{o1} = 5$ kΩ, $A_{ISC1} = -60$; $R_{i2} = 1.4$ kΩ, $R_{o2} = 6$ kΩ, $A_{ISC2} = -75$. $R_s = 10$ kΩ, and $R_L = 1$ kΩ. We want to determine the overall current gain (the ratio of i_o to i_s).

Solution. We start at the input and work our way toward the output. First,

$$i_1 = i_s \frac{R_s}{R_s + R_{i1}}$$

next,

$$i_2 = A_{ISC1} i_1 \frac{R_{o1}}{R_{o1} + R_{i2}}$$

then,

$$i_o = A_{ISC2} i_2 \frac{R_{o2}}{R_{o2} + R_L}$$

Now we are ready to put the separate parts together to determine the overall gain.

$$A_I = \frac{i_o}{i_s} = \frac{i_o}{i_2} \times \frac{i_2}{i_1} \times \frac{i_1}{i_s} \qquad (10.12)$$

The three separate parts of A_I can be identified from the three preceding

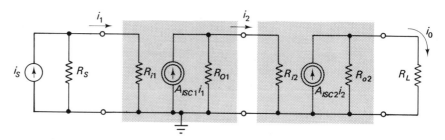

Figure 10.8 Cascading of two current amplifiers.

equations. Thus,

$$\frac{i_o}{i_2} = A_{ISC2} \frac{R_{o2}}{R_{o2} + R_L}$$

$$\frac{i_2}{i_1} = A_{ISC1} \frac{R_{o1}}{R_{o1} + R_{i2}}$$

$$\frac{i_1}{i_s} = \frac{R_s}{R_s + R_{i1}}$$

Using the above expressions in Eq. (10.12) yields the desired current gain:

$$A_I = (A_{ISC1})(A_{ISC2}) \left(\frac{R_s}{R_s + R_{i1}}\right) \left(\frac{R_{o1}}{R_{o1} + R_{i2}}\right) \left(\frac{R_{o2}}{R_{o2} + R_L}\right) \quad (10.13)$$

If we replace the symbols by the numbers for this particular example, we obtain the overall current gain.

$$A_I \cong (-60)(-75)(0.91)(0.78)(0.86) \cong 2747$$

Note one interesting aspect in this example. We could rewrite the gain expression of Equation (10.13) in a slightly different form:

$$A_I = \left(\frac{R_s}{R_s + R_{i1}}\right) \left[(A_{ISC1})(A_{ISC2}) \left(\frac{R_{o1}}{R_{o1} + R_{i2}}\right)\right] \left(\frac{R_{o2}}{R_{o2} + R_L}\right) \quad (10.14)$$

If we now recognize that the quantity inside the square brackets is the effective short-circuit current gain of the cascaded pair of amplifiers, we can draw an equivalent circuit combining the two amplifiers into one, as shown in Figure 10.9.

We could have arrived at this conclusion directly as well. We can now write

$$A_{ISC} = A_{ISC1} A_{ISC2} \frac{R_{o1}}{R_{o1} + R_{i2}} \quad (10.15)$$

where A_{ISC} denotes the combined short-circuit current gain for the cascade of two amplifiers.

The new single-stage equivalent amplifier shown in Figure 10.9 has the input impedance of the first stage and the output impedance of the second stage.

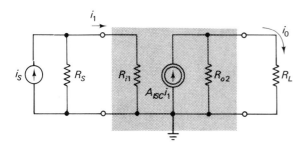

Figure 10.9 Single amplifier representation of the two-stage amplifier of Figure 10.8.

10.5.2 Voltage Gain

When two voltage amplifiers are cascaded as illustrated in Figure 10.10, a similar procedure is used to determine the properties of the equivalent single-stage amplifier. We shall only list the results here, leaving the details as an exercise.

$$A_{VOC} = A_{VOC1} A_{VOC2} \frac{R_{i2}}{R_{i2} + R_{o1}} \quad (10.16)$$

where A_{VOC} stands for the combined open-circuit voltage gain for the cascade of two amplifiers.

The new single-stage equivalent amplifier, shown in Figure 10.11, has the input impedance of the first stage and the output impedance of the second stage.

10.5.3 Bandwidth

The cascading of amplifiers obviously increases the gain. At the same time that it increases the gain, however, it also affects the bandwidth of an amplifier.

Suppose that a single amplifier stage has a gain of 20 dB (gain magnitude of $\times$ 10) and an upper cutoff frequency of 100 kHz. If we start cascading additional amplifier stages to it, the combined amplifier will have a larger gain and an upper cutoff frequency which is lower. For example, if we cascade identical stages, each with a gain of 20 dB, the resulting amplifier will have the gain response curves shown in Figure 10.12, where n is the number of similar stages cascaded.

Note two things. First, the 3 dB point becomes successively lower the more stages that we cascade. Second, the gain falls off more rapidly as the number of

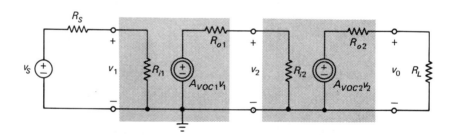

Figure 10.10 Cascading of two voltage amplifiers.

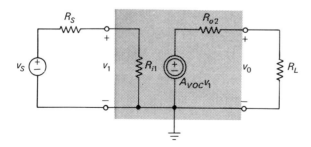

Figure 10.11 Single amplifier representation of the two-stage amplifier of Figure 10.10.

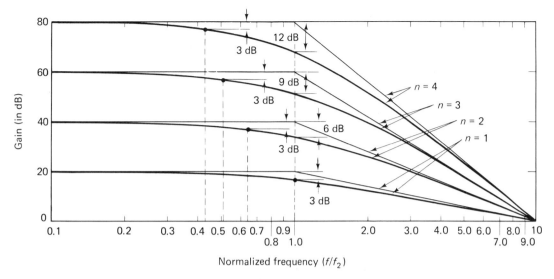

Figure 10.12 Bandwidth reduction in cascaded amplifiers.

stages in cascade is increased. If a single amplifier gain falls off at a rate of 20 dB/decade (6 dB/octave), then the cascaded amplifier gain will fall off at a rate of $20n$ dB/decade ($6n$ dB/octave), where n is the number of stages in the cascade. The results of Figure 10.12 are summarized in Table 10.1.

In a similar manner, the lower cutoff frequency of a number of identical amplifier stages increases as the number of stages increases. This increase is also summarized in Table 10.1. The procedure for determining gain and cutoff frequencies is illustrated in Example 10.5.

Example 10.5

Suppose that we construct an amplifier containing three identical stages, each with a gain of 20 dB, an upper cutoff frequency of 10 kHz, and a lower cutoff frequency of 50 Hz. We want to determine the gain and upper and lower cutoff frequencies of the cascaded amplifier.

TABLE 10-1 UPPER AND LOWER CUTOFF FREQUENCIES FOR A CASCADED AMPLIFIER (ONLY VALID FOR IDENTICAL STAGES)

Lower cutoff frequency (of cascaded amplifier)	n (number of states in cascade)	Upper cutoff frequency (of cascaded amplifier)
1.00 f_1*	1	1.00 f_2†
1.56 f_1	2	0.64 f_2
1.96 f_1	3	0.51 f_2
2.28 f_1	4	0.43 f_2
2.56 f_1	5	0.39 f_2

* f_1 denotes the lower cutoff frequency of each stage.
† f_2 denotes the upper cutoff frequency of each stage.

Solution. We can determine the gain by adding the individual dB gains, which results in 60 dB for the cascaded amplifier.

We can calculate the upper cutoff frequency by looking up the shrinkage factor for $n = 3$ in Table 10.1 to get 0.51. Therefore, the new upper cutoff frequency is $(0.51)(10 \text{ kHz}) \cong 5.1$ kHz. Similarly, we obtain the lower cutoff frequency as $(1.96)(50 \text{ Hz}) \cong 98$ Hz. Thus, the three-stage cascaded amplifier has a bandwidth that is roughly half of the single-stage amplifier.

REVIEW QUESTIONS

1. How is the input impedance of an amplifier defined?
2. What is the practical significance of the input impedance?
3. How is the output impedance of an amplifier defined?
4. What is the practical significance of the output impedance of an amplifier?
5. What is meant by *real* gain? By *apparent* gain?
6. What are the conditions that yield the highest possible current gain? Explain.
7. What are the conditions that yield the highest possible voltage gain? Explain.
8. What are the universal equivalent circuits that apply equally well for any amplifier? What components do they contain?
9. What is meant by the term *loading* as applied to an amplifier?
10. What are the possible ways in which an amplifier may be loaded?
11. In order to maximize the current gain, what is the desired condition for the input impedance? For the output impedance?
12. In order to maximize the voltage gain, what is the desired condition for the input impedance? For the output impedance?
13. What is meant by the term *impedance matching* as applied to an amplifier?
14. Why is impedance matching important? When is it important?
15. What is meant by the term *cascading* as applied to an amplifier?
16. What is the effect of cascading a number of amplifiers on the gain? On the input impedance? On the output impedance?
17. What happens to the gain when two amplifiers are cascaded? What happens to the gain as expressed in decibels?
18. What is the effect of cascading a number of amplifiers on the upper cutoff frequency?
19. What is the effect of cascading a number of amplifiers on the lower cutoff frequency?
20. When a number of amplifiers are cascaded, what is the effect on the bandwidth of the overall amplifier?

PROBLEMS

*1. The amplifier in Fig. 9.1 has the following circuit values: $R_1 = 56$ kΩ, $R_2 = 22$ kΩ, and $R_C = 6.8$ kΩ. The transistor parameters are: $h_{ie} = 2$ kΩ, $h_{fe} = 75$, and $1/h_{oe} = 40$ kΩ. Determine the input impedance, the output impedance, the short-circuit current gain, and the open-circuit voltage gain.

2. Using the results of Problem 1, determine the actual circuit current and voltage gains.

3. The FET amplifier shown in Fig. 9.4 has the following circuit values: $R_{G1} = 2.2$ MΩ, $R_{G2} = 1$ MΩ, and $R_D = 18$ kΩ. The FET parameters are r_{gs} infinite, $g_m = 2000$ μS, and $r_d = 100$ kΩ. Determine the amplifier input and output impedances, the short-circuit current gain, and the open-circuit voltage gain.

4. Using the results of Problem 3, determine the actual circuit voltage gain.

5. Determine the output current and voltage in Problem 1, if the input is driven by a voltage of 1 mV with a source impedance of 10 kΩ.

*6. Determine the output voltage in Problem 3, if the input is a 100 μV signal with a 100 kΩ source impedance.

7. Repeat Problem 6 if the source impedance is 1 MΩ.

8. If the amplifiers of Problems 1 and 3 are cascaded (the FET stage first), determine the input impedance, the output impedance, and the open-circuit voltage gain for the cascaded combination.

9. For the cascaded amplifier of Problem 8, what is the largest source impedance that may be applied without causing significant loading of the source? Explain.

10. A single-stage amplifier with a gain of 28 dB and upper and lower cutoff frequencies of 50 kHz and 10 Hz, respectively, is available. We wish to construct an amplifier containing two cascaded stages. It will have a gain of 60 dB; its upper and lower frequencies will be 0.64 and 1.56 times the respective cutoff frequencies of the available amplifier stage. Specify the parameters of the second stage that must be used.

*11. Three identical amplifiers are cascaded. The parameters for each stage are: gain = 23, upper cutoff frequency = 1.1 MHz, and lower cutoff frequency = 40 Hz. Determine the gain (in decibels) and the upper and lower cutoff frequencies of the cascaded amplifier.

12. On semilogarithmic graph paper, sketch the results of Problem 11. Show also the frequency response of a single stage.

Chapter 11

Tuned Amplifiers

In communication circuits (such as radio and television), in instrumentation, and in other applications, amplifiers are needed that amplify signals of only certain predetermined frequencies. Such amplifiers are called *tuned*, or *frequency-selective*, amplifiers. For example, in a radio receiver, we need to be able to select one station from all the available stations; that is, we need to tune to a desired station. The amplifier, therefore, must pass signals of the desired frequency and reject all other frequencies. Rejecting undesired signals in many cases is as important as amplifying desired signals.

Figure 11.1 shows the gain versus frequency characteristics of an ideal tuned amplifier. The gain is zero for all frequencies below f_1, becomes very high for frequencies in the desired pass band (between f_1 and f_2), and is again zero for all frequencies above f_2. The difference between the upper and lower cutoff frequencies is the *bandwidth* (*BW*). The *center frequency*, f_c, is either arithmetically or geometrically on a logarithmic scale the average of f_1 and f_2. An ideal tuned amplifier does not exist in the real world. Practical tuned amplifiers have characteristics that at best only approximate those of the ideal tuned amplifier shown in Figure 11.1.

11.1 SINGLE-TUNED AMPLIFIERS

Single-tuned BJT and FET amplifiers are illustrated in Figures 11.2 and 11.3, respectively. These circuits resemble the single-stage amplifiers discussed in Chapter 9. However, the resistive load of the single-stage amplifier is replaced by a *tuned*, or *tank*, circuit containing C and L. In both circuits, the three resistors establish the dc bias conditions. Capacitors C_1 and C_2 provide dc isolation between the source and load, respectively; C_B is the bypass capacitor, as discussed in Section 9.2. All the capacitors (with the exception of C) are large enough to be effective short circuits at the frequencies we are interested in, so they will not enter into the analysis.

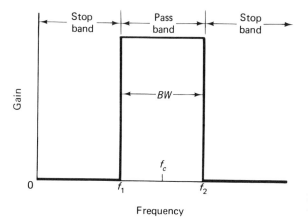

Figure 11.1 Characteristics of an ideal tuned amplifier.

Using the generalized form of the voltage amplifier equivalent circuit developed in Chapter 10, we can draw a single equivalent circuit for both the BJT and FET amplifiers, as indicated in Figure 11.4. In the case of the BJT amplifier, R_i is the parallel combination of h_{ie}, R_1, and R_2. A_{VOC} is given by

$$A_{VOC} = -\frac{h_{fe}}{h_{ie}h_{oe}} \tag{11.1}$$

and R_o is $1/h_{oe}$.

In the case of the FET amplifier, R_i is the parallel combination of R_{G1} and R_{G2}. A_{VOC} is given by

$$A_{VOC} = -g_m r_d = -\mu \tag{11.2}$$

and R_o is r_d.

In the equivalent circuit of Figure 11.4, the inductor is represented by an

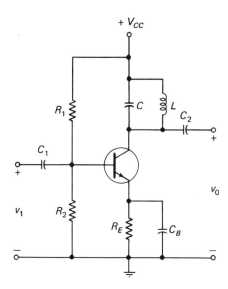

Figure 11.2 Single-tuned BJT amplifier.

Sec. 11.1   Single-Tuned Amplifiers

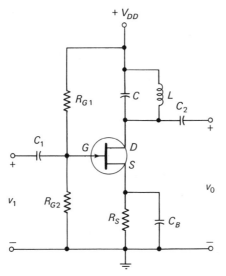

Figure 11.3 Single-tuned FET amplifier.

ideal inductor in series with a resistance R, which accounts for the nonzero dc resistance of the actual coil. Note that the impedance (Z_L) of the resonant circuit load in Figure 11.4 is the parallel combination of capacitor C in parallel with the series connection of inductor L and resistor R. Thus,

$$Z_L = \frac{\frac{1}{j\omega C}(R + j\omega L)}{\frac{1}{j\omega C} + (R + j\omega L)} \quad (11.3)$$

The *resonant frequency* (f_o) is given by

$$f_o = \frac{\omega_o}{2\pi} = \frac{1}{2\pi\sqrt{LC}} \quad (11.4)$$

The resonant frequency is the frequency at which the LC circuit responds with maximum amplitude.

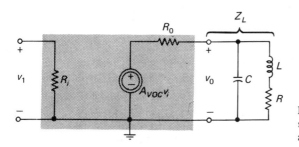

Figure 11.4 Equivalent circuit for single-tuned amplifiers of Figures 11.2 and 11.3.

11.1.1 Quality Factor (Q)

At this time, we must define another important parameter, called the *quality* factor Q:

$$Q_o = \frac{2\pi f_o L}{R} = \frac{1}{2\pi f_o RC} \tag{11.5}$$

The Q of the coil is a unitless measure of the quality of the coil. It is almost invariably designated at a specific frequency. (Note: Q_o is specified at f_o.) So, if we know the inductance and the Q at a certain frequency, we can determine the effective series resistance, R, of the coil. This is illustrated in Example 11.1.

Example 11.1

Assume that an IF radio coil has 1.0 mH inductance with a Q of 100 at 455 kHz. We want to determine the effective series resistance of the coil.

Solution. We can solve for R from Eq. (11.5):

$$R = \frac{2\pi f_o L}{Q_o} \cong \frac{(2\pi)(455 \times 10^3)(0.1 \times 10^{-3})}{10^2} \Omega \cong 28.6 \Omega$$

For convenience and brevity, we use a new variable here: δ. It defines the normalized frequency deviation:

$$\delta = \frac{f - f_o}{f_o} \tag{11.6}$$

Using the definitions of Equations (11.4), (11.5), and (11.6), we rearrange the expression for the load impedance from Equation (11.3) to read*:

$$Z_L = \frac{R Q_o^2}{1 + j2\delta Q_o} \tag{11.7}$$

Now we evaluate the load impedance Z_L at resonance, where $f = f_o$ and therefore $\delta = 0$. The equation for Z_L is then:

$$Z_{L(\text{resonance})} = R Q_o^2 \equiv R_{res} \tag{11.8}$$

Note that the load is purely resistive at resonance; this must be true because of the definition of resonance. Moreover, even though the inductor's effective resistance (R) is usually quite low, the load impedance at resonance, R_{res}, offers a very high resistance for typical values of Q.

At resonance, the tuned load circuit of Figure 11.4 may be replaced with an equivalent approximate load, as shown in Figure 11.5. Here we see that the R of Figure 11.4 is now represented as a parallel equivalent resistance, R_{res}, which is

---

*There are certain approximations involved in the step between Eqs. (11.3) and (11.7). These approximations are called high-Q approximations and are valid near the resonant frequency f_o if $Q \gg 1$.

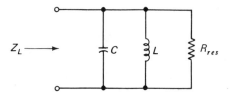

Figure 11.5 Approximate equivalent circuit of the load near resonance.

determined by Equation (11.8). At resonance, the impedance of the load (Z_L) of Figure 11.5 is equal to R_{res} because the capacitive susceptance ($j1/X_C$) and the inductor susceptance ($-j1/X_L$) are equal but opposite in phase and add to zero thus:

$$Z_L = \frac{1}{1/R_{res} + (j1/X_C - j1/X_L)} = R_{res}$$

To see the effect of Q on the load impedance, we show a plot in Figure 11.6 of the normalized load impedance as a function of δ for different values of Q. Observe that the higher the Q, the more selective (the narrower) the curve around $\delta = 0$ (which corresponds to $f = f_o$). Note also that the curves are symmetrical about $\delta = 0$ when δ is very small.

Returning to the amplifier of Figure 11.4, we can write the voltage gain A_V (which is v_0/v_1):

$$V_0 = A_{VOC} V_1 \frac{Z_L}{Z_L + R_o}$$

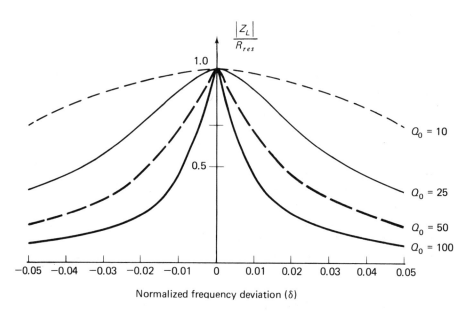

Figure 11.6 Variation of Z_L for different Q_o.

and since $A_V = V_0/V_1$ then

$$A_V = A_{VOC} \frac{Z_L}{Z_L + R_o} \quad (11.9)$$

If we substitute Equation (11.7) and make use of Equation (11.8) in Equation (11.9), after some manipulation, the gain can be rewritten

$$A_V = A_{VOC} \left(\frac{R_P}{R_o}\right)\left(\frac{1}{1 + j2\delta Q_e}\right) \quad (11.10)$$

where the parallel resistance R_P and the loaded or effective Q (Q_e) have been defined as

$$R_P = \frac{R_o R_{res}}{R_o + R_{res}} \quad (11.11)$$

$$Q_e = Q_o \frac{R_o}{R_o + R_{res}} = Q_o \frac{R_P}{R_{res}} \quad (11.12)$$

There are a few important points worth considering here. The parallel resistance R_P must always be smaller than either R_o or R_{res} because it is the parallel combination of these two resistors. The ratio R_P/R_{res} in Equation (11.12), therefore, is always less than 1. Furthermore, the *effective Q* or *loaded Q* (labeled Q_e) must at all times be smaller than the unloaded or circuit Q (labeled Q_o). We say that the output of the amplifier loads down the tuned circuit and degrades the Q (i.e., lowers it).

11.1.2 Resonant Amplifier Performance

Let us next examine the performance of the resonant amplifier. At resonance, as we have already seen, δ is zero and the amplifier gain A_{Vres} is

$$A_{Vres} = A_{VOC} \frac{R_P}{R_o} \quad (11.13)$$

This gain is the highest possible for the resonant amplifier. However, because R_P is smaller than R_o, the resonant amplifier gain is smaller than the open-circuit voltage gain. The gain response is the plot of the gain as a function of frequency. It has the same shape and characteristics as the impedance plot of Figure 11.6.

We can determine both the cutoff frequencies (3 dB frequencies) and the bandwidth by determining the value of δ, for which the gain is 0.707 of the resonant gain. Equation (11.10) tells us that this occurs when $\delta = \pm 1/2Q_e$. From this fact, remembering the definition of δ, we can write

$$f_1 = f_o\left(1 - \frac{1}{2Q_e}\right) \quad \text{and} \quad f_2 = f_o\left(1 + \frac{1}{2Q_e}\right) \quad (11.14)$$

Knowing the cutoff frequencies, f_1 and f_2, we can determine the 3 dB bandwidth as follows:

$$BW_{3\ dB} \cong f_2 - f_1 = \frac{f_o}{Q_e} \tag{11.15}$$

We specify here that it is the 3 dB bandwidth because other bandwidths are also used to specify the performance of tuned amplifiers. For example, an ideal amplifier (as shown in Figure 11.1) has a gain characteristic that falls and rises with infinite slope. It is customary to define a parameter, which specifies the steepness of the slopes in the characteristics of an amplifier. This parameter is sometimes called the *selectivity* of an amplifier. One possible definition for the selectivity is

$$S \cong \frac{60\ \text{dB bandwidth}}{3\ \text{db bandwidth}} \tag{11.16}$$

By the 60 dB bandwidth, we mean the bandwidth at the point where the gain is down 60 dB from its resonant value. It can be shown that for the single-tuned amplifier, the 60 dB bandwidth is given by $(1{,}000\ f_o/Q_e)$. The selectivity of a single-tuned amplifier is then 1,000—a constant that is not dependent on the resonant gain or the center frequency or either the loaded or unloaded Q. If we used the same definition for the selectivity of an ideal tuned amplifier, we would obviously get $S = 1$, because the passband of an ideal amplifier is constant.

In practice, we would like as small a numerical value as possible for the selectivity. In other words, we would like the amplifier to be *very* or *highly selective*. The procedure for this is illustrated in Example 11.2.

Example 11.2

A single-tuned BJT amplifier of the type shown in Fig. 11.2 is constructed to operate as an IF amplifier in an AM superheterodyne receiver. The center frequency is to be 455 kHz. The transistor parameters are: $h_{ie} = 2$ kΩ, $h_{fe} = 50$, and $h_{oe} = 10$ µS. The inductor of 1 mH with a Q_o of 100 at 455 kHz will be used in the tuned circuit. Determine the capacitor value, the resonant voltage gain, the cutoff frequencies, and the bandwidth.

Solution. First calculate C from Eq. (11.4):

$$C = \frac{1}{(2\pi f_o)^2 L}\ \text{F} \cong \frac{1}{(6.28 \times 4.55 \times 10^5)^2 (10^{-3})}\ \text{F} \cong 122\ \text{pF}$$

Just as in Ex. 11.1, we calculate the series inductor resistance to obtain $R = 28.6\ \Omega$. From Eq. (11.8) we find R_{res}. Thus,

$$R_{res} = RQ_o^2 \cong 28.6 \times 10^4\ \Omega \cong 286\ \text{k}\Omega$$

We now turn our attention to the amplifier. With the parameters given:

$$R_o = \frac{1}{h_{oe}} = 100\ \text{k}\Omega$$

And from Eq. 11.1:

$$A_{VOC} = -\frac{h_{fe}}{h_{ie}h_{oe}} = -\frac{50}{(2)(0.01)} \cong -2500$$

With R_o and R_{res} known, we calculate that R_P from Eq. (11.11) is approximately 74 kΩ. We can now calculate the resonant gain from Eq. (11.13):

$$A_{Vres} = A_{VOC}\frac{R_P}{R_o} = (-2500)\frac{(74)}{(100)} \cong -1850$$

We can similarly calculate the loaded Q, Q_e, from Eq. (11.12):

$$Q_e = Q_o\frac{R_P}{R_{res}} = (100)\frac{(74)}{(286)} \cong 25.9$$

We next calculate the bandwidth from Eq. (11.15):

$$BW_{3\,dB} = \frac{f_o}{Q_e} \cong \frac{455}{25.9}\text{ kHz} \cong 17.6\text{ kHz}$$

To calculate the cutoff frequencies, it is best to rewrite Eq. (11.14) as

$$f_1 = f_o - \frac{BW}{2} \cong 455 - 8.8 \cong 446.2 \text{ kHz}$$

$$f_2 = f_o + \frac{BW}{2} \cong 455 + 8.8 \cong 463.8 \text{ kHz}$$

This completes the problem.

11.1.3 Adjusting the Amplifier

In practice, it is often necessary to adjust the tuned circuit for exactly the right center frequency. We can make this adjustment by tuning either the inductor or the capacitor, depending on which is variable. We can tune the inductor by turning the ferrite slug inside the coil. In the case of the capacitor, a small trimmer capacitor with a screw adjustment may be used. You must be careful to consider transistor internal capacitances. Our example and analysis have dealt with a "high-frequency transistor" where the internal capacitance is very low. In the previous example, however, if the amplifier stage were cascaded to an identical stage that typically might have an input shunting (internal) capacitance of a few hundred picofarads, the calculations would not hold.

Bandwidth adjustments in the single-tuned amplifier are also possible. In order to increase the bandwidth (it cannot be decreased), the effective Q needs to be decreased. So we add a parallel compensating resistor R_x, as shown in Figure 11.7. As a result, the R_P is modified; it now is the parallel combination of R_{res}, R_o, and R_x. All other equations remain the same. The procedure for this is illustrated in Example 11.3.

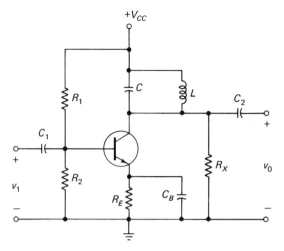

Figure 11.7 Modification of the circuit to increase the bandwidth by adding R_x.

Example 11.3

Let us assume that we want to change the performance of the amplifier in Ex. 11.2 in such a way that it has a bandwidth of 25 kHz. The compensating resistor value has to be found.

Solution. From Eq. (11.15):

$$Q_e = \frac{f_o}{BW} = \frac{455}{25} \cong 18.2$$

From Eq. (11.12):

$$R_P = R_{res} \frac{Q_e}{Q_o} = (286)\frac{18.2}{100} \text{ k}\Omega \cong 52 \text{ k}\Omega$$

$$\frac{1}{R_P} = \frac{1}{R_x} + \frac{1}{R_o} + \frac{1}{R_{res}}$$

Therefore, with R_P as calculated here (52 kΩ) and R_o and R_{res} the same as in the previous example, we can determine:

$$\frac{1}{R_x} = \frac{1}{R_P} - \frac{1}{R_o} - \frac{1}{R_{res}} = \frac{1}{52} - \frac{1}{100} - \frac{1}{286} \text{ mS} = 5.73 \text{ mS}$$

$$\therefore R_x = 174 \text{ k}\Omega$$

The only manner in which the bandwidth could be narrowed is by replacing the inductor with another one having a higher Q at the same frequency.

11.2 COUPLING OF TUNED AMPLIFIERS

When we cascade amplifiers, the input impedance of the second amplifier appears as part of the load to the first stage. In the case of tuned amplifiers, this procedure presents some special problems, especially when BJT amplifiers are involved. To see the full consequences, let us consider the amplifier in Example 11.2. Suppose there is another identical amplifier cascaded to it. In effect, we would place the input impedance (in this case about 2 kΩ) of the second stage in parallel with the output of the first stage, the equivalent of placing R_x as we did in Example 11.3. However, the input impedance of a BJT amplifier in the tuned circuit is extremely low, and the loading would destroy any selectivity of the amplifier. (It would in the case mentioned given an effective Q of less than 1.) We could not even truthfully call such a cascaded amplifier a tuned amplifier.

To prevent this loading effect in BJT amplifiers, we have three basic methods available to us. In the first method, we can use an impedance matching transformer, as illustrated in Figure 11.8. The transformer raises the effective impedance seen by the tuned circuit and thus prevents undesirable loading and bandwidth broadening that would otherwise result. Although a successful method, the interstage matching transformer is not used frequently, because other more efficient methods offer the same benefits at lower cost.

A more popular scheme for minimizing, or in some cases eliminating, the loading effect is shown in Figure 11.9. It uses tapped inductors. The total inductance in the output of the first stage is $L_1 + L_2 + 2M$ (where M is the mutual inductance between the two windings). If the number of turns in L_2 is sufficiently smaller than the number of turns in L_1, the low input impedance of the second stage is stepped up, and loading is minimized.

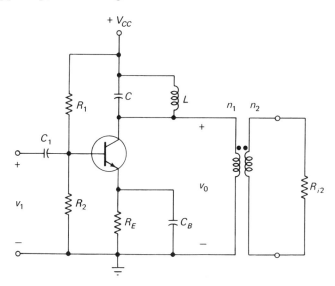

Figure 11.8 Impedance matching with an interstage transformer ($n_1 \gg n_2$).

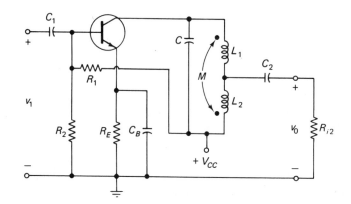

Figure 11.9 Impedance matching with tapped inductors.

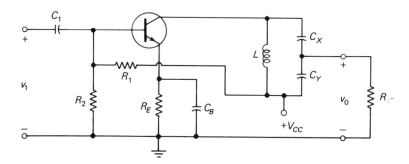

Figure 11.10 Impedance matching with tapped capacitors.

The third method is similar, but it uses tapped capacitors, as indicated in Figure 11.10.

In the case of FET tuned amplifiers, stages can be cascaded directly without loading problems because of the extremely high input impedance of FETs, either JFETs or MOSFETs. Because of this property, the FET is the device of choice for radio frequency circuits.

11.3 DOUBLE-TUNED AMPLIFIERS

In many applications the selectivity of a single-tuned amplifier stage is insufficient. In these cases, two or more tuned amplifier stages may be cascaded, as indicated in Figure 11.11, or one amplifier stage may contain two tuned circuits, as shown in Figure 11.12. In either case, two possibilities exist. The tuned circuits may be adjusted to the same center frequency (called *synchronous tuning*), or they may be adjusted to either side of the desired center frequency (called *stagger tuning*).

11.3.1 Synchronously-Tuned Amplifiers

Let us consider the circuit of Figure 11.11. Assume that the amplifiers are identical; that is, they have the same parameters and are tuned to the same frequency.

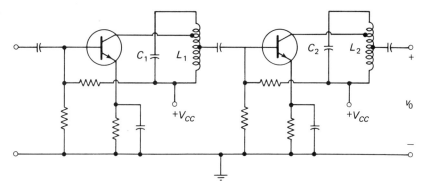

Figure 11.11 Double-tuned two-stage amplifier.

The results are tabulated (for up to five cascaded stages) in Table 11.1 and shown graphically (for up to three cascaded stages) in Figure 11.13. Note that as the number of stages is increased, the bandwidth of the overall amplifier is reduced and the selectivity is improved. Improvement in the selectivity is especially marked (by a factor of over 30) when we go from a single stage to two stages, so two stages are sufficient in most applications.

11.3.2 Stagger-Tuned Amplifiers

In order to achieve a stagger-tuned arrangement, we need to tune each of the tuned circuits depicted in either Figure 11.11 or Figure 11.12 slightly to one side of the desired center frequency. For example, we could tune L_1C_1 to slightly below the desired center frequency, with L_2C_2 slightly above the desired frequency. In order to achieve the optimum response (see Figure 11.14), the tuning is critical. As shown in Figure 11.14, if the two tuned circuits are tuned too far away from the desired center frequency, the response has a hump or dip in it. On the other hand, if the two tuned circuits are tuned too close to the center frequency, there is a loss in the selectivity possible with critical tuning. When the two tuned circuits are tuned too far away, we say that they are *loosely coupled*,

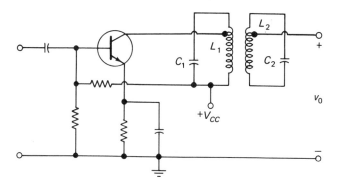

Figure 11.12 Double-tuned one-stage amplifier.

TABLE 11.1 BANDWIDTH REDUCTION IN CASCADED TUNED AMPLIFIERS

n (Number of identical amplifier stages in cascade)	BW_n (3 dB bandwidth of overall n-stage amplifier)	S (Selectivity, see Sec. 11.2)
1	BW*	1000.
2	$0.64\ BW$	31.6
3	$0.51\ BW$	10.
4	$0.43\ BW$	5.5
5	$0.39\ BW$	3.8

* BW denotes the 3 dB bandwidth (f_0/Q_c) of each individual stage, with all stages assumed to be identical.

or *undercoupled*. When the tuned circuits are tuned too closely, we say that they are *overcoupled*.

For *critical coupling*, assume that the overall bandwidth is known (labeled BW_t for total bandwidth). Then the two tuned circuits should be adjusted as follows:

$$f_{o1} = f_c - 0.35\ BW_t \tag{11.17}$$

$$f_{o2} = f_c + 0.35\ BW_t \tag{11.18}$$

$$BW_1 = BW_2 = 0.7\ BW_t \tag{11.19}$$

where f_{o1} and f_{o2} are the resonant frequencies of the two tuned circuits; BW_1

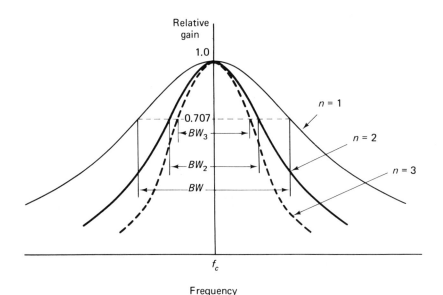

Figure 11.13 Response of multiple, synchronously tuned, cascaded (single-tuned) amplifiers showing the increased selectivity with an increase in the number of stages.

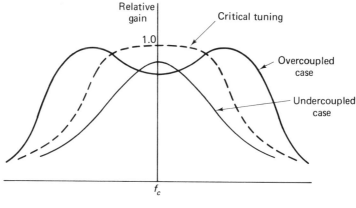

Figure 11.14 Possible cases for the response of a two-stage stagger-tuned amplifier.

and BW_2 are the 3 dB bandwidths of the two tuned circuits, respectively; and f_c and BW_t are the desired center frequency and bandwidth, respectively.

Note that one resonant circuit is tuned exactly the same amount below the center frequency as the other resonant circuit is above the center frequency. Obviously, the center frequency will be exactly between the two resonant frequencies of the two tuned circuits.

Practical adjustments of stagger-tuned amplifiers are quite tricky and involve iterative trial-and-error procedures. Tuning one of the two tuned circuits changes the overall bandwidth and the center frequency. The design procedure is illustrated in Example 11.4.

Example 11.4

We wish to design a two-stage stagger-tuned IF amplifier with a bandwidth of 10 kHz and a center frequency of 455 kHz. Specify the amplifier parameters.

Solution. For critical tuning, we use Eqs. (11.17) through (11.19):

$$f_{o1} = 455 - (0.35)(10) \cong 451.5 \text{ kHz}$$

$$f_{o2} = 455 + (0.35)(10) \cong 458.5 \text{ kHz}$$

$$BW_1 = BW_2 = (0.7)(10) \cong 7.0 \text{ kHz}$$

We can also proceed to calculate the loaded Qs for each stage:

$$Q_{e1} = \frac{f_{o1}}{BW_1} \cong \frac{451.5}{7.0} \cong 64.5$$

$$Q_{e2} = \frac{f_{o2}}{BW_2} \cong \frac{458.5}{7.0} \cong 65.5$$

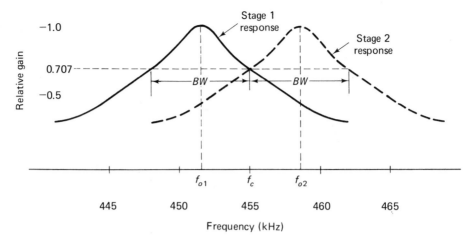

Figure 11.15 Individual response curves for stagger-tuned stages.

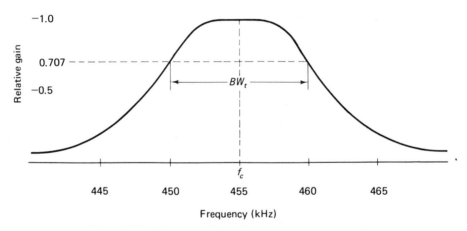

Figure 11.16 Response of overall IF stagger-tuned amplifier.

Note that the two effective (loaded) Qs are not exactly the same.

For the preceding numbers, the individual stage responses are shown in Fig. 11.15. The response of the overall amplifier is shown in Fig. 11.16. Note that the selectivity for the stagger-tuned amplifier is better (that is, the skirt of the response is steeper) than for the single-tuned amplifier.

REVIEW QUESTIONS

1. What is a tuned amplifier? Describe it in comparison to other types of amplifiers.
2. What are the characteristics of an ideal tuned amplifier?

3. What is meant by center frequency? By bandwidth?
4. What is the difference between single-tuned and double-tuned amplifiers?
5. What are the circuit components that distinguish a tuned amplifier from any other amplifier?
6. What is the significance of a tank circuit in a tuned amplifier? (Discuss in terms of its function.)
7. What is the quality factor Q? Of what is it a function and what does it indicate?
8. What is the relative magnitude of the impedance of a tuned circuit at resonance with respect to its magnitude away from resonance?
9. What are the conditions for maximum gain in a tuned amplifier? Why?
10. What is the sensitivity of a tuned amplifier? How is it defined?
11. What would be the sensitivity of an ideal tuned amplifier if it existed?
12. What are the different methods of double-tuning an amplifier?
13. Describe what is meant by synchronous tuning.
14. What is meant by stagger tuning?
15. What is the effect of resistive loading on a tuned circuit? (Discuss in terms of the response.)
16. What is meant by critical coupling as applied to a double-tuned amplifier?

PROBLEMS

1. A coil of 1 mH has an equivalent series resistance of 1 Ω. Determine its Q at the following frequencies: 10, 100, 455, and 560 kHz. At what frequency is the Q equal to 100?
*2. The coil specified in Ex. 11.1 is to be used in a tuned amplifier with a center frequency of 910 kHz. Determine the circuit Q.
3. Repeat Problem 2 if the tank circuit is loaded by an equivalent parallel resistance of 60 kΩ.
4. The single-tuned BJT amplifier of Ex. 11.2 is to be modified to provide a 3 dB bandwidth of 10 kHz by using a different inductor. With all other circuit values the same, can this modification be accomplished? If so, what are the parameters of the coil needed?
5. A tuned circuit uses a 200 pF capacitor and a 2.5 mH inductor. At its resonant frequency, it offers an equivalent shunt resistance of 100 kΩ. Determine the Q_o, f_o, and BW.
*6. If the tuned circuit in Problem 5 is used at the output of an FET amplifier (illustrated in Fig. 11.3), which has $r_{ds} = 100$ kΩ and $g_m = 500$ μS? Determine the resonant gain, the effective Q, and the bandwidth.
7. What are the ways in which the resonant gain of the amplifier in Problem 6 could be increased? Explain.
8. The bandwidth of a tuned amplifier depends on the center frequency. If the capacitor in Problem 6 is varied from 200 to 400 pF (in steps of 50 pF), with

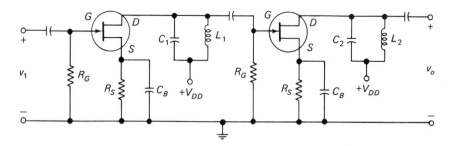

Figure 11.17 Circuit for Problem 10.

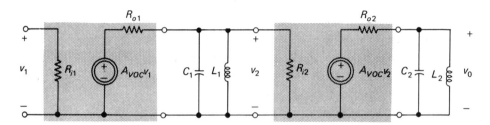

Figure 11.18 Equivalent circuit for Problem 10.

all other circuit values constant, make a plot of the bandwidth as a function of the resonant frequency.

9. A double-tuned amplifier is formed by cascading two single-tuned amplifiers. Each has a resonant gain of 26 dB, a center frequency of 1 MHz, and an effective Q of 50. What is the response of this synchronously tuned amplifier? Sketch it.

10. A two-stage FET amplifier is shown in Fig. 11.17, with its equivalent circuit given in Fig. 11.18. We want to make it into an IF amplifier ($f_o = 455$ kHz). The two tank circuits are to be synchronously tuned. The circuit values are: $R_G = 1$ MΩ, $L_1 = L_2 = 10$ μH with $Q = 150$ at 455 kHz. Identical FETs are used; r_{gs} is infinite; $r_{ds} = 100$ kΩ; and $g_m = 500$ μS. Determine the resonant gain, the capacitance values needed ($C_1 = C_2$), and the bandwidth.

*11. A stagger-tuned amplifier is constructed with the desired bandwidth of 6 MHz and a center frequency of 60 MHz. Determine the parameters (f_o, BW, and Q_e) for each stage.

12. Repeat Problem 10 if $C_1 \neq C_2$; that is, the amplifier is stagger-tuned with critical coupling. $L_1 = L_2 = 0.1$ mH with $Q = 250$ at 455 kHz if total bandwidth is 3.03 kHz.

Chapter 12

Power Amplifiers

In this chapter we study power amplifiers, which are characterized by two distinctive properties: (1) large signals and (2) nonlinear operation. As we discussed in Chapter 10, when the signal swing around the operating point is large, small-signal analysis is no longer valid. We must use other means. We shall introduce here graphical techniques that can be used to analyze power amplifiers.

12.1 CLASSES OF POWER AMPLIFIERS

We will begin our study of power amplifiers by examining the different classes of operation. The first class has already been discussed in terms of all small-signal amplifiers: namely, class A. As shown in Figure 12.1(a), a class-A amplifier has an output signal that is present during the complete input-signal cycle. All linear, and some power, amplifiers are operated as class-A amplifiers.

In a class-B amplifier, the input signal is actually half-wave rectified, as indicated in Figure 12.1(c). A class-B amplifier usually consists of two transistors operated in tandem, as we shall see in a later section.

A class-AB amplifier, as the name suggests, is a compromise between a class-A and class-B amplifier. The output waveshape resembles the input waveshape for almost the complete cycle, as shown in Figure 12.1(b).

The class-C amplifier waveshapes are illustrated in Figure 12.1(d). A class-C amplifier usually has a tuned circuit as a load and is characterized by very high power efficiencies. Its applications include radio and TV transmitters.

12.2 SERIES-FED CLASS-A AMPLIFIERS

Figure 12.2 demonstrates how the simple fixed-bias circuit discussed in Chapter 4 can be used for a power amplifier. We determine the dc Q-point by drawing the load line and then calculating the base current from the circuit values. This procedure is illustrated in Example 12.1.

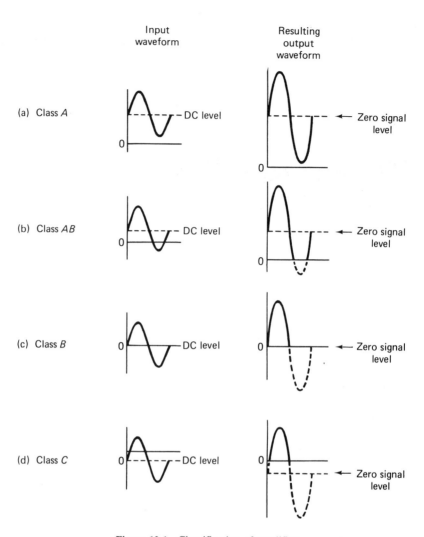

Figure 12.1 Classification of amplifiers.

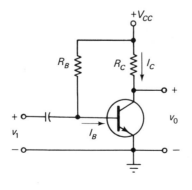

Figure 12.2 Series-fed class-A amplifier.

Example 12.1

A silicon *NPN* transistor TIP29 is used in the amplifier circuit of Fig. 12.2 with $V_{CC} = 10$ V, $R_C = 4\ \Omega$, and $R_B = 470\ \Omega$. The output characteristics for the transistor may be displayed on a Transistor Curve Tracer and the characteristics will look like those shown in Fig. 12.3. We want to determine the dc Q-point.

Solution. To plot the load line of Fig. 12.3, we observe that when no collector current flows, the collector voltage is equal to the supply voltage. Thus, we have point $P1$ as (10 V, 0 A). Point $P2$ (6 V, 1 A) is determined by noting that when the collector current is 1 A, the voltage drop across R_C is 4 V. The collector voltage (V_{CE}) must be $V_{CO} - V_{RC}$ or $10 - 4 = 6$ V. The load line is then drawn by connecting points $P1$ and $P2$, as indicated on Fig. 12.3. We calculate the base current in the 470 Ω resistor by assuming a voltage drop of about 0.6 V between base and emitter. (It is a forward-biased junction of a silicon transistor.) Thus,

$$I_B = \frac{V_{CC} - V_{BE}}{R_B} \cong \frac{10 - 0.6}{0.47}\ \text{mA} \cong 20\ \text{mA}$$

The Q-point is now established where the load line intersects the $I_B = 20$ mA characteristic curve. The Q-point is at $I_C \cong 0.65$ A and $V_{CE} \cong 7.4$ V.

From the previous example it should be evident that we determine the dc Q-point for a power amplifier the same way as we did for small-signal amplifiers in Chapter 4.

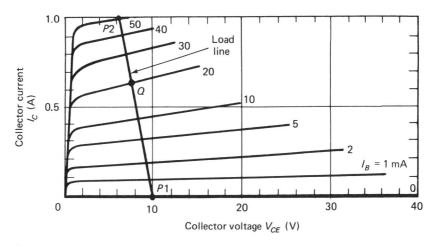

Figure 12.3 Typical power transistor output characteristics drawing the load line for Example 12.1.

Example 12.2

A sinusoidal ac signal is applied to the amplifier described in Ex. 12.1. It causes a 20 mA peak (40 mA peak-to-peak) base current. We want to determine the collector current waveshape.

Solution. The first step is to determine graphically the relationship between the base and collector currents. We make a table of two columns: one for I_B, one for I_C. We fill in the I_B column to correspond to values for which we have characteristic curves. In our example, we would choose 0, 1, 2, 5, 10, . . . , mA for the I_B column. For these specific values of I_B, we look up the intersections of the characteristic curves (on the output characteristics) with the load line; then we read off the corresponding values of I_C. As an example, the characteristic curve for I_B of 10 mA intersects the load line at approximately 0.45 A for I_C. The rest of the table is completed in a similar manner. The points from the table are then plotted and connected by a smooth curve, as shown in Fig. 12.4. This curve is the *current transfer* curve.

The next step is to make a plot of a sinusoidal waveshape for I_B below the transfer curve, as indicated in Fig. 12.5. (When you plot a sinusoidal waveshape, it should be sufficient to plot the nine points A through I, as shown.) The important thing to remember is to line up the zero for the waveshape with the dc value of I_B (in this case, 20 mA). Because, in a manner of speaking, the ac part of the waveshape "rides" on top of the dc level, this procedure must be followed. Once the complete base current signal is plotted below the transfer curve as shown in Fig. 12.5, a second time scale is drawn to the right of the transfer curve for the collector current.

To obtain the collector current waveshape, we use the following construction: With a straight edge, project the point A on the base current waveshape directly upwards until it intersects the transfer curve (also point A). Now project this point directly to the right until it intersects the $\omega t = 0$ point

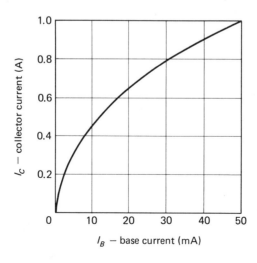

Figure 12.4 Current transfer curve for Example 12.2.

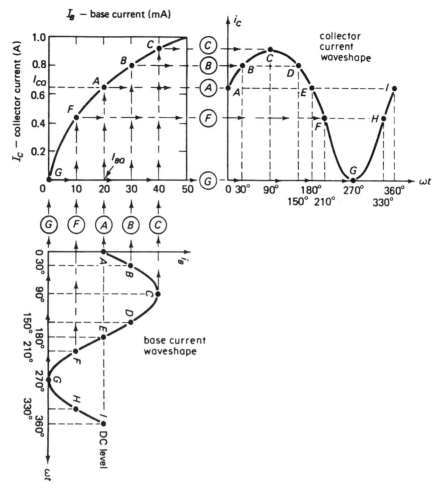

Figure 12.5 Constructing the collector current waveshape using the transfer characteristic (see Example 12.2).

on the collector current scale. This point is also labeled A. The time reference (value of ωt) on both the base current and collector current scales must be observed. In a similar manner, project point B from the base current graph upwards to the transfer curve, then to the right to the point $\omega t = 30$ on the collector current graph, also labeled point B. To complete the construction, proceed with the projections of points C, D, E, F, G, H and I. Note that, in reality, points D, E, H, and I do not need to be projected. They give the same intercepts on the transfer curve as points B, A, F, and A, respectively. Therefore, they may be plotted on the I_c graph without projection, still observing the time reference.

Once the points A through I are obtained on the graph to the right, the collector current waveshape may be sketched in by connecting the points with a smooth curve. We note that the collector current is 0.9 A peak-to-peak and that the waveshape is not an exact reproduction of the input waveshape. In other words, the output waveshape is distorted.

It is imperative to realize that the points A through I in the previous example were not chosen at random. Points A, E, and I were chosen where the sinusoid is zero—at 0°, 180°, and 360°, respectively. Points B and D are located where the sinusoid is one-half of its positive peak value—at 30° and 150°, respectively. Points F and H indicate where the sinusoid is one-half of the negative peak value—at 210° and 330°, respectively. Lastly, points C and G were chosen where the sinusoid has its positive and negative peaks—at 90° and 270°, respectively. It is most important to observe these choices, especially in the calculation of harmonic distortion.

12.3 POWER EFFICIENCY AND DISSIPATION

In a power amplifier, unlike a small-signal amplifier, we are interested in the average ac power delivered to the load rather than in voltage or current gain. Moreover, we want to see how efficiently we are delivering power to the load. Lastly, we must concern ourselves with the power dissipated in the transistor to ensure that it is not excessive. Excessive power would not only cause permanent damage to the transistor but also make it useless.

12.3.1 Power Efficiency

The power efficiency of an amplifier is given by the ratio between the ac power delivered to the load and the dc power that the power supply delivers:

$$\% \text{ efficiency} = \frac{P_{o(ac)}}{P_{i(dc)}} \times 100 \tag{12.1}$$

The dc input power is calculated as the product of the power supply voltage (V_{CC}) and the average current delivered (approximately given by I_{CQ}). Thus,

$$P_{dc} = V_{CC} I_{CQ} \tag{12.2}$$

The ac power delivered to the load is the product of the rms values of the output voltage and currents. The rms value for any sinusoidal signal may be found by dividing the peak-to-peak value by $2\sqrt{2}$. Thus,

$$P_{ac} = \frac{(V_{p-p})(I_{p-p})}{(2\sqrt{2})(2\sqrt{2})} = \frac{(V_{p-p})(I_{p-p})}{8} \tag{12.3}$$

where V_{p-p} and I_{p-p} are the output peak-to-peak voltage and current, respectively. In the case where the load is resistive, we can express the voltage in terms of the current:

$$V_{p-p} = I_{p-p} R_C \tag{12.4}$$

and rewrite the ac power equation accordingly:

$$P_{ac} = \frac{(I_{p-p})^2 R_C}{8} \qquad (12.5)$$

With the preceding equations, it is possible to determine the ac output power, the dc input power, and the power efficiency for any given power amplifier.

Efficiency in power amplifiers is important because power levels are in the watt range; whereas in small-signal amplifiers the power involved does not exceed a few hundred milliwatts at most. Calculation of power efficiency is illustrated in Example 12.3.

Example 12.3

Calculate the ac and dc power as well as the power efficiency for the series-fed class-A amplifier discussed in Ex. 12.1 and 12.2.

Solution. The Q-point collector current was determined in Ex. 12.1 to be 0.65 A. Using Eq. (12.2) we get

$$P_{dc} = (0.65)(10) \text{ W} \cong 6.5 \text{ W}$$

From Fig. 12.5 and Ex. 12.2 we see that the peak-to-peak collector current is 0.9 A. This fact, together with an R_C of 4 in Eq. (12.5), gives us

$$P_{ac} = \frac{(0.9)^2(4)}{8} \text{ W} \cong 0.405 \text{ W}$$

The power efficiency is now calculated from Eq. (12.1):

$$\% \text{ efficiency} = \frac{(0.405)}{(6.5)} \times 100 \cong 6.23\%$$

As we can readily see, the power efficiency of a series-fed class-A amplifier is typically very low, making this circuit rather impractical for high-power applications.

As a sidelight to the preceding example note that the power efficiency of a series-fed class-A amplifier has 25% as the absolute maximum, again making this type of amplifier a poor choice where power efficiency is a serious consideration.

12.3.2 Transistor Maximum Power Dissipation

How do we choose a transistor to use in a given situation? First, we are usually limited to those transistors that are already in stock. Then we must decide on the specific transistor that would be suitable for a certain application. The answer is usually contained in the manufacturers' data sheets. Without implying that it is the best, we have chosen a Texas Instruments TIP29 silicon *NPN* transistor as being representative for the examples here. Full data sheets are given in Appendix A.

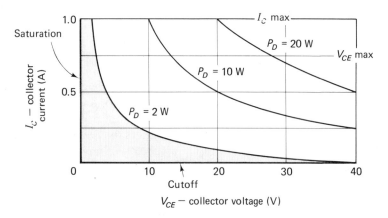

Figure 12.6 Permissible region of operation (shaded area) for a typical power transistor (TIP29) without a heat sink. The maximum power capability increases with the use of a heat sink.

The first step in determining the suitability of a transistor for a certain application is to establish its permissible region of operation. The permissible region of operation for a TIP29 transistor is shown as the shaded area in Figure 12.6. It is determined by the saturation and cutoff regions, as shown, and by three quantities specified by the manufacturer. These quantities are (1) the absolute maximum collector current, (2) the absolute maximum collector voltage, and (3) the continuous device power dissipation. For a TIP29, the manufacturer lists these as 1 A, 40 V, and 2 W, respectively. The manufacturer also supplies two power ratings. One that is usually specified as "continuous device dissipation at (or below) 25°C free-air temperature" is the maximum power rating *without a heat sink*. The second rating, "continuous device dissipation at (or below) 25°C case temperature," is the maximum power rating with additional cooling provided by a fan or a heat sink, or both.

Under normal operation, the emitter junction is forward biased and thus has a voltage of less than 1 V across it. The collector junction, on the other hand, usually has a much higher voltage across it. With the current through the two junctions approximately the same, the collector junction must be dissipating essentially all of the power given out by the transistor. If we make that approximation, then the total power dissipated in the transistor is given by:

$$P_D = I_C V_{CE} \tag{12.6}$$

We can plot this equation for $P_D = 2$ W in Figure 12.6 by choosing values of either I_C or V_{CE} and calculating the other.

The manufacturer also specifies the "operating collector junction temperature range." For the TIP29 this range is listed as $-65°C$ to $150°C$. The more important quantity here is the upper limit. Typically, the maximum collector junction (or simply junction) temperature is 150 to 200°C for silicon transistors and 75 to 100°C for germanium transistors. This maximum junction temperature should not be exceeded; it determines the amount of power that can safely be dissipated with a given heat sink mounted on the transistor.

12.3.3 Derating Curve

Manufacturers usually provide power dissipation derating information or curves of the type depicted in Figure 12.7. The relationship between the junction temperature, ambient temperature, and the total power dissipated is

$$T_J - T_A = \theta_{JA} P_D \qquad (12.7)$$

where T_J is the collector junction temperature in degrees Celcius (°C), T_A is the ambient or room temperature in degrees Celcius (°C), and θ_{JA} is the junction-to-ambient thermal resistance in degrees Celcius per watt (°C/W). Without a heat sink, θ_{JA} is a property of the transistor and indicates the increase in the junction temperature above the ambient temperature for each additional watt of power dissipated in the transistor. For example, suppose that we dissipate 1 W in a TIP29 transistor, which has θ_{JA} listed as 62.5°C/W. Then the junction temperature will rise 62.5°C above the ambient temperature, or will be 87.5°C. This temperature is well below the 150°C maximum, so safe operation would result. If, however, we tried to dissipate 3 W (without a heat sink), the transistor would probably burn out because the junction temperature would reach upwards of 200°C, which is well in excess of the 150°C maximum for this transistor.

12.3.4 Heat Sink Application

With the transistor mounted on a heat sink, the power dissipation at the collector junction inside the transistor can be increased without exceeding the maximum junction temperature. The transistor may be mounted directly to the heat sink or

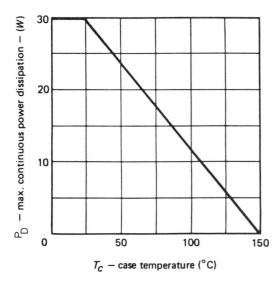

Figure 12.7 Power dissipation derating curve for the TIP29 silicon power transistor.

it may be electrically insulated from the heat sink with a thermally conductive mica (or other material) washer. With a mica washer between the transistor case and the heat sink, and with the ambient temperature assumed to be 25°C, then Equation (12.7) may be rewritten as

$$T_J - 25 = P_T(\theta_{JC} + \theta_{CS} + \theta_{SA}) \tag{12.8}$$

where the single junction-to-ambient thermal resistance has been replaced by three separate thermal resistances: θ_{JC} is the thermal resistance between the junction and the case of the transistor; θ_{CS} is the thermal resistance between the case and the heat sink (i.e., the thermal resistance of the mica washer); and θ_{SA} is the thermal resistance between the heat sink and the air around it (i.e., the thermal resistance of the heat sink). If no washer is used, then $\theta_{CS} \approx 0$ and θ_{CS} and θ_{SA} are replaced by θ_{CA}, as pictured in Figure 12.8. θ_{CA} is the thermal resistance of the heat sink by itself.

Figure 12.8 is used to help you visualize the operation of the heat transfer. Here we show an electrical analog for Equation (12.8) wherein the total power is analogous to current, the different temperatures are analogous to voltages, and thermal resistances are analogous to electrical resistances. The procedure for this operation is illustrated in Example 12.4.

Example 12.4

Without a heat sink, the TIP29 transistor (case type TO-220) can safely dissipate 2 W when the ambient air temperature is ≤25°C. Its thermal resistances are listed in the data specification sheet as $\theta_{JC} = 4.17°C/W$ and $\theta_{JA} = 62.5°C/W$. Let us determine the thermal resistance of the heat sink needed to safely dissipate (a) 10 watts and (b) 20 watts of power. A mica washer coated with silicon thermal compound to improve heat transfer is used to electrically insulate the transistor from the metal heat sink. Its thermal resistance is determined from Table 12.1 as 0.5°C/W.

Solution. (a) For a 10 W power dissipation with T_J equal to a maximum of 150°C and $T_A = 25°C$, the θ_{JA} is calculated to be

$$\theta_{JA} = \frac{T_J - T_A}{P_D} = \frac{150 - 25}{10} = 12.5°C/W$$

The heat sink thermal resistance θ_{SA} is found next:

$$\theta_{SA} = \theta_{JA} - \theta_{JC} - \theta_{CS} = 12.5 - 4.17 - 0.5 = 7.8°C/W$$

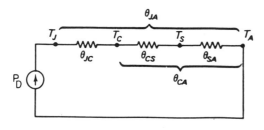

Figure 12.8 Electrical analog for determining power dissipation and temperature for power transistors.

TABLE 12.1 THERMAL RESISTANCE CASE TO SINK θ_{CS} (°C/W) W/Si THERMAL COMPOUND

	Case type			
Material	TO-36	TO-3	TO-220	TO-8
Mica	0.4	0.4	0.5	0.5
Teflon	0.8	0.8	1.0	1.0
Anodized Al	0.4	0.4	0.5	0.5
Metal-to-metal	0.1	0.2	0.3	0.3

As a matter of interest, we can determine the temperatures T_C and T_S in Fig. 12.8 to be 108°C and 98°C, respectively. There is then a 42° temperature differential between the junction and the transistor case, a 5° differential across the mica washer (case to sink), and a 78° differential across the heat sink (sink to ambient). Thus, the effect of the heat sink is to reduce the total temperature differential between the collector junction of the transistor and its case, or to help remove the heat from the junction.

(b) We proceed in a similar manner to calculate the thermal resistance of the heat sink needed to dissipate 20 W. We obtain

$$\theta_{JA} = \frac{T_J - T_A}{P_D} = \frac{150 - 25}{20} = 6.25°C/W$$

Thus, the heat sink thermal resistance must then be

$$\theta_{SA} = 6.25 - 4.17 - 0.5 = 1.58°C/W$$

Observation. Heat sink specifications are available from manufacturers of heat sinks. Some examples of the available specifications may be found in Appendix A. As a general rule, the lower the thermal resistance, the larger the heat sink.

Figure 12.6 plots the power limits of the 10 and 20 W device dissipation (with the heat sink) as determined in the preceding example. The permissible region of operation, therefore, can be increased with the use of proper heat sinks.

We must mention one important fact in connection with maximum power considerations. You have to take care when mounting and installing power transistors in a case so that there is sufficient ventilation and air circulation around the transistor and its heat sink (if used). This is to ensure that the air in contact with the case or heat sink will be at the specified ambient air temperature and not some higher value, which would invalidate all the calculations carried out to determine the proper heat sink.

One additional specification in the manufacturer's data sheet needs further clarification: the "peak collector current," which usually has an accompanying note. This quantity is not the same as the maximum collector current discussed earlier. For example, in the TIP29 data sheet, the manufacturer specifies "peak collector current ... 3 A" with a note that "this value applies for $t_w \leq 0.3$ ms, duty cycle ≤ 10%." This specification means that the collector current may go as high as 3 A for a maximum time of 0.3 millisecond (ms) or for less than 10% of one cycle, whichever is less. Thus, if the signal frequency is 1 kHz, the collector current could go as high as 3 A for a time not to exceed 0.1 ms during each cycle.

12.4 HARMONIC DISTORTION

As we noted earlier (see Figure 12.5), nonlinear operation results in distortion of the output waveshape. From the graphic analysis discussed in Section 12.2, we can determine the harmonic content and distortion in the output waveshape. We can express the output waveshape in the form of a truncated series:

$$I_C = M_0 + M_1 \cos(\omega t) + M_2 \cos(2\omega t) + M_3 \cos(3\omega t) + M_4 \cos(4\omega t) \quad (12.9)$$

where ω is the frequency of the input signal (called the *fundamental frequency*); M_1 is the amplitude of the fundamental component of the output wave; M_2, M_3, and M_4 are the amplitudes of the second, third, and fourth harmonics, respectively. These M coefficients can be determined from the graphic analysis:

$$M_0 = \frac{1}{6}(I_M + I_m) + \frac{1}{3}(I_1 + I_2) - I_Q$$

$$M_1 = \frac{1}{3}(I_M - I_m) + \frac{1}{3}(I_1 - I_2)$$

$$M_2 = \frac{1}{4}(I_M + I_m) - \frac{1}{2}I_Q \quad (12.10)$$

$$M_3 = \frac{1}{6}(I_M - I_m) - \frac{1}{3}(I_1 - I_2)$$

$$M_4 = \frac{1}{12}(I_M + I_m) - \frac{1}{3}(I_1 + I_2) + \frac{1}{2}I_Q$$

The symbols have the following definitions:

I_Q—no-signal collector current (point A, Fig. 12.5)

I_M—peak collector current (point C, Fig. 12.5)

I_m—minimum collector current (point G, Fig. 12.5)

I_1—half-peak collector current (point B, Fig. 12.5)

I_2—half-minimum collector current (point F, Fig. 12.5)

Once the M coefficients are determined, the harmonic content of the output wave is known and the amount of distortion may be calculated as follows:

$$D_2 \cong \left|\frac{M_2}{M_1}\right| \times 100\%$$

$$D_3 \cong \left|\frac{M_3}{M_1}\right| \times 100\% \qquad (12.11)$$

$$D_4 \cong \left|\frac{M_4}{M_1}\right| \times 100\%$$

where D_2, D_3, and D_4 are the second, third, and fourth harmonic distortions in percentage, respectively. We find the per cent total harmonic distortion (THD), D_T, from the following equation:

$$D_T = \sqrt{D_2{}^2 + D_3{}^2 + D_4{}^2} \qquad (12.12)$$

These calculations are best illustrated in Example 12.5.

Example 12.5

For the series-fed class-A amplifier described in Ex. 12.2, we want to determine the harmonic content and distortion in the output waveshape.

Solution. The numbers to be used in Eq. (12.10) are determined by inspecting the output waveshape shown in Fig. 12.5. They are

$I_Q \cong 0.65$ A   (from point A on Fig. 12.5)

$I_M \cong 0.9$ A   (from point C on Fig. 12.5)

$I_m \cong 0.0$ A   (from point G on Fig. 12.5)

$I_1 \cong 0.8$ A   (from point B on Fig. 12.5)

$I_2 \cong 0.45$ A   (from point F on Fig. 12.5)

We then insert these values into Eq. (12.10) to find the M coefficients:

$$M_0 = \frac{1}{6}(0.9) + \frac{1}{3}(0.8 + 0.45) - 0.65 \cong -0.083 \text{ A} \cong -83 \text{ mA}$$

$$M_1 = \frac{1}{3}(0.9) + \frac{1}{3}(0.8 - 0.45) \cong 0.42 \text{ A}$$

$$M_2 = \frac{1}{4}(0.9) - \frac{1}{2}(0.65) \cong -0.1 \text{ A}$$

$$M_3 = \frac{1}{6}(0.9) - \frac{1}{3}(0.8 - 0.45) \cong 0.03 \text{ A} \cong 30 \text{ mA}$$

$$M_4 = \frac{1}{12}(0.9) - \frac{1}{3}(0.8 + 0.45) + \frac{1}{2}(0.65) \cong -0.017 \text{ A} \cong -17 \text{ mA}$$

The separate and total harmonic distortion (THD) can now be calculated:

$$D_2 = \frac{0.1}{0.42} \times 100\% \cong 24\%$$

$$D_3 = \frac{0.03}{0.42} \times 100\% \cong 7\%$$

$$D_4 = \frac{0.017}{0.42} \times 100\% \cong 4\%$$

$$D_T = \sqrt{(24)^2 + (7)^2 + (4)^2} \cong 25\%$$

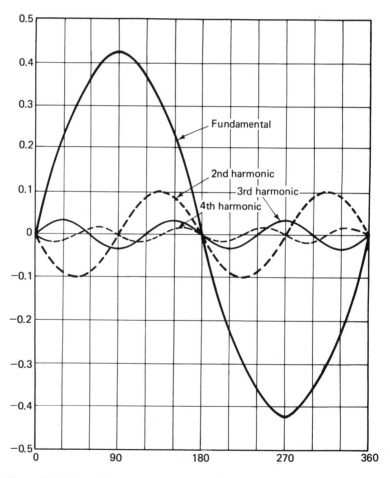

Figure 12.9 Approximate harmonic content of the collector waveshape of Figure 12.5. (The phase relationship among the harmonics is not shown.)

Obviously, harmonic distortion is high (as we might have expected from the actual waveshape). In essence, the total harmonic distortion is mainly due to the second harmonic.

To see the meaning of these harmonic calculations, you must plot the individual waveshapes, (which add up to produce the collector current of Figure 12.5) for the numbers obtained in Example 12.5 (see Figure 12.9).

In practice, harmonic distortion analysis may be carried out either (1) by observing the actual waveshape on an oscilloscope and proceeding as just outlined or (2) by using a special instrument called a *wave*, or *distortion analyzer* that gives the harmonic amplitudes directly. Another method involves using a *spectrum analyzer* to make the measurements.

12.5 SINGLE-ENDED CLASS-A AMPLIFIERS

As we have seen in the previous sections, the series-fed class-A amplifier performs poorly, both in terms of power efficiency and distortion. Consequently, it is not commonly used, but it does serve to illustrate some of the basic graphic techniques. Let us next examine the transformer-coupled single-ended class-A amplifier, whose typical circuit is illustrated in Figure 12.10.

The analysis of the circuit shown in Figure 12.10 is quite similar to that of the series-fed amplifier, with two important differences. First, the dc load line of the transformer-coupled amplifier has a slope proportional to the negative recip-

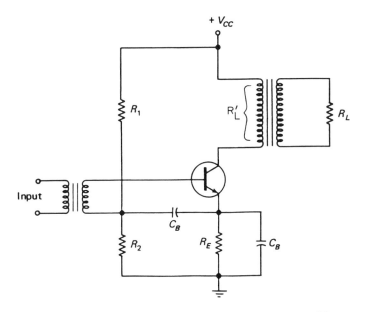

Figure 12.10 Transformer-coupled single-ended class-A amplifier.

rocal of $R_E + R$, where R is the effective series resistance of the primary (input winding) of the transformer. The other difference is that the ac load line of the transformer-coupled amplifier is not the same as the dc load line; the ac load line is governed by the turns ratio n and the load resistance R_L. The effective or reflected ac load of the amplifier is given by

$$R'_L = \frac{1}{n^2} R_L \qquad (12.13)$$

where the transformer turns ratio is defined in terms of the number of turns in the primary (N_1) and secondary (N_2) winding:

$$n = \frac{N_2}{N_1} \qquad (12.14)$$

The transformer is used as an impedance-matching device to match the actual load, R_L, to the output of the transistor. The actual load might typically be 4 or 8 Ω (for a speaker). By using the transformer, we can increase the load of the transistor to a few hundred or even a few thousand ohms.

In analyzing the transformer-coupled circuit, we may proceed to draw the dc load line and establish the Q-point as we did for the series-fed amplifier. The ac load line is then plotted to pass through the Q-point and to have the slope of $-1/R'_L$. This plot is shown on a typical set of output characteristics in Figure 12.11.

From that point on, the graphical analysis parallels the one used in the previous sections. However, we use the ac load line, which is different from the dc load line in this case.

Typical amplifier performance comparisons of the series-fed vs. the transformer-coupled amplifiers, we shall see, yield somewhat lower distortion and somewhat higher power efficiency for the transformer-coupled amplifier. The absolute maximum power efficiency for the transformer-coupled amplifier is 50%, as compared to 25% for the series-fed amplifier.

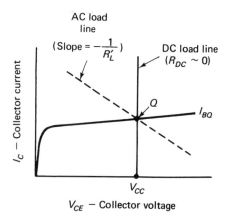

Figure 12.11 ac and dc load lines for transformer-coupled class-A amplifier.

12.6 TRANSFORMER-COUPLED PUSH-PULL AMPLIFIERS

A typical transformer-coupled push-pull amplifier circuit is depicted in Figure 12.12. We can explain its operation by drawing waveshapes, as shown. If the input signal is sinusoidal, the input transformer applies it to the two transistors. But the two base signals are 180° out of phase. Let us consider that the two transistors are biased so that each will operate in class B (with R_1 removed and R_2 shorted). When the base of $Q1$ goes positive, it conducts and amplifies the input signal. At the same time, the base of $Q2$ is being driven negative, and $Q2$ is cut off. Thus, the output is being supplied by $Q1$. During the other half of the cycle, the base of $Q1$ is driven negative, while the base of $Q2$ is driven positive. The situation is exactly reversed, with $Q1$ being cut off and $Q2$ conducting. The collector signals are coupled through the output transformer to the load.

The name for the amplifier (push-pull) is indicative of circuit operation. One transistor is conducting while the other one is off, and vice versa.

The waveshapes shown along with the push-pull circuit in Figure 12.12 are for class-AB operation. Before we go on to class-AB operation, however, let us examine the characteristics of class-B amplifiers. Then you will understand the need for class AB.

12.6.1 Class B Operation

Let us assume that the input signal is from a very low impedance source, so that the input to the transistors is a voltage. A set of voltage-to-current transfer char-

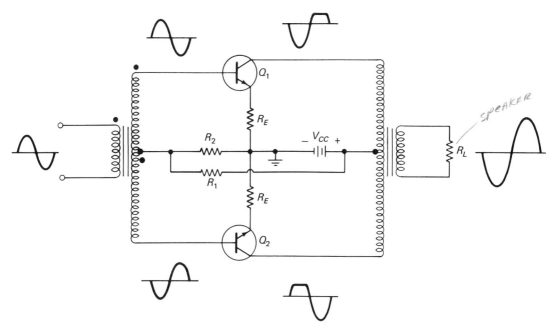

Figure 12.12 Transformer-coupled push-pull amplifier (shown with waveshapes for class-AB operation).

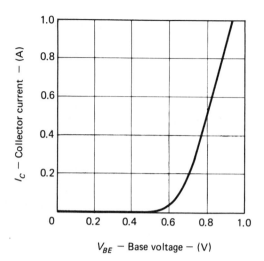

Figure 12.13 Typical voltage transfer curve for a power transistor.

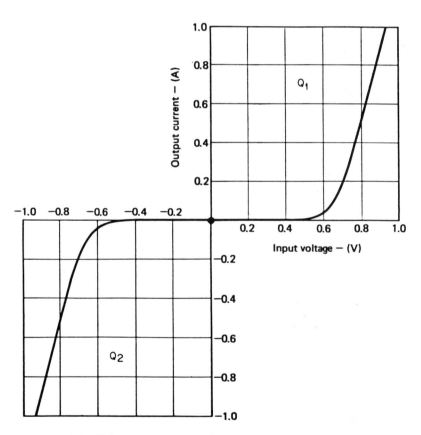

Figure 12.14 Voltage-to-current transfer curve for a class-B push-pull (or complementary-symmetry) amplifier.

acteristics for a typical transistor is shown in Figure 12.13. We can also assume that the transistors used in the push-pull amplifier are matched and have exactly the same properties. The composite transfer curve for the push-pull amplifier as a whole is obtained (in class-B operation) by placing the individual transistor transfer curves as indicated in Figure 12.14. The horizontal axis (which was V_{BE} for the individual transistors) now becomes the input voltage axis for the composite device: the push-pull amplifier. Above the horizontal axis is the collector current for $Q1$, below is the collector current for $Q2$. The output current is the algebraic difference of the two collector currents, thus accounting for the inversion of the $Q2$ characteristic. The vertical axis is then the output current for the push-pull amplifier.

One of the advantages of the push-pull amplifier is that the second and fourth harmonics of the two transistors are exactly in phase and cancel in the output transformer. The output current contains only the third harmonic. We, therefore, would expect very low distortion from a push-pull amplifier. However, this is not the case. Consider Example 12.6.

Example 12.6

The transformer-coupled amplifier in Fig. 12.12 is operated in class-B. The voltage-current transfer curve for the composite device is illustrated in Fig. 12.14. The input voltage is a 0.9 V peak sinusoid with zero dc component, as shown in Fig. 12.15(a). Determine the output waveform.

Solution. We use the same graphic techniques as in Ex. 12.2; that is, we sketch the input wave below the transfer characteristics and project it up to the characteristics and across to the right to get the output waveshape. Although this construction is not shown, the result (i.e., the output waveshape) appears in Fig. 12.15(b). Note the extreme flattening near the origin; this characteristic is called *crossover distortion*.

Although the second and fourth harmonics are absent for a class-B push-

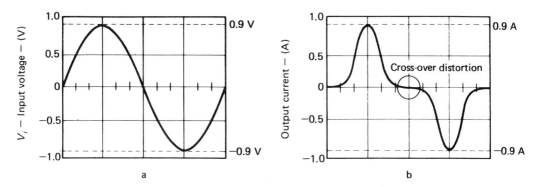

Figure 12.15 Class-B push-pull amplifier: (a) input waveshape and (b) output waveshape.

pull amplifier, distortion is not low. There is a very large third harmonic component.

12.6.2 Class AB Operation

To eliminate some of this distortion, we operate the transistors in class AB by biasing slightly. The dc voltage across R_2 of Figure 12.12 is allowed to exceed the dc voltage across R_E. For example, let us assume that the base-emitter voltage of each transistor is raised to, say, 0.65 V at the operating point. The composite characteristics for this case (class AB) are found by matching the individual characteristics at the V_{BE} operating-point value; we obtain 0.65 V, as shown in Figure 12.16. To determine the composite curve, we calculate the algebraic sum of the individual curves at a particular voltage. The result is indicated by the dashed curve in Figure 12.16. Note that the class-AB composite curve is nearly a straight line; in any case, it is much closer to a straight line than the transfer curve for the class-B amplifier in Figure 12.14.

By biasing the bases even more positive, we can make the composite transfer

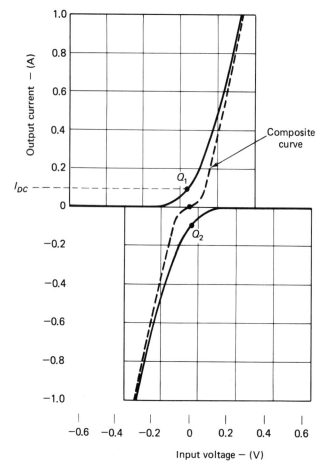

Figure 12.16 Transfer characteristics for class-AB operation.

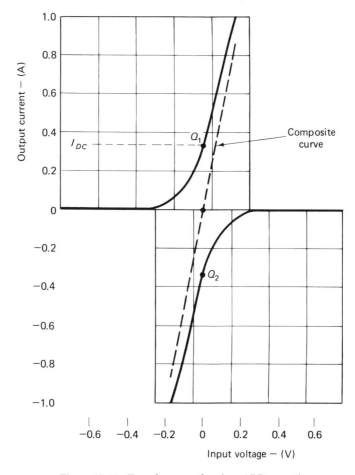

Figure 12.17 Transfer curve for class-ABB operation.

curve almost perfectly linear, as shown in Figure 12.17. This special case of class AB is sometimes called class ABB. The linearity of the transfer curve indicates that distortion will be extremely small. However, this improvement is not completely free. As we can see in Figure 12.17, the Q-point is fairly high; that is, there is a fairly large dc (idling) current which decreases the power efficiency. In general, class-B amplifiers have the best power efficiency, but large distortion, while class-AB amplifiers have fairly good (i.e., high) power efficiency and very low distortion. The choice is a trade-off between power efficiency and harmonic distortion.

We must mention another difference between class-B and class-AB amplifiers. From Figure 12.17, it becomes evident that only a 0.2 V peak input voltage is necessary to cause approximately the same output current as the 0.8 V peak signal in the class-B amplifier. Thus, the gain is increased when the transistors are biased upwards toward class-A operation. In fact, even higher gain and lower harmonic distortion are possible when the amplifier is operated in class A. The transfer curve for class-A operation is shown in Figure 12.18. Class-A operation is very low in power efficiency—so low, in fact, that it is uneconomical.

Among the four operating points (Figures 12.14, 12.15, 12.17, and 12.18), perhaps the best compromise is shown in Figure 12.16. The power efficiency is quite good, and the distortion is low enough that it can be brought within acceptable levels for use in stereo amplifiers through negative feedback. We shall discuss feedback in detail in Chapter 13.

Transformer-coupled push-pull amplifiers can offer very high-quality performance as the final stage in audio and servo amplifiers. However, they have one major drawback. They need large, heavy, and very expensive transformers, both at the input and output. A circuit that replaces the input transformer, together with the ensuing waveforms, is depicted in Figure 12.19. The operation of this circuit is identical to that already discussed, with the exception that the driver stage is a class-A operated transistor.

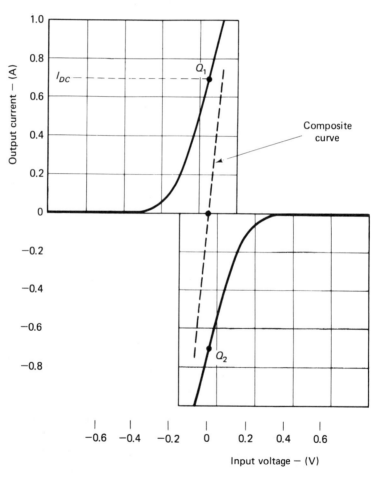

Figure 12.18 Transfer curve for class-A operation.

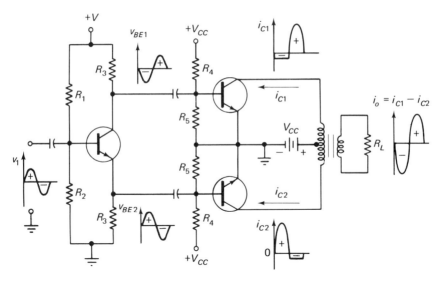

Figure 12.19 Split-load phase-inverter input circuit for a push-pull amplifier.

12.7 OTHER PUSH-PULL AMPLIFIERS

Many schemes exist for eliminating the need for the transformers at both the input and output. Basically, such schemes fall into two categories. One encompasses push-pull amplifiers, which use the same type of transistors for the output stage, i.e., either both *NPN* or both *PNP*. The second system includes complementary-symmetry amplifiers where two different types of transistors are used in the output stage.

Figure 12.20 gives an example of these two different types of transformerless push-pull amplifier output stages. Their operation is almost identical to that of the transformer-coupled push-pull amplifier, with a few exceptions. First, the amplifiers in Figure 12.20 are dc-coupled to the load; that is, there may be a direct

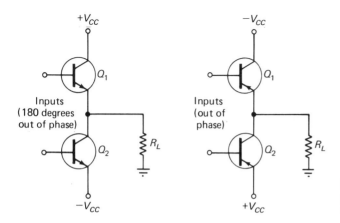

Figure 12.20 Basic transformerless push-pull configuration with (a) *NPN* and (b) *PNP* transistors.

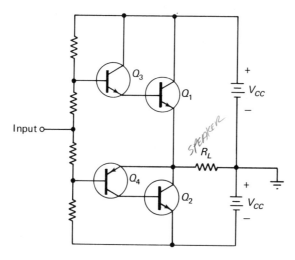

Figure 12.21 Transformerless push-pull amplifier circuit.

current through the load if any imbalance exists between the two transistors. Secondly, these amplifiers require a dual (+ and −) supply, whereas the transformer-coupled amplifiers only require a single supply.

The inputs to the output stage still need signals which are 180° out of phase, just like the transformer-coupled amplifier. We can furnish such signals by an input (center-tapped) driver transformer or by the driver circuit shown in Figure 12.21. Driver transistors Q3 and Q4 are of complementary types, one a *PNP* and the other an *NPN*, thus providing outputs which are out of phase while having a common input.

The transformerless push-pull amplifier circuit illustrated in Figure 12.22 needs only one power supply to operate. It may be a more desirable version for certain applications. Note that in this circuit the load is capacitively coupled to the amplifier.

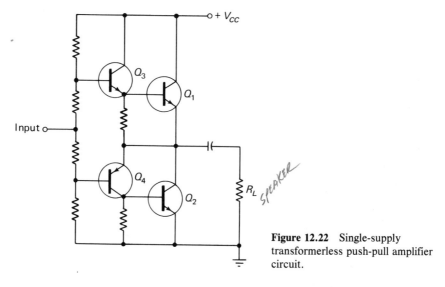

Figure 12.22 Single-supply transformerless push-pull amplifier circuit.

12.8 COMPLEMENTARY-SYMMETRY AMPLIFIERS

Push-pull operation may be accomplished by using complementary transistors (one *NPN* and one *PNP*), as depicted in Figure 12.23. The use of complementary matched transistors in the output stage eliminates the need for two out-of-phase input signals.

The circuit operates as follows: When the input signal is positive, $Q1$ is biased on and amplifies the input signals while $Q2$ is essentially cut off. This action gives

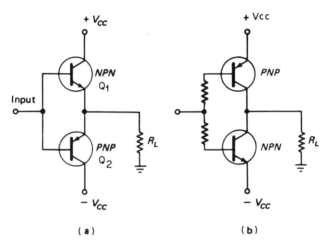

Figure 12.23 Basic complementary-symmetry configurations.

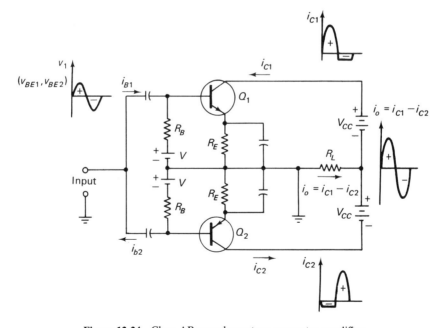

Figure 12.24 Class-AB complementary-symmetry amplifier.

Sec. 12.8   Complementary-Symmetry Amplifiers   233

the positive output signal. When the input is negative, Q1 is cut off and Q2 is biased on, so that the negative part of the output signal is provided by Q2.

Class-AB complementary-symmetry amplifiers are shown in Figures 12.24 and 12.25. In Figure 12.24, resistors R_B together with power supplies V provide the AB-base bias. Resistors R_E provide current limiting at the output (when the transistors saturate) as well as bias stability. The waveshapes, assuming a sinusoidal signal, are shown.

The power amplifier pictured in Figure 12.25(a) uses diodes D_1 and D_2 to set the amplifier into class-AB operation by slightly forward biasing the output stage, thereby removing the crossover distortion.

Because the audio power amplifier of Figure 12.25(a) is designed to develop >30 watts in the load (R_L), a large drive current is needed. The drive current for the *output* transistors Q5 and Q6 is provided by the *driver* transistors Q3 and Q4.

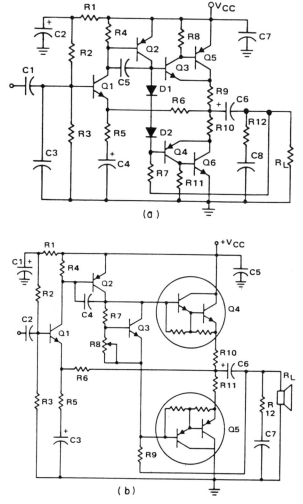

Figure 12.25 Complete schematic diagram of a complementary-symmetry power amplifier: (a) using power transistors in the output and (b) using Darlington transistors in the output. (*Courtesy of Motorola, Inc.*)

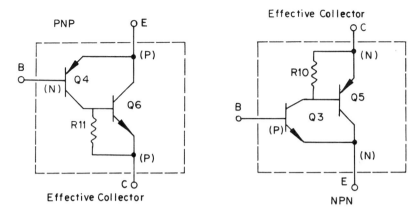

Figure 12.26 Composite high-gain power transistors: (a) *PNP* and (b) *NPN*.

Transistors $Q4$ and $Q6$ form a *composite PNP* transistor while transistors $Q3$ and $Q5$ form a *composite NPN* transistor.

The results of connecting $Q4$–$Q6$ and $Q3$–$Q5$ together are shown in Figure 12.26. In Figure 12.26(a) we see that the collector current of the *PNP* driver transistor ($Q4$) becomes the base current of the *NPN* power output transistor ($Q6$). $Q6$ is operated as an emitter-follower and provides current gain without inversion. Since the emitter of the *NPN* output transistor ($Q6$) acts as the "*effective*" collector of the composite *PNP* transistor, then the circuit is equivalent to a high-gain, high-power *PNP* transistor.

A capacitor [$C6$ in Figure 12.25(a) and (b)] is used to couple the amplifier output when a single supply ($+V_{CC}$) is used. The single supply voltage of $+V_{CC}$ is divided so that the dc voltage at the *effective emitters* of $Q5$ and $Q6$ is equal to $+V_{CC}/2$. This voltage is set by the voltage divider formed between $R1 + R2$ and $R3$. Since the emitter voltage of Q_1 is approximately equal to the base voltage, then the voltage at the junction of $R9$ and $R10$ is approximately equal to $+V_{CC}/2$ volts if both V_{BE} and the small V_{R6} drop is neglected.

12.9 IC POWER AMPLIFIERS

To complete our study of power amplifiers, we will look at integrated circuit amplifiers. IC power amplifiers are available in two popular circuit configurations, *duals* and *monos*. Dual audio power amplifiers are used in stereo phonographs, tape players, recorders, and AM/FM receivers. Mono audio power amplifiers were designed primarily for automotive applications.

12.9.1 Dual IC Power Amplifiers

The LM378 is an example of a dual IC amplifier. It is a dual 4-watt audio amplifier that has self-centered biasing and operates from a single supply of up to 35 V. Its total harmonic distortion (THD) is a respectable 0.1% at 2 watts. Figure 12.27

pictures the schematic diagram of the LM378. Here we see a complementary class-AB power amplifier consisting of a Darlington *NPN* emitter follower (made up of $Q12$ and $Q13$) and a composite *PNP* emitter follower (made up of $Q9$, $Q14$, and $Q15$). The LM378 has both thermal protection and current limiting. The 1.2 A current limiter is provided by transistors $Q9$ and $Q10$.

The LM378, as well as similar devices, may have its output power *boosted* by the addition of a simple complementary-symmetry booster circuit, as shown in Figure 12.28(a). As noted in Figure 12.28(b), the THD of the power amplifier is less than 2% at 10 watts output (per channel) with a frequency response of 50 Hz to 30 kHz.

12.9.2 Mono IC Power Amplifier

The LM383 8-watt audio power amplifier is an example of a mono IC power amplifier. This device is packaged in a five-lead TO-220 case to facilitate heat-sinking. As with most IC power amplifiers, the LM383 has both current limiting and thermal shutdown circuitry. Because this device was specifically designed for automotive application, its typical supply voltage is 14.4 volts. The LM383 has a maximum supply voltage of 20 V with a shutdown provision for supply voltages greater than the maximum. In order to achieve rated power levels at 14.4 V, high currents (up to 3.5 A) must pass through low impedance loads ($<4\ \Omega$). Example 12.7 explores the basic operating circuit of the LM383.

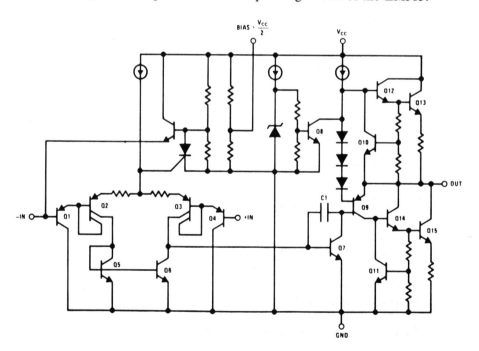

Figure 12.27 A simplifier schematic diagram of the LM378 dual 4-watt power amplifier showing the design features of the amplifier. (*Courtesy of National Semiconductor Corporation*)

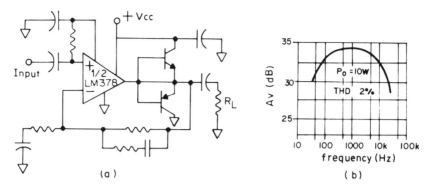

Figure 12.28 The LM378 with a booster circuit to provide 10 watts into a 4-ohm load. (*Circuit design courtesy of National Semiconductor Corporation*)

Example 12.7

Using the LM383, describe the selection of the circuit components and heat sink for the amplifier ($A_v = 100$) pictured in Fig. 12.29. Also, determine the THD for output power of 3, 5, and 8 watts for the load impedance and supply voltage specified in Fig. 12.29.

Solution. Although the input impedance is specified at 150 kΩ, a fairly large ($C_1 = 10$ µF) capacitor is used at the input. This size capacitor is not needed to ensure low frequency response but, rather, to prevent speaker "pop" at turn ON.

To ensure good supply voltage ripple rejection (up to 40 dB) in the output signal, the network of R_1 and R_2 is selected with low values of resistance and C_2 is set to 470 µF.

Capacitor C_4 (0.2 µF) is used to decouple the amplifier from the supply line and it must be installed at the $+V_{CC}$ pin of the circuit board. Capacitor

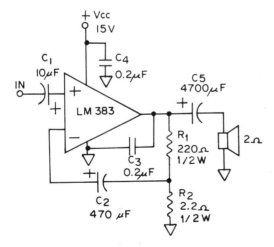

Figure 12.29 The LM383 mono IC power amplifier circuit used with Example 12.7.

Sec. 12.9   IC Power Amplifiers   237

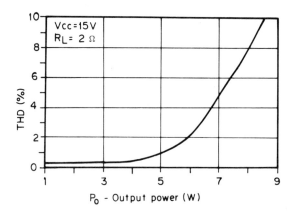

Figure 12.30 Total harmonic distortion (THD) vs. power out of the LM383.

C_3 is used to prevent *parasitic oscillations* and it is installed between the V_{out} and ground pins next to the IC. The value of C_3 is specified by the manufacturer as 0.2 μF. To ensure good low-frequency response and to block dc from entering the speaker, C_5 is selected as 4700 μF.

As previously mentioned, R_1 and R_2 have low values of resistance. A typical pick for R_1 is 220 Ω. Once R_1 is selected, then R_2 is determined as follows:

$$A_V = 1 + R_1/R_2 \quad (12.15)$$

With A_V specified as 100 and R_1 = 220 Ω, R_2 is computed to be ≈2.2 Ω.

The LM383 has a 4°C/W junction to case thermal resistance (θ_{JC} = 4°C/W) and a maximum operating temperature of 150°C (T_j = 150°C). The device dissipation is 6 W for a $+V_{CC}$ = 15 V and an output power of 8 watts into a 2 Ω load. The thermal impedance of the heat sink (θ_{SA}) is determined with the aid of Table 12.1 and Eq. (12.7) where $\theta_{JA} = \theta_{jc} + \theta_{CS} + \theta_{SA}$. We will assume an elevated ambient temperature (T_A) of 40°C. Thus,

$$T_J - T_A = (\theta_{JC} + \theta_{CS} + \theta_{SA})P_D$$

$$150 - 40 = (4 + 0.3 + \theta_{SA})6$$

$$\theta_{SA} = \frac{110}{6} - 4.3 = 14°C/W$$

Because of heat sink availability and to provide a margin of safety, a 12°C/W sink is selected.

From Fig. 12.30, the THD is determined to be 0.2% at 3 W, 2% at 6 W, and 8% at 8 W.

12.10 SUMMARY

There is no "best" power amplifier circuit since most of the circuits have some good and some bad points. The specific application should be the deciding factor, as well as the availability or nonavailability of a dual power supply and of matched complementary transistors, etc.

The design of discrete component power amplifiers is by no means an easy task. There are trade-offs between output power, power efficiency, and harmonic distortion. We can analyze power amplifiers most easily by using graphical techniques. Remember, however, that these techniques are at best only good approximations. Moreover, an analysis carried out for one transistor, or a pair of transistors, may not hold for another transistor of the same type or another pair of similar transistors. However, the circuits included in this chapter, transformer-coupled and transformerless push-pull, as well as the complementary-symmetry amplifiers, will give you a good start in the design of power amplifiers.

REVIEW QUESTIONS

1. In what way are power amplifiers different from small-signal amplifiers?
2. What is the main method of analyzing power amplifiers?
3. What is a class-A power amplifier? Discuss in terms of the input and output signals.
4. What is a class-B amplifier?
5. What is a class-AB amplifier?
6. What gives rise to harmonic distortion in a power amplifier?
7. What is meant by the efficiency in a power amplifier?
8. Why is the efficiency of a power amplifier important, whereas that of a small-signal amplifier is not?
9. What is the role of a transformer used in the output of a power amplifier?
10. In a series-fed class-A power amplifier, what is the relationship between the output power and harmonic distortion?
11. What is the maximum power efficiency possible from a class-A amplifier?
12. What are the advantages and disadvantages of a push-pull amplifier over the single-ended amplifier?
13. What classes of amplifiers are operated in push-pull? Why these classes and not others?
14. What is the action of the two transistors used in a push-pull amplifier?
15. The input stage of a push-pull amplifier may have a center-tapped transformer. Why is this transformer necessary?
16. What other circuits may be used to feed a push-pull amplifier besides transformer-coupled? Explain.
17. What are the differences and similarities in the performance of class-B and class-AB push-pull amplifiers?
18. What are the advantages and disadvantages of a complementary-symmetry amplifier as compared to a push-pull amplifier?
19. What types of transistors must be used in a complementary-symmetry amplifier? Why?

20. What are the advantages and disadvantages of a quasi-complementary-symmetry amplifier over a regular complementary-symmetry amplifier?

PROBLEMS

1. A TIP29B silicon *NPN* transistor (see Appendix A) is to be used in a power amplifier. Determine and sketch its permissible region of operation.
2. Repeat Problem 1 for the TIP29C transistor.
3. The TIP29B transistor is used in conjunction with a mica washer and heat sink, which have a combined thermal resistance of 5°C/W. Determine the maximum power that can be dissipated without exceeding the maximum junction temperature.
*4. What thermal resistance of the heat sink will be needed if a TIP29A transistor is to dissipate 15 W? (Assume that no mica washer is used.)
5. Repeat Problem 4 for a TIP29C transistor and a mica washer with a thermal resistance of 1.2°C/W.
6. What is the approximate operating junction temperature of the transistors in Problems 4 and 5?

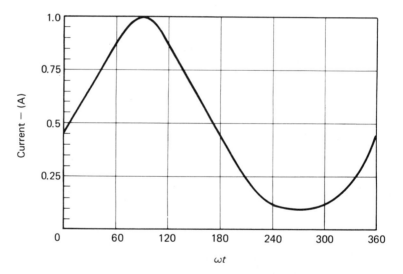

Figure 12.31 Circuit for Problem 7.

7. A class-A amplifier driving a 4 Ω load produces the current waveshape shown in Fig. 12.31. If the dc power supplied is 4 W, determine (a) output power, (b) total harmonic distortion, and (c) power efficiency.
8. For the load line and Q-point as shown in Fig. 12.3, the input signal is 20 mA peak-to-peak. Determine (a) the output waveshape, (b) output power, (c) total harmonic distortion, and (d) power efficiency.

*9. Repeat Problem 8 if the load resistance is 8 Ω. (Note that a new load line must be drawn.)
10. Repeat Ex. 12.6 if the input voltage is 1.6 V peak-to-peak.
11. If the load in Problem 10 is 16 Ω, determine the output power.
12. A complementary-symmetry amplifier has the composite transfer characteristics shown in Fig. 12.17. The input voltage is 0.3 V peak-to-peak. Sketch the output (current) waveshape. What is the power delivered into an 8 Ω load?

Chapter 13

Negative Feedback Amplifiers

In this chapter we introduce negative feedback amplifiers and discuss their basic properties. We shall also be concerned with the advantages and disadvantages of negative feedback amplifiers as compared to the basic nonfeedback type of amplifier.

Two types of feedback are possible. The first is called *negative* feedback: A part or all of the output signal (voltage or current) is diverted and applied at the input so as to subtract from the input signal. In this manner, the apparent input signal to the original amplifier is reduced, and thus the output signal is accordingly reduced. Negative feedback amplifiers are characterized as having lower gain than the same amplifier without feedback.

The second type of feedback is called *positive* feedback: A part or all of the output signal is applied to the input so as to add to it. Positive feedback is undesirable in amplifiers because it usually causes the amplifier to be unstable and oscillate. However, this property is used fully in oscillator circuits. In this chapter we shall confine the discussion to negative feedback amplifiers. Chapter 15 will treat the positive feedback amplifier, or oscillator.

13.1 GENERAL FEEDBACK CONCEPTS

We first saw the advantage of using feedback constructively in the discussion of Q-point stability in Chapter 4. Feedback is an extremely useful tool in numerous applications, particularly in control systems. Control systems encompass all circuits where the output is sensed and used to control or correct the input to provide the desired output. Other uses of feedback include sensing the output, then comparing it to some reference signal, and finally controlling the input (and thus the output) in accordance with the difference between the output and reference signal. Specifically, negative feedback in an amplifier can be used to

1. Stabilize the gain (either voltage or current)

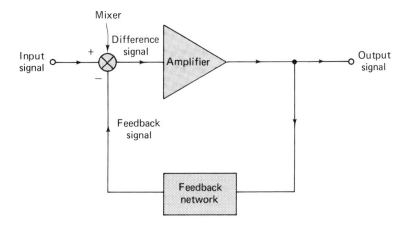

Figure 13.1 General block diagram of a feedback amplifier.

2. Achieve more linear operation
3. Broaden the bandwidth
4. Lower or raise input impedance
5. Lower or raise output impedance
6. Reduce the noise in the amplifier
7. Reduce thermal effects

By stabilizing the gain, we mean making the gain less dependent on the specific device parameters. Linearity of operation is important for all amplifiers, but an improvement in the linearity (i.e., lower distortion) is especially important in power amplifiers, discussed in the previous chapter. Items 3, 4, and 5 are self-explanatory. The noise (spurious electrical signals generated within the amplifier) is especially bothersome in amplifiers with extremely small signal levels. In these cases, negative feedback can be used to decrease the amount of noise in the amplifier. We have already covered the reduction in thermal effects in our discussion on stability in Chapter 4.

We shall classify types of feedback according to the action of the feedback on the gain. Two types are *current feedback* and *voltage feedback* circuits; they are characterized by a decrease in gain. Two other feedback types, termed *shunt* and *series feedback* circuits, will also be treated.

The basic feedback amplifier block diagram is shown in Figure 13.1, with signal paths as shown. The signal at any point in Figure 13.1 could be either a voltage or a current, depending on the type of performance desired.

13.2 VOLTAGE-FEEDBACK AMPLIFIERS

Referring to Figure 13.1, we see that when all the signals are voltages, the circuit is a voltage-feedback amplifier. The general form of a voltage-feedback amplifier is shown in Figure 13.2. Negative feedback is accomplished by causing a part of the output voltage to subtract from the input voltage.

13.2.1 Voltage Gain

In Figure 13.2, the output voltage V_o appears across both the external load R_L and the feedback network. The reverse voltage gain of the feedback network β_v is defined as

$$\beta_v = \frac{V_f}{V_o} \tag{13.1}$$

(Note: The β as defined here is *not* the same as the transistor dc β defined previously and should not be confused with it.)

The open-circuit voltage gain of the amplifier A_v is defined by

$$A_v = \frac{V_o}{V_1} \tag{13.2}$$

Setting the sum of the voltages at the input to zero as indicated in Figure 13.2, we obtain

$$V_s = V_1 + V_f \tag{13.3}$$

The open-circuit gain of the feedback amplifier A_{vf} is given by

$$A_{vf} = \frac{V_o}{V_s} = \frac{V_o}{V_1 + V_f} = \frac{\frac{V_o}{V_1}}{1 + \frac{V_f}{V_1}} \tag{13.4}$$

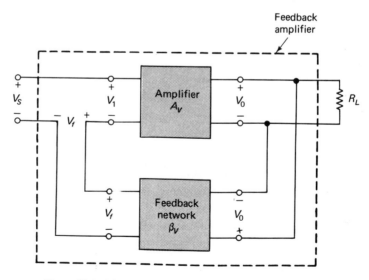

Figure 13.2 Block diagram for a voltage-feedback amplifier.

From Equation (13.1) we see that $V_f = \beta_v V_o$. Also noting that $A_v = V_o/V_1$, we obtain

$$A_{vf} = \frac{A_v}{1 + \beta_v A_v} \tag{13.5}$$

We define negative feedback for $(1 + \beta_v A_v)$ as being greater than 1 and positive feedback for $(1 + \beta_v A_v)$ as being less than 1.

Usually, $|A_v|$ is much larger than 1, so that we can approximate:

$$A_{vf} \cong \frac{1}{\beta_v} \tag{13.6}$$

13.2.2 Input Resistance

The input resistance for the feedback amplifier is defined as the ratio of V_s to I_1. Substituting Equation (13.1) for V_f into Equation (13.3), we obtain

$$V_s = V_1 + \beta_v V_o = V_1(1 + \beta_v A_v) \tag{13.7}$$

We can substitute $V_1 = R_i I_1$. Thus,

$$R_{if} = \frac{V_s}{I_1} = R_i(1 + \beta_v A_v) \tag{13.8}$$

When negative voltage feedback is used, the input resistance is increased.

13.2.3 Output Resistance

Assuming that the current drawn off by the feedback network in Figure 13.2 is negligibly small, we can write

$$V_o = A_v V_1 - I_o R_o \tag{13.9}$$

Substituting for V_1 from Equation (13.3), we have

$$V_o = A_v V_s - A_v V_f - I_o R_o \tag{13.10}$$

We then rearrange the equation:

$$V_o(1 + \beta_v A_v) = A_v V_s - I_o R_o \tag{13.11}$$

Dividing both sides by $(1 + \beta_v A_v)$, we obtain

$$V_o = A_{vf} V_s - I_o \frac{R_o}{1 + \beta_v A_v} \tag{13.12}$$

We solve for the feedback amplifier output resistance by setting $V_s = 0$. Thus,

$$R_{of} = \frac{V_o}{-I_o} = \frac{R_o}{1 + \beta_v A_v} \tag{13.13}$$

When the feedback is negative, the output resistance with feedback is lower than the output resistance without feedback.

13.2.4 Equivalent Circuit

An examination of Equation (13.12) suggests an equivalent circuit for the output of the feedback amplifier. The complete equivalent circuit for the feedback amplifier is given in Figure 13.3. The procedure for determining the parameters of a voltage-feedback amplifier is illustrated in Example 13.1.

Example 13.1

The amplifier shown in Fig. 13.4 is a voltage-feedback amplifier. The feedback network consists of voltage-divider resistors R_9 and R_{10}. The amplifier without feedback has the following parameters: $A_v = 100$, $R_i = 2$ kΩ, and $R_o = 5$ kΩ. We want to determine the parameters of the amplifier with feedback.

Solution. The feedback factor β_v is calculated from the resistor ratio:

$$\beta_v = \frac{R_{10}}{R_{10} + R_9} = \frac{0.1}{0.1 + 2.2} = \frac{1}{23}$$

We next find the amount of feedback:

$$1 + \beta_v A_v = 1 + \frac{100}{23} = 5.35$$

The voltage-feedback amplifier parameters can now be calculated as follows:

$$R_{if} = R_i(1 + \beta_v A_v) \cong (2)(5.35) \text{ k}\Omega \cong 10.7 \text{ k}\Omega \tag{13.8}$$

$$R_{of} = \frac{R_o}{1 + \beta_v A_v} \cong \frac{5}{5.35} \text{ k}\Omega \cong 0.935 \text{ k}\Omega \cong 93 \text{ }\Omega \tag{13.13}$$

$$A_{Vf} = \frac{A_v}{1 + \beta_v A_v} \cong \frac{100}{5.35} \cong 18.7 \tag{13.5}$$

Note that by using the approximation in Eq. (13.6) we obtain $A_{Vf} \cong 23$, which is a crude approximation in this case. The approximation becomes usable when $\beta_v A_v$ is greater than 10.

The calculation of the gain, input impedance, and output impedance for the amplifier without feedback must be carried out with care, because the feedback network cannot be ignored completely. For input calculations, the amplifier in

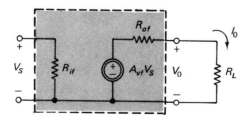

Figure 13.3 Equivalent circuit for the voltage-feedback amplifier.

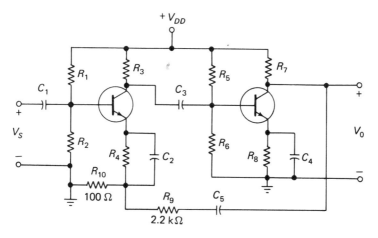

Figure 13.4 Example of a voltage-feedback amplifier.

Figure 13.4 without feedback must be considered as having $V_o = 0$ (output short-circuited). For output calculations, the amplifier must be considered as having $V_f = 0$ (in this case, R_{10} shorted). However, the input impedance in this example is considered without the parallel combination of R_1 and R_2. The total input impedance contains these two resistors.

13.3 CURRENT-FEEDBACK AMPLIFIERS

When all the signals in Figure 13.1 are currents, the circuit is a current-feedback amplifier. Such a block diagram is depicted in Figure 13.5. Negative feedback is accomplished by causing the output current to subtract from the input current, as shown.

13.3.1 Current Gain

In Figure 13.5, the output current I_o is supplied to the load R_L and to the feedback network. The reverse current gain of the feedback network, β_I, is defined as

$$\beta_I = \frac{I_f}{I_o} \qquad (13.14)$$

Therefore, when the output current I_o flows into the feedback network, the component that reaches the input of the amplifier is

$$I_f = \beta_I I_o \qquad (13.15)$$

The amplifier input current I_1 is given by

$$I_1 = I_s - I_f = I_s - \beta_I I_o \qquad (13.16)$$

The feedback amplifier input current is I_s and can be obtained from Equation (13.16):

$$I_s = I_1 + \beta_I I_o \qquad (13.17)$$

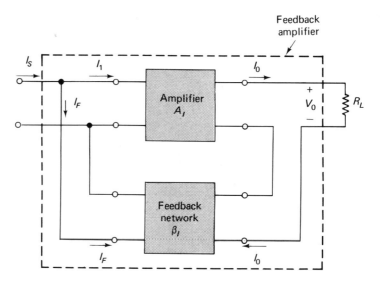

Figure 13.5 Block diagram for a current-feedback amplifier.

If the forward short-circuit current gain of the amplifier is

$$A_I = \frac{I_o}{I_1} \tag{13.18}$$

we can write

$$I_o = A_I I_1 \tag{13.19}$$

The short-circuit current gain of the feedback amplifier A_{If} is given by the ratio of I_o to I_s. Using Equations (13.17) and (13.19) we obtain

$$A_{If} = \frac{I_o}{I_s} = \frac{A_I I_1}{I_1 + \beta_I I_o} \tag{13.20}$$

Dividing the numerator and denominator of Equation (13.20) by I_1 gives us

$$A_{If} = \frac{A_I}{1 + \beta_I A_I} \tag{13.21}$$

This equation relates the short-circuit current gain of the feedback amplifier, A_{If}, to the short-circuit current gain of the amplifier without feedback, A_I.

We can approximate the current gain of the feedback amplifier if we note that usually $|A_I|$ is much larger than 1. Thus, if we divide both the numerator and the denominator of Equation (13.21) by A_I, we have

$$A_{If} \cong \frac{1}{\beta_I} \tag{13.22}$$

Consequently, the short-circuit current gain of the feedback amplifier can be made independent of the device parameters and dependent only on the feedback network component values.

13.3.2 Input Resistance

The input resistance of the feedback amplifier R_{if} is defined as the ratio of V_s to I_s, where V_s is the voltage across the input terminals in Figure 13.5.

$$R_{if} = \frac{V_s}{I_s} = \frac{I_1 R_i}{I_f + I_1} = \frac{R_i}{1 + \dfrac{I_f}{I_1}} \tag{13.23}$$

Noting that $I_f = \beta_I I_o$, we have

$$R_{if} = \frac{R_i}{1 + \beta_I A_I} \tag{13.24}$$

When the feedback is negative, $(1 + \beta_I A_I)$ is greater than 1 and the input resistance is lowered as a result of the feedback.

13.3.3 Output Resistance

According to Figure 13.5, the output resistance of the feedback amplifier is defined as the ratio of V_o to $-I_o$ for the condition $I_s = 0$. If we assume that the voltage developed across the feedback network in the output loop is negligibly small when compared to either V_o or the voltage across R_o, then we can say that the voltage across R_o is approximately equal to V_o:

$$V_o = (A_I I_1 - I_o) R_o \tag{13.25}$$

Substituting Equation (13.16) for I_1, we have

$$V_o = (A_I I_s - \beta_I A_I I_o - I_o) R_o \tag{13.26}$$

We can now factor out $(1 + \beta_I A_I)$:

$$V_o = \left[\left(\frac{A_I}{1 + \beta_I A_I} \right) I_s - I_o \right] R_o (1 + \beta_I A_I) \tag{13.27}$$

When $I_s = 0$, we can solve for the output resistance of the feedback amplifier:

$$R_{of} = \frac{V_o}{-I_o} = R_o (1 + \beta_I A_I) \tag{13.28}$$

Thus, we see that the effect of negative current feedback is to increase the output resistance.

13.3.4 Equivalent Circuit

We have noted the effect of negative current feedback on the current gain, input resistance, and output resistance. We can summarize these factors in an equivalent circuit for the feedback amplifier. In Equation (13.27), we recognize the coefficient of I_s as A_{If}. If we use the definition of R_{of}, we can write

$$V_o = (A_{If} I_s - I_o) R_{of} \tag{13.29}$$

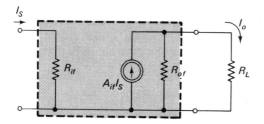

Figure 13.6 Equivalent circuit for the current-feedback amplifier.

This equation suggests an equivalent circuit of the output with a current generator $A_{If}I_s$ and output resistance R_{of}. The input current is I_s and the input resistance is R_{if}. This feedback amplifier equivalent circuit is shown in Figure 13.6. The procedure for determining the parameters of a current-feedback amplifier is illustrated in Example 13.2.

Example 13.2

The circuit shown in Figure 13.7 is a current-feedback amplifier. Without feedback, the amplifier parameters are: $A_I = 800$, $R_i = 1$ kΩ, and $R_o = 10$ kΩ. Feedback is applied through the feedback network consisting of R_8 and R_9 (220Ω and 4.7 kΩ, respectively). We want to determine the amplifier parameters with feedback.

Solution. The feedback factor β_I is found from the resistor ratio:

$$\beta_I \cong \frac{R_8}{R_8 + R_9} \cong \frac{0.22}{0.22 + 4.7} \cong \frac{1}{22.4}$$

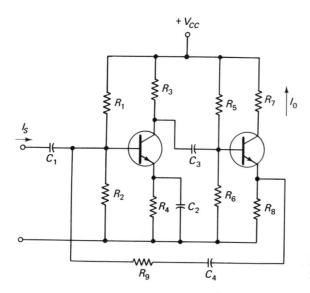

Figure 13.7 Example of a current-feedback amplifier.

250 Negative Feedback Amplifiers Chap. 13

We next calculate the amount of feedback:

$$1 + \beta_I A_I \cong 1 + \frac{800}{22.4} \cong 36.7$$

The current-feedback amplifier parameters can be calculated as follows:

$$R_{if} = \frac{R_i}{1 + \beta_I A_I} \cong \frac{1000}{36.7}\,\Omega \cong 27\,\Omega \tag{13.24}$$

$$R_{of} = R_o(1 + \beta_I A_I) \cong (10)(36.7)\text{ k}\Omega \cong 367\text{ k}\Omega \tag{13.28}$$

$$A_{If} = \frac{A_I}{1 + \beta_I A_I} \cong \frac{800}{36.7} \cong 21.8 \tag{13.21}$$

The approximation for the current gain given in Equation (13.22) is quite valid in this case because $\beta_I A_I$ is greater than 10. Using the approximation, we get $A_{If} = 22.4$, which compares quite favorably with the value just calculated. This equation means that in the current-feedback amplifier in this example the current gain is insensitive to the transistor parameters and depends on the values of the feedback resistors R_8 and R_9.

Determining the amplifier parameters without feedback should be done with care. If we want to determine the input parameters, the output current should be zero (output open-circuited at the second emitter in Figure 13.7). When we calculate the output parameters, the input current should be zero (input open-circuited at the first base). In this way, feedback is eliminated, although the loading of the feedback circuit on the amplifier without feedback is taken into account.

13.4 EFFECT OF FEEDBACK ON FREQUENCY RESPONSE

As we have seen in the previous two sections, feedback changes the gain and input and output impedances of an amplifier. As might be expected, it also modifies the frequency response of an amplifier. The following discussion is applicable for the current-gain response of the current-feedback amplifier and the voltage-gain response of the voltage-feedback amplifier. We shall make no distinction between the two responses in this section.

An amplifier without feedback has lower and upper 3 dB frequencies labeled f_1 and f_2, respectively. The same amplifier with voltage feedback will have lower and upper 3 dB frequencies (labeled f_{1f} and f_{2f}, respectively) given by

$$f_{1f} = \frac{f_1}{1 + \beta A} \tag{13.30}$$

$$f_{2f} = f_2(1 + \beta A) \tag{13.31}$$

where β and A would have the appropriate subscripts (I or V) depending on whether it was a current-feedback or voltage-feedback amplifier. The effect of the feedback is to *decrease* the lower 3 dB frequency and to *raise* the upper 3 dB frequency.

The bandwidth of the amplifier with voltage feedback is modified accordingly. If we assume that the lower 3 dB frequency is quite small compared to the upper 3 dB frequency, the bandwidth with feedback is given by:

$$BW_f \cong BW(1 + \beta A) \tag{13.32}$$

The effect of feedback on the amplifier frequency response is illustrated in Example 13.3.

Example 13.3

An amplifier (without feedback) has a voltage gain of 1,000 and lower and upper 3 dB frequencies of 100 Hz and 100 kHz, respectively. It is made into a feedback amplifier having 20 dB of feedback. We want to determine the frequency response of the feedback amplifier.

Solution. The frequency response of the amplifier is shown in Fig. 13.8. The amount of feedback is

$$\text{dB of feedback} = 20 \log |1 + \beta A| = 20 \text{ dB}$$

Therefore,

$$1 + \beta A = 10$$

With feedback, the gain of the amplifier is

$$A_{Vf} = \frac{1000}{10} = 100 \text{ or } 40 \text{ dB}$$

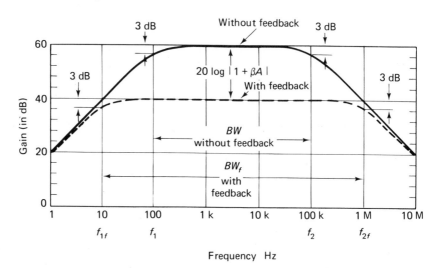

Figure 13.8 Effect of feedback on the frequency response of an amplifier.

The lower and upper 3 dB frequencies are

$$f_{1f} = \frac{100}{10} \text{ Hz} = 10 \text{ Hz}$$

$$f_{2f} = (100)(10) \text{ kHz} = 1 \text{ MHz}$$

These results are plotted in Fig. 13.8. Note that the increase in bandwidth is always the same as the decrease in gain. In this case, the bandwidth is increased tenfold, whereas the gain is decreased tenfold.

13.5 SERIES- AND SHUNT-FEEDBACK AMPLIFIERS

Block diagrams for series-feedback and shunt-feedback amplifiers are shown in Figures 13.9 and 13.10, respectively. In a series-feedback circuit, the output current is sampled and fed back to the input as a voltage. This connection is sometimes called a *voltage-series feedback* amplifier.

13.5.1 Series-Feedback Amplifiers

The series-feedback amplifier modifies the effective transconductance of the basic amplifier without feedback. Any amplifier can be represented by the equivalent circuit shown in Figure 13.11. A typical example might be a single-stage FET amplifier.

The transconductance of the amplifier without feedback is modified, once

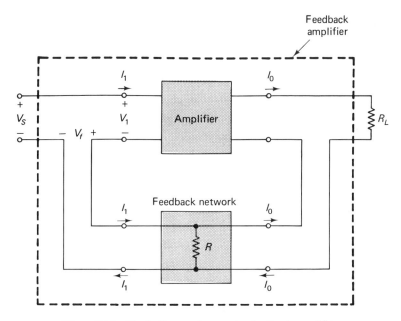

Figure 13.9 Block diagram for a series-feedback amplifier.

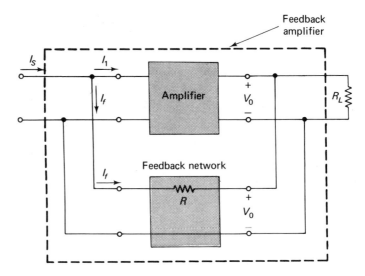

Figure 13.10 Block diagram for a shunt-feedback amplifier.

series-feedback is applied, according to the following equation:

$$G_{mf} = \frac{G_m}{1 + \beta_m G_m} \qquad (13.33)$$

where G_{mf} and G_m are the respective transconductances, with and without feedback, and β_m is defined as

$$\beta_m = \frac{V_f}{I_o} \qquad (13.34)$$

and has units of conductance.

Table 13.1 summarizes the effect of the series-feedback circuit on the voltage and current gains and gives the input and output impedances. The procedure for determining the parameters of a series-feedback amplifier is illustrated in Example 13.4.

Example 13.4

The circuit in Fig. 13.12 is an example of a series-feedback amplifier. The circuit values are: $R_1 = R_2 = 100 \text{ k}\Omega$, $R_C = 2.2 \text{ k}\Omega$, and $R_E = 1 \text{ k}\Omega$. The

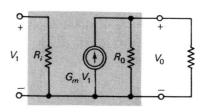

Figure 13.11 A transconductance amplifier.

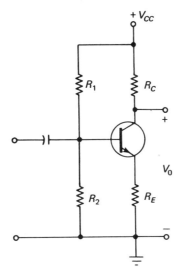

Figure 13.12 Example of a series-feedback amplifier.

transistor parameters are: $h_{ie} = 1 \text{ k}\Omega$, $h_{fe} = 100$, with h_{re} and h_{oe} negligible. We want to find the parameters for the amplifier as shown.

Solution. For the amplifier without feedback (R_E bypassed by a capacitor):

$$R_i = h_{ie} \cong 1 \text{ k}\Omega$$

$$A_I = -h_{fe} \cong -100$$

$$A_V \cong -h_{fe}\frac{R_C}{h_{ie}} \cong -100\,\frac{2.2}{1} \cong -220$$

We can assume that the output impedance is infinite. From Table 13.1, for

TABLE 13.1 SUMMARY OF EQUATIONS FOR DIFFERENT FEEDBACK CIRCUITS

Type of feedback	A_{Vf} (open circuit)	A_{If} (short circuit)	R_{if}	R_{of}
Voltage (Fig. 13.2)	$\dfrac{A_V}{1 + \beta_V A_V}$	A_I	$R_i(1 + \beta_V A_V)$	$\dfrac{R_o}{1 + \beta_V A_V}$
Current (Fig. 13.5)	A_V	$\dfrac{A_I}{1 + \beta_I A_I}$	$\dfrac{R_i}{1 + \beta_I A_I}$	$R_o(1 + \beta_I A_I)$
Series (Fig. 13.9)	$\sim -\dfrac{R_c{}^*}{R_m}$	A_I	$R_i + R_m(1 - A_I)$	$R_o + R_m(1 - A_V)$
Shunt (Fig. 13.10)	A_V	$\sim -\dfrac{R_c{}^\dagger}{R_m}$	$\left(\dfrac{R_m}{1 - A_V}\right) \| (R_i)$	$(R_o) \| \left(\dfrac{A_V R_m}{A_V - 1}\right)$

* For $A_V R_m \gg (R_o + R_i)$.

† For $A_I \gg \left(1 - \dfrac{R_i}{R_m}\right)$.

the amplifier with series feedback, where $R_m = R_E = 1\text{ k}\Omega$, we see that

$$R_{if} = R_i + R_m(1 - A_I) \cong 1 + 1(1 + 100)\text{ k}\Omega = 102\text{ k}\Omega$$

$$R_{of} = R_o + R_m(1 - A_V) \cong \infty$$

$$A_{vf} \approx -\frac{R_C}{R_m} \cong -\frac{2.2}{1} = -2.2$$

$$A_{If} = A_I \cong -100$$

∴ With feedback: R_{in} is increased, R_{out} is increased, voltage gain is decreased.

The preceding results can also be obtained through simple circuit analysis by drawing the equivalent circuit of the amplifier shown.

13.5.2 Shunt-Feedback Amplifiers

The block diagram in Figure 13.10 shows the connection for a shunt-feedback amplifier. The output voltage is sampled by the feedback network and fed back to the input in the form of a current.

We can choose to represent an amplifier (with or without feedback) in the form of a transresistance amplifier, as depicted in Figure 13.13. Once shunt feedback is applied, the transresistance of the amplifier is modified according to

$$R_{mf} = \frac{R_m}{1 + \beta_r R_m} \qquad (13.35)$$

where R_{mf} and R_m are the transresistances of the amplifier, with and without feedback, respectively. The feedback factor is defined as

$$\beta_r = \frac{I_f}{V_o} \qquad (13.36)$$

Table 13.1 lists the voltage and current gains as well as the input and output impedance transformations for the shunt-feedback amplifier illustrated in Figure 13.10. The procedure for determining the parameters of a shunt-feedback amplifier is illustrated in Example 13.5.

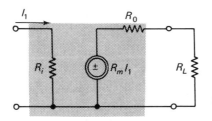

Figure 13.13 A transresistance amplifier.

Example 13.5

The circuit in Fig. 13.14 is a simple example of a shunt-feedback amplifier. The circuit values are: $R_C = 3.3$ kΩ and $R_B = 56$ kΩ. The transistor parameters are: $h_{ie} = 1.5$ kΩ, $h_{fe} = 75$; h_{re} and h_{oe} are negligible. We want to determine the parameters for the amplifier shown in Fig. 13.14.

Solution. For the amplifier without feedback:

$$R_i \cong h_{ie} \cong 1.5 \text{ k}\Omega$$

$$A_V \cong -h_{fe}\frac{R_C}{h_{ie}} \cong -165$$

$$A_I \cong -h_{fe} \cong -75$$

The output impedance is assumed to be infinite.

From Table 13.1 we can determine the parameters of the amplifier with feedback where $R_m = R_B = 56$ kΩ. Thus,

$$R_{if} = \frac{R_m}{1 - A_V} \parallel R_i \cong \left\| \frac{56 \text{ k}\Omega}{1 + 165} \right\| 1.5 \text{ k}\Omega \cong 275 \Omega$$

$$R_{of} = R_o \parallel \frac{A_V R_m}{A_V - 1} \cong R_B \cong 56 \text{ k}\Omega$$

$$A_{Vf} = A_V \approx -165$$

$$A_{If} \approx -\frac{R_C}{R_m} \cong -\frac{3.3}{56} \approx -0.06$$

∴ With feedback: R_i is decreased, R_o is decreased, current gain is decreased.

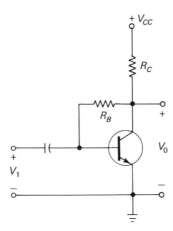

Figure 13.14 Example of a shunt-feedback amplifier.

TABLE 13.2 SUMMARY OF THE EFFECTS OF DIFFERENT FEEDBACK TYPES ON CIRCUIT PERFORMANCE

Type of feedback	Effect on			
	Gain		Impedance	
	Voltage†	Current‡	Input	Output
Voltage (Fig. 13.2)	Reduced	Unchanged	Increased	Decreased
Current (Fig. 13.5)	Unchanged	Reduced	Decreased	Increased
Series (Fig. 13.9)	*	Unchanged	Increased	Increased
Shunt (Fig. 13.10)	Unchanged	*	(a) Decreased for R_i large (b) Unchanged for R_i small	(a) Decreased for R_o large (b) Unchanged for R_o small

* See Table 13.1.
† Open-circuit voltage gain.
‡ Short-circuit current gain.

The preceding values may also be obtained by more conventional circuit analysis after the small-signal equivalent circuit for the amplifier in Fig. 13.14 is drawn.

The effect of different types of feedback on circuit performance is summarized in Table 13.2.

13.6 EFFECT OF FEEDBACK ON NONLINEAR DISTORTION AND NOISE

As one result of negative feedback, nonlinear distortion in amplifiers is decreased. In particular, the amplitude of the distorted signal is reduced (for a discussion of harmonics, see Chapter 12):

$$D_f = \frac{D}{1 + \beta A} \tag{13.37}$$

where β and A refer to voltage feedback and D_f and D are the amplitudes of the distortion signals with and without feedback.

Because all signals in a negative feedback amplifier are fed back to the input, their own amplitude at the output is reduced. Furthermore, distortion signals as well as internally generated noise signals are also reduced in amplitude.

REVIEW QUESTIONS

1. What is *negative* feedback?
2. What is *positive* feedback?
3. What type of feedback (positive or negative) is used in a feedback amplifier? Why?
4. What is the effect of negative feedback on the amplifier gain stability?
5. What is the effect of negative feedback on the amplifier bandwidth?
6. What is the effect of negative feedback on the nonlinearity and noise in an amplifier?
7. In a voltage feedback circuit, how is the voltage gain stabilized against transistor parameter variations? Explain.
8. What is the effect of negative voltage feedback on the amplifier input resistance? On the output resistance?
9. In a current feedback circuit, how is the current gain stabilized against transistor parameter variations? Explain.
10. What is the effect of negative current feedback on the amplifier input resistance? On the output resistance?
11. What is the effect of negative feedback on the frequency response of an amplifier?
12. What does shunt feedback modify?
13. What does series feedback modify?
14. What are the basic advantages of a negative feedback amplifier over one using no feedback?

PROBLEMS

1. Suppose that we construct a voltage feedback amplifier, using an amplifier with: $A_V = 200$, $R_i = 100$ kΩ, and $R_o = 10$ kΩ. Determine the feedback factor β to give a gain (with feedback) of 26 dB. Also calculate the input and output resistances of the feedback amplifier.
*2. An amplifier is characterized by $A_{Isc} = 80$, $R_i = 1.8$ kΩ, and $R_o = 40$ kΩ. What type of feedback must be used to achieve an input resistance of 500Ω? Determine the current gain and output resistance of the amplifier with feedback driving a 1 kΩ load.
3. The feedback amplifier shown in Fig. 13.12 is constructed with $R_1 = R_2 = 100$ kΩ, $R_C = 1.2$ kΩ, and $R_E = 680$ Ω. The transistor parameters are: $h_{ie} = 1$ kΩ, $h_{fe} = 85$, and $1/h_{oe} = 55$ kΩ. Determine the input and output resistances as well as the voltage gain.
4. The feedback amplifier shown in Fig. 13.14 is constructed with $R_B = 220$ kΩ and $R_C = 5.6$ kΩ. Transistor parameters are the same as in Problem 3. De-

termine the input and output resistances and the current gain if a 1 kΩ load is placed from collector to ground.

5. An amplifier with a gain of 66 dB and upper and lower cutoff frequencies of 100 kHz and 100 Hz, respectively, is made into a feedback amplifier. We want bandwidth of 2 MHz. Specify the amount of feedback needed and the new cutoff frequencies.
6. Repeat Problem 5 if a gain of 20 (magnitude) is desired.
7. Repeat Problem 3 if the output voltage is taken across R_E.
8. Repeat Ex. 13.2 if the gain without feedback is changed to 1,000. Make a comparison of your answers with those in Ex. 13.2 and make a conclusion.
*9. If the open-loop (no feedback) gain of the amplifier in Ex. 13.1 changes by 5%, how much does the gain with feedback change? How could this change be minimized still further?
10. If the resistors used in the feedback network in Ex. 13.2 have a tolerance (i.e., an uncertainty in the value) of ±10%, what is the uncertainty of the gain with feedback?
11. Determine the values of resistors R_8 and R_9 in Ex. 13.2 to provide a current gain of 40. What are the new input and output resistance values?
12. If the input voltage in Ex. 13.4 is 100 mV, and $R_C = R_E = 1$ kΩ, determine the output voltage across R_C. Also determine the voltage across R_E. How could we use this circuit?

Chapter 14

Differential and Operational Amplifiers

We introduce a very important group of amplifier circuits in this chapter. Differential and operational amplifiers are dc-coupled amplifiers that are used as building blocks in a multitude of applications. They have become important since the cost of integrated circuits (IC) that use these devices has decreased. Because of the popularity of ICs, we shall emphasize IC differential and operational amplifiers.

14.1 EMITTER-FOLLOWER CIRCUITS

The emitter-follower circuit was introduced in Chapter 3; here we want to re-emphasize its usefulness and apply it in more complicated circuits. Figure 14.1 shows an emitter-follower circuit with two power supplies. This connection allows the input voltage to be at or near ground; or, if so desired, the output voltage could be maintained at or near ground potential. These conditions may be necessary in a dc-coupled amplifier consisting of many stages of amplification.

In some applications, the input impedance of the simple emitter-follower circuit depicted in Figure 14.1 may be insufficient even though it is high (up to 50 kΩ). For these high impedance applications, the emitter-follower circuit shown in Figure 14.2 is used where the two transistors are interconnected as indicated in Figure 14.3. Here we see the two collectors are connected together while the emitter of one is connected to the base of the other; this interconnection constitutes what is called a *Darlington configuration*.

The Darlington emitter-follower configuration of Figure 14.2 is analyzed by combining the Darlington pair into one composite device. In fact, Darlington-connected transistor pairs are available in a single, three-lead package. The ac small-signal equivalent circuit for the Darlington configuration is illustrated in Figure 14.4. If the Darlington configuration is to be used in an emitter-follower configuration, h_{oe1} usually cannot be neglected because of the high effective input resistance at $B2$.

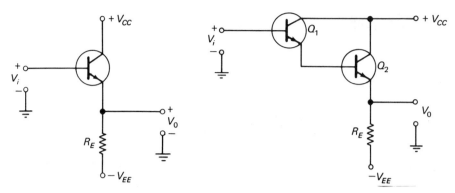

Figure 14.1 Emitter-follower with two power supplies.

Figure 14.2 Darlington emitter-follower.

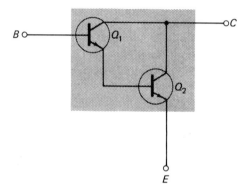

Figure 14.3 Darlington transistor configuration.

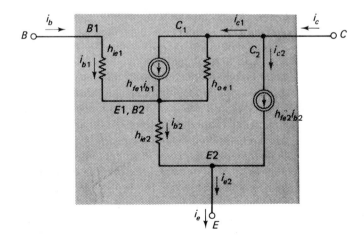

Figure 14.4 Small-signal equivalent circuit for the Darlington circuit.

262   Differential and Operational Amplifiers   Chap. 14

The small-signal equivalent circuit for the Darlington emitter-follower is shown in Figure 14.5. Note that at $B2$ (which is the same point as $E1$), the impedance to ground is

$$R_{i2} = h_{ie} + (h_{fe} + 1)R_E \cong h_{fe}R_E \tag{14.1}$$

There is current division between R_{i2} and h_{oe}, so that i_{b2} is given by

$$i_{b2} = (h_{fe} + 1)i_{b1} \frac{1}{1 + h_{oe}R_{i2}} \tag{14.2}$$

The output current (through R_E) is i_e, that is,

$$i_e = (h_{fe} + 1)i_{b2} \tag{14.3}$$

We can calculate the current gain (i_e/i_{b1}) from Equations (14.2) and (14.3):

$$A_I = \frac{(h_{fe} + 1)^2}{1 + h_{fe}h_{oe}R_E} \tag{14.4}$$

The input impedance may also be calculated. Note that R_{i2} is the effective load on transistor 1:

$$R_i = \frac{h_{fe}^2 R_E}{1 + h_{fe}h_{oe}R_E} \tag{14.5}$$

Although the voltage gain of the Darlington emitter-follower circuit is just slightly less than 1, it may be assumed to be 1 for all practical considerations. The procedure for determining the parameters of an emitter-follower circuit is illustrated in Example 14.1.

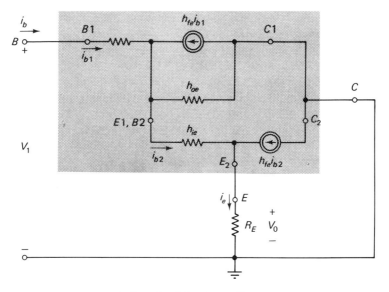

Figure 14.5 Equivalent circuit for the Darlington emitter-follower.

Sec. 14.1 Emitter-Follower Circuits

Example 14.1

The Darlington emitter-follower circuit shown in Fig. 14.2 is constructed with transistors having the following parameters: $h_{ie} = 1$ kΩ, $h_{fe} = 60$, and $h_{oe} = 1/40$ kΩ. We want to determine the amplifier parameters for R_E of 100 Ω and 1 kΩ.

Solution. We calculate the input impedance for the two values of R_E using Eq. (14.5) as follows:

$$R_i = \frac{(60)^2 R_E}{1 + \frac{60 R_E}{40 \text{ k}}}$$

For $R_E = 100$ Ω, this calculation is 313 kΩ. When $R_E = 1$ kΩ, it is 1.44 MΩ. The current gain is determined for the two values of R_E using Eq. (14.4) as follows:

$$A_I = \frac{(60 + 1)^2}{1 + \frac{60 R_E}{40 \text{ k}}}$$

For $R_E = 100$ Ω, this calculation is 3236. When $R_E = 1$ kΩ, it becomes 1488.

From this example we can see the effect of R_E on both the input impedance and current gain as well as the increase in input impedance over a comparable single-transistor emitter-follower circuit.

Both the emitter-follower and Darlington emitter-follower circuits are used in applications where a high-input impedance, low-output impedance, and high-current gain are needed.

14.2 DIFFERENTIAL AMPLIFIERS

Differential amplifiers are characterized by the use of two matched transistors, having two possible inputs and two possible outputs. They are usually dc-coupled and require two power supply voltages, one positive and one negative.

14.2.1 Basic Differential Amplifiers

The basic differential amplifier circuit is illustrated in Figure 14.6. It features two matched transistors with the emitters tied together. The input signals V_1 and V_2 are applied to the two bases, with two possible outputs taken at the two collectors.

Ideally, R_E should be infinite in order to provide infinite input impedance and to make the circuit operate as an ideal differential amplifier (DIFF AMP). The characteristics of an ideal DIFF AMP are as follows: (1) for identical inputs, the output is zero; and (2) for different inputs, the output is proportional to the difference between the inputs. You can see the need for infinite R_E by considering

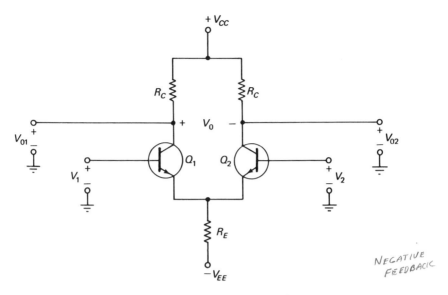

Figure 14.6 Basic differential amplifier.

the equivalent circuit in Figure 14.7. If R_E were indeed infinite, it would not draw current and i_{e1} would equal $-i_{e2}$. Under these conditions i_{c1} would equal $-i_{c2}$ and V_{o1} would equal $-V_{o2}$. Thus, the differential output V_o would be zero.

Because $Q1$ and $Q2$ must have dc bias, the size of R_E is limited. If R_E is increased, the negative supply voltage must also be increased in order to maintain the same dc bias current for the two transistors.

14.2.2 Improved DIFF AMP Circuits

Let us study the circuit shown in Figure 14.8, where the resistor has been replaced with an ideal constant current source. This arrangement provides the dc bias current needed for Q_1 and Q_2 yet it still maintains an infinite resistance between the two emitters and ground.

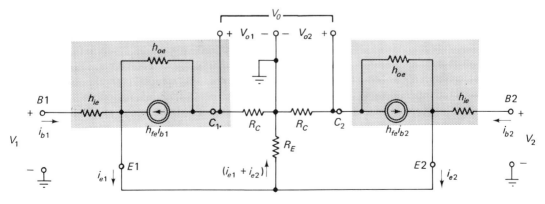

Figure 14.7 DIFF AMP equivalent circuit.

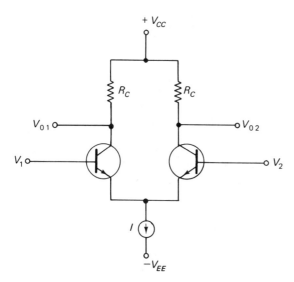

Figure 14.8 DIFF AMP with a constant current source.

In practice, a current generator can be approximated by a transistor, as shown in Figure 14.9. The current I is controlled by bias resistors R_1, R_2, and R_3. Diodes $D1$ and $D2$ act to regulate the temperature of transistor $Q3$. If we assume forward diode voltages of 0.6 V and neglect the base current of $Q3$, the base voltage of $Q3$ with respect to ground is

$$V_{B3} = -\frac{(V_{EE} - 1.2)R_1}{R_1 + R_2} \tag{14.6}$$

The bias current I may be calculated from the voltage across R_3:

$$V_{R3} = V_{B3} - V_{BE3} - V_{EE} = IR_3 \tag{14.7}$$

If R_1 and R_2 are equal, then

$$I = \tfrac{1}{2}V_{EE}/R_3 \tag{14.8}$$

The bias current I is, therefore, independent of the base-emitter voltage, which is quite sensitive to temperature; instead, it is a constant determined only by the supply voltage and R_3.

An alternate method of obtaining a temperature-stable bias current is shown in Figure 14.10. A Zener diode is used to provide the bias to the current source transistor $Q3$. The series resistor R_Z is a current-limiting resistor used to keep the current through the Zener diode to a safe limit. If the Zener diode temperature characteristics match those of the base-emitter diode of $Q3$, then the bias current I will be insensitive to temperature and given by

$$I = \frac{V_Z - V_{BE3}}{R_3} \tag{14.9}$$

where V_Z is the Zener diode (reference) voltage.

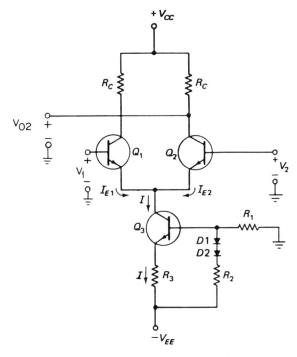

Figure 14.9 A practical DIFF AMP circuit.

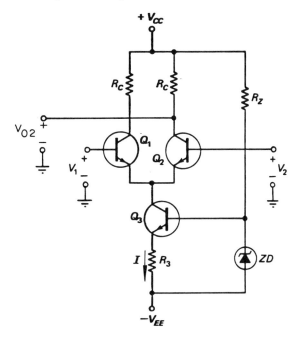

Figure 14.10 Zener-referenced current source in a DIFF AMP.

The ac performance of the two improved DIFF AMP circuits illustrated in Figures 14.9 and 14.10 is quite similar. The small-signal equivalent circuit of Figure 14.7 may be used with R_E replaced by the output impedance of $Q3$, as shown in Figure 14.11. Note that the base of $Q3$ has no ac drive; therefore, all of its terminal currents have zero alternating current, and its ac equivalent circuit contains only the output impedance in the collector circuit, h_{oe}.

Difference gain. Let us consider the DIFF AMP circuit of either Figure 14.9 or 14.10. If we apply input signals V_1 and V_2 (shown in Figure 14.12) in such a way that

$$V_1 = -V_2 = \frac{V_i}{2} \tag{14.10}$$

the resulting output voltage (V_o) divided by V_i is called the *difference gain*, A_d. Thus,

$$A_d \cong \frac{V_o}{V_i} = \frac{V_o}{V_1 - V_2} \tag{14.11}$$

where V_o of Figure 14.12 is V_{o2} of Figure 14.11. From the DIFF AMP equivalent circuit in Figure 14.11, we can find that the *single-ended* difference gain is given by:

$$A_d \cong -\frac{h_{fe}R_C}{2h_{ie}} \tag{14.12}$$

The difference gain for this amplifier is not very high. The *differential output gain* is $-2A_d$ of Equation (14.12).

Common-mode gain. When the input signals applied to the DIFF AMP are the same, $V_1 = V_2 = V_i$, as shown in Figure 14.13, the output ideally should be zero, since there is no difference in the input signals. In an actual amplifier,

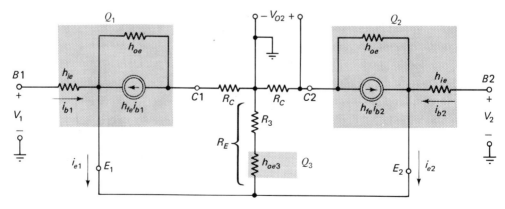

Figure 14.11 An ac small-signal equivalent circuit for improved DIFF AMP (Figures 14.9 and 14.10).

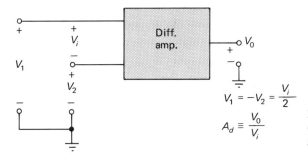

Figure 14.12 Configuration for determining DIFF AMP difference gain, A_d.

however, this is not the case. Consequently, we define the common-mode gain as the ratio of the output to the input when the two inputs are equal. Thus,

$$A_c \cong \frac{V_o}{V_i} = \frac{V_o}{\frac{1}{2}(V_1 + V_2)} \quad (14.13)$$

It can be shown from the equivalent circuit that the common-mode gain is given by

$$A_c \cong -\frac{1}{2} R_C/R_E \quad (14.14)$$

where $R_E = R_3 + 1/h_{oe}$. The common-mode gain appears to be quite small if R_E is made very large.

Common-mode rejection ratio. The common-mode rejection ratio (*CMRR*) is a figure of merit for both DIFF AMPs and OP AMPs—a numerical indication of quality. It is defined as the ratio of the difference gain to the common-mode gain:

$$CMRR \cong \left| \frac{A_d}{A_c} \right| \quad (14.15)$$

The *CMRR* measures the amplifier's ability to reject signals that are unwanted but are common to both inputs. Such signals might be noise or other undesirable signals. The larger the *CMRR*, the closer the proportion of the DIFF AMP output will be to the difference between the two input signals. The output voltage can,

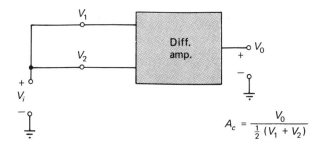

Figure 14.13 Configuration for determining DIFF AMP common-mode gain, A_c.

in general, be expressed in terms of the *CMRR*:

$$V_o = A_d V_d \left(1 + \frac{1}{CMRR}\frac{V_c}{V_d}\right) \tag{14.16}$$

where V_d is the difference input signal ($V_1 - V_2$), and V_c is the common-mode input signal $\frac{1}{2}(V_1 + V_2)$.

For the two DIFF AMP circuits shown in Figures 14.9 and 14.10, the *CMRR* is obtained by dividing Equation (14.12) by Equation (14.14):

$$CMRR \cong \frac{h_{fe}R_E}{h_{ie}} \tag{14.17}$$

This analysis is illustrated in Example 14.2.

Example 14.2

The DIFF AMP circuit in Fig. 14.10 is constructed with $R_C = 100$ kΩ, $R_3 = 50$ kΩ, $R_Z = 1.8$ kΩ, $V_{CC} = V_{EE} = 12$ V. All the transistors may be assumed to be identical with $\beta = 100$, $h_{ie} = 5$ kΩ, $h_{fe} = 50$, and $h_{oe} = 1/200$ kΩ. The Zener diode is a 1N754 (6.8 V) type. We want to determine (1) the dc voltages and currents in the DIFF AMP, (2) the difference and common-mode gains in the circuit, and (3) the *CMRR*.

Solution. Because $\beta = 100$, we may safely neglect all base currents in comparison to either emitter or collector currents. From Eq. (14.9):

$$I \cong \frac{6.8 - 0.6}{50} \text{ mA} \cong 124 \text{ } \mu\text{A}$$

If transistors $Q1$ and $Q2$ are identical, then I divides equally between the two emitters:

$$I_{E1} = I_{E2} = \frac{I}{2} \cong 62 \text{ } \mu\text{A}$$

If we assume that no dc input is being applied to either input terminal (inputs grounded), then $V_{B1} = V_{B2} = 0$ V. The emitters of both $Q1$ and $Q2$ must then be at -0.6 V with respect to ground.

Because β is large, we can say that the collector and emitter currents in $Q1$, $Q2$, and $Q3$ are equal. The voltage drops across the collector resistors are

$$V_{RC} \cong (62 \text{ } \mu\text{A})(100 \text{ k}\Omega) \cong 6.2 \text{ V}$$

The collectors of $Q1$ and $Q2$ must then be 6.2 V below V_{CC}, or at $12 - 6.2 = 5.8$ V with respect to ground. The collector-to-emitter voltages for the three transistors are

$$V_{CE1} = V_{CE2} = 5.8 - (-0.6) = 6.4 \text{ V}$$

$$V_{CE3} = -0.6 + 12 - (124 \text{ } \mu\text{A})(50 \text{ k}\Omega) \cong 5.2 \text{ V}$$

The difference and common-mode gains may be calculated from Eqs. (14.12) and (14.14), respectively, noting that R_E is $1/h_{oe}$ in series with R_3:

$$A_d \approx -\frac{h_{fe}R_C}{2h_{ie}} = -\frac{50(100)}{2(5)} \cong -500$$

$$A_C \approx -1/2R_C/R_E = -\frac{100}{2(250)} \cong -0.20$$

The *CMRR* can now be determined as

$$CMRR = \frac{500}{0.20} \cong 2500 \text{ or } 20\log(2500) = 68 \text{ dB}$$

In the previous example, if we had applied two input signals, one of 2 mV and the other of 5 mV, we would have expected an output voltage given by Equation (14.16) of

$$V_o = (-500)(3) \text{ mV} \left[1 + \frac{7/2}{(200)(3)}\right] = -1500(1.0005) \text{ mV} = -1.5007 \text{ V}$$

This number is (for all practical purposes) exactly -500 times the *difference input signal* of 3 mV. If transistor $Q3$ had had a larger output impedance (i.e., a lower h_{oe}), the *CMRR* would have been even higher and the error in the output would have been even lower than the 0.05% of Example 14.2.

14.3 INTEGRATED CIRCUIT DIFF AMPS

Because integrated circuit (IC) DIFF AMPs can be manufactured with matched transistors on the same substrate at low cost, they are widely used in many circuit applications. An example of a commercially available IC DIFF AMP is the RCA CA3000, whose circuit diagram is shown in Figure 14.14. Some of the applications listed by the manufacturer are RC-coupled feedback amplifier, crystal oscillator, modulator, Schmitt trigger, comparator, and sense amplifier.

The circuit diagram is a little more complicated than the configurations examined previously. However, we can recognize transistors $Q1$ and $Q2$ connected in the differential mode; transistor $Q3$, together with its resistors and diodes $D1$ and $D2$, acts as the constant current source. Transistors $Q4$ and $Q5$ are connected in the emitter-follower configuration and are the input transistors.

The manufacturer lists the following parameters for the CA3000:

$$A_d = 32 \text{ dB} \quad \text{(magnitude} \approx 40\text{)}$$

$$CMRR = 98 \text{ dB} \quad \text{(magnitude} \approx 80\,000\text{)}$$

$$R_i = 195 \text{ k}\Omega$$

From these values we can get some idea as to how this IC DIFF AMP compares with the discrete component DIFF AMP of Example 14.2. For example, we can

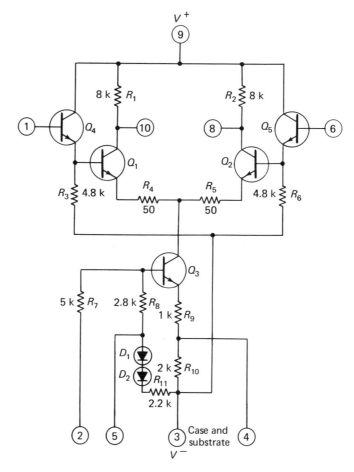

Figure 14.14 An example of an integrated circuit (IC) differential amplifier (RCA CA3000).

estimate the common-mode gain for the CA3000 as $A_c \approx A_d/CMRR = 40/80\,000 = 5 \times 10^{-4}$.

14.4 MONOLITHIC OPERATIONAL AMPLIFIERS

Modern fabrication techniques and volume production have lowered the cost of high-quality monolithic (IC) operational amplifiers to a point where the operational amplifier has become the most important linear integrated circuit.

By using an external feedback network, the operational amplifier can be adapted to many different applications. The same operational amplifier may be used to perform mathematical functions (see Chapter 19), shape waves, or amplify signals by simply changing the components in the feedback network. This capability makes the operational amplifier (OP AMP) by far the most versatile circuit configuration used for linear ICs.

The modern IC OP AMP is so versatile because it is an excellent approximation of the *ideal* operational amplifier. Table 14.1 compares ideal parameters to those of two popular OP AMP designs, the bipolar 741 and the FET input LF351.

14.4.1 Internal Operation

Figure 14.15 pictures the schematic diagram of the classic 741 monolithic OP AMP. Here, we see the connection of the two voltage supplies—one positive and one negative with respect to common (ground). The two supplies provide for dc coupling of the signals at both the input terminals as well as the output terminal with an allowable voltage swing above and below ground.

Bias circuit. Bias current for all the amplifying transistors is derived from the current through the 39 kΩ resistor in the center of the circuit. This current is scaled down and *mirrored* by $Q10$. The same current passes through $Q9$ and is *mirrored* by $Q8$ to provide the constant current bias to the differential amplifier of $Q1$ and $Q2$. $Q5$ and $Q6$ form a current mirror to equally divide the current from $Q8$, thereby providing the differentiated transistors with equal currents.

Amplifier circuit. The differential input signal is amplified by $Q1$ and $Q2$ followed by $Q3$ and $Q4$. The signal is then sent on to the Darlington amplifier of $Q16$ and $Q17$ for further processing. The signal then passes through the dc-level translator, $Q18$, which acts like a coupling capacitor in that it removes the dc component from the signal. The signal is output through the complementary emitter-follower transistors $Q14$ and $Q20$ to provide a low output impedance as well as a large output current capability. The stability of the amplifier is ensured by the 30 pF *compensation* capacitor connected from the output to the input of the Darlington emitter-follower. Figure 14.16 pictures the block diagram of a typical IC operational amplifier.

To summarize, IC operational amplifiers, like the 741, are high-gain direct coupled differential amplifiers with very high input impedance, very low output impedance, and an open-loop gain in excess of 100 dB (100 000).

TABLE 14.1 IDEAL VS. PRACTICAL OP AMP

Parameter	Ideal	Bipolar	FET input
Gain	∞	200 000*	100 000*
Input R	∞	2 MΩ	1 TΩ
Output R	0	10–80Ω**	1–50Ω**
CMRR	∞	90 dB†	100 dB†
Gain bandwidth product	∞	1.5 MHz	4 MHz

* Open loop voltage gain (A_{VOL}).

** Output impedence depends on the closed loop gain and the bandwidth.

† At 1 kHz.

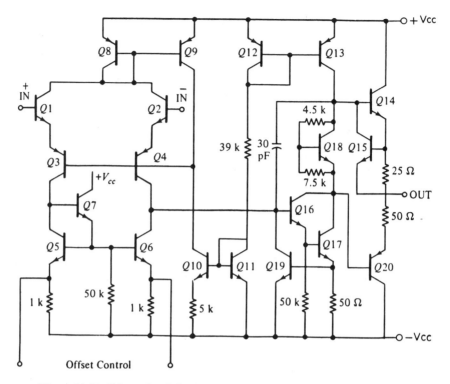

Figure 14.15 Schematic of the 741 type of internally compensated integrated circuit (IC) operational amplifier (OP AMP).

14.4.2 Definition of OP AMP Terms

Large signal voltage gain (A_{VOL}). This is the ratio of output voltage (maximum swing) to the difference in voltage between the two input terminals required to produce the output. Note that this is the open-loop gain of the OP AMP without any feedback, so it is the largest gain possible. Any (negative) feedback produces closed loop gains lower than this value. Typically, this parameter is in excess of 100 000. (It might also be specified as 100 V/mV).

Input offset voltage (V_{OS}). The voltage by which the two input terminals must be offset in order to cause zero output voltage. It is more practically thought of as the uncompensated voltage resulting in the output voltage when the OP AMP

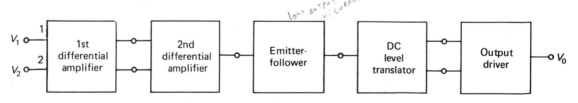

Figure 14.16 Block diagram of a typical IC operational amplifier.

is operated as a voltage follower with the input grounded. This offset results from the mismatch in the base-emitter characteristics of the two input transistors. Typically, the input offset voltage is a few mV.

Input voltage range (V_I). The maximum positive and negative voltage that may be applied to either input terminal and maintain proper bias of the input transistors. This parameter is defined for specific power supply voltages; typically, the input terminals cannot be brought closer than within one or two V_{BE} of the supply voltage.

Output voltage swing (V_o). The maximum excursion (positive and negative) of the output voltage specified for specific power supply voltages. Like the input voltage range, the output voltage swing is limited by the need to maintain bias on the current sources and amplifying transistors internal to the OP AMP. The output voltage can typically swing to within one or two V_{BE} of either positive or negative supply voltage (depending on the specific circuit configuration).

Input bias current (I_B). The actual dc base current required by each of the two input transistors. (Sometimes, it is defined as the average of the two base currents.) Note that this current enters the input terminals for *NPN* input transistors, whereas for *PNP* input transistors it leaves the input. Typically, the input bias current is 80 nA, significantly less for super-β input transistor OP AMPs (less than 1 nA), and even smaller for FET input OP AMPs (less than 100 pA).

Input offset current (I_{OS}). The difference between the two input base currents. It is caused by the mismatch in the base-emitter characteristics of the input transistors. Typically, the input offset current is less than the magnitude of the input bias current.

Supply current (I_S). The current taken from the power supply to operate the OP AMP with no load and zero volts at the output.

Input resistance (R_{IN}). The effective resistance seen at either input terminal with the other terminal grounded. This parameter is defined for the open-loop amplifier. Typically, this value is in excess of 1 MΩ and may be as high as 10^{12} Ω for FET input OP AMPs.

Output resistance (R_o). The effective (small-signal) resistance that appears in series with the output (assuming that the output has been modelled by an ideal voltage generator). Typically, this value is less than 50 Ω.

Common-mode rejection ratio (*CMRR*). Usually defined as the ratio of the difference gain to the common-mode gain, and usually expressed in dB, typically in excess of 80 dB. It is an important parameter since it gives an indication of how closely a given OP AMP resembles an ideal OP AMP.

Power supply rejection ratio (PSRR). The ratio of input offset voltage to the change in power supply voltage producing the input offset voltage. As a result of a change in the supply voltage, the master bias current, and therefore all bias currents, changes; specifically the bias on the input differential pair is changed. The *PSRR* is typically greater than 80 dB. *PSRR* is sometimes specified in μV/V, meaning that the stated number of μV change results from a 1 V change in supply voltage. For example, a *PSRR* of 100 dB is the same as one specified as 10 μV/V.

Slew rate (SR). The internally set limit in the rate of change in the output voltage with a large-amplitude step function applied to the input. Typically 0.5 V/μs for bipolar and 13 V/μs for FET input OP AMPs.

Gain-bandwidth product (GBP). The product of the closed loop gain and closed loop bandwidth resulting in the gain-bandwidth product—a constant. That is to say, a noninverting OP AMP with a 4 MHz GBP would yield an X100 amplifier with a bandwidth of 40 kHz, or an X10 amplifier with a 400 kHz bandwidth, or an X1 amplifier with a 4 MHz bandwidth, etc. Equation (4.43) of Chapter 4 relates gain to bandwidth.

14.5 BASIC AMPLIFIER CONFIGURATIONS

The complete operational amplifier circuit is represented by the symbol pictured in Figure 14.17(a). Here we see the inverting input (designated −) and the non-inverting input (designated +) along with the output, which is referenced to ground (common connection). The two supply leads are shown to emphasize that both + and − voltage (in relation to common) must be applied to the OP AMP. In subsequent drawings, we will assume proper bias is present and these connections will not be shown. The equivalent circuit for the OP AMP is shown in Figure

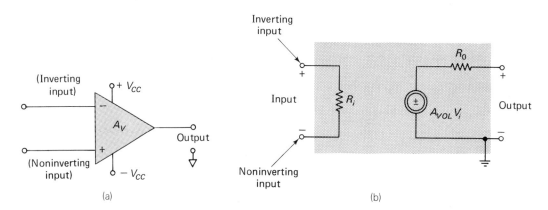

Figure 14.17 OP AMP: (a) symbol with the voltage supplies indicated and (b) equivalent circuit.

14.17(b). Notice that the input resistance is pictured between the two inputs and not from one of the inputs to ground; however, the output is referenced to ground (common). As indicated in Table 14.1, R_i is large in the practical amplifier and is assumed to be infinite in the ideal amplifier; R_o is very small in the practical amplifier and is assumed to be zero in the ideal amplifier. A_{VOL} is large in the practical amplifier and is assumed to be infinite in the ideal amplifier.

14.5.1 Inverting Amplifiers

If we assume an ideal OP AMP, then the inverting amplifier has its gain set by the external negative feedback components, as shown in Figure 14.18.

Virtual ground. An unusual condition called a *virtual ground* exists at the inverting node of the OP AMP. The virtual ground is unlike a typical ground in that the inverting node is at ground potential even though there is no physical connection between the inverting node and ground. Because the voltage is zero at the inverting terminal (negative terminal) even though the impedance at the terminal is not zero, no current flows into the inverting terminal.

With no current drawn between the input terminals, then the output must source or sink current to force the difference between the inputs to remain at zero. Because of the virtual ground, the inverting terminal (negative terminal) of an OP AMP with feedback acts like a dead short to voltage and an open to current.

Closed-loop gain. The concept of a virtual ground leads to the following definitions for the inverting OP AMP of Figure 14.18. Since the node at the inverting terminal is shorted to voltage, then

$$I_1 = V_i/R_1 \tag{14.18}$$

where R_1 includes the source resistance (impedance) as well as additional feedback elements (resistors), V_i is the source voltage. Also,

$$I_f = -V_o/R_f \tag{14.19}$$

where R_f is the feedback resistor and V_o is the output voltage. R_L is large enough so its effects are negligible.

Because of virtual ground at the inverting terminal, the inverter node looks

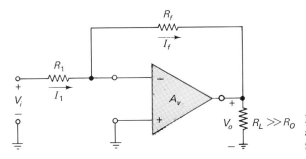

Figure 14.18 The basic inverting amplifier uses the OP AMP with shunt feedback.

like an open to current. Thus, $I_1 = I_f$ and from Equations (14.18) and (14.19),

$$V_i/R_1 = -V_o/R_f \quad \text{and}$$

$$\frac{V_o}{V_i} = -\frac{R_f}{R_1} \qquad (14.20)$$

Equation (14.20) is the expression for closed-loop gain in an inverting feedback OP AMP. This equation is very important; it tells us that the gain is dependent only on the impedances (resistances) used in the feedback circuit and not on any amplifier parameters as long as (1) the open-loop gain of the amplifier is large, (2) its input impedance is very large, and (3) its output impedance is very small.

The input impedance of the inverting amplifier with feedback is equal to R_1 and the closed-loop bandwidth is equal to the gain-bandwidth product divided by one plus the absolute value of the closed loop gain. Thus,

$$BW_{3dB} = \frac{GBP}{1 + \dfrac{R_f}{R_1}} \qquad (14.21)$$

The complete inverting amplifier, along with R_3 to minimize the offset voltage error due to bias current, is shown in Figure 14.19. R_3 is selected to be equal to the parallel combination of R_1 and R_f. Thus,

$$R_3 = 1/(1/R_1 + 1/R_f) \qquad (14.22)$$

Offset voltage is mainly a concern when amplifying dc voltage signals; it is not as important in ac coupled applications.

14.5.2 Noninverting Amplifiers

The complete noninverting OP AMP circuit is pictured in Figure 14.20. The principle difference between the inverting and noninverting amplifier is that the output and input are in phase, the input signal is applied to the noninverting terminal, and the input impedance of the amplifier with feedback is very, very high

$$\left[\approx \text{input } Z \times A_{VOL} \Big/ \left(\frac{R_f}{R_1 + R_f} \right) \right].$$

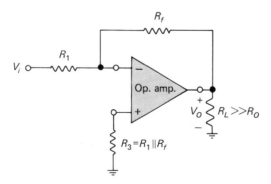

Figure 14.19 The inverting amplifier with bias current compensation (R_3).

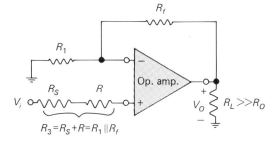

Figure 14.20 The noninverting amplifier with bias current compensation (R_3) for low values of source impedance (R_s).

You should note that the source resistance (impedance) R_S is not in the feedback component R_1, as was the case in the inverting amplifier.

Since the voltage across the input terminals is zero, due to the virtual ground, then V_i will appear at the junction of R_1 and R_f. The current through R_1 is V_i/R_1. Since the inverting input draws no current, then $I_1 = I_f$ and V_o is equal to the drop across both R_f and R_1. To summarize,

$$(1) \quad V_o = I_1 R_1 + I_f R_f \qquad (14.23)$$

$$(2) \quad V_i = I_1 R_1 \qquad (14.24)$$

$$(3) \quad I_1 = I_f \qquad (14.25)$$

Setting up the gain proportion from Equations (14.23) and (14.24) results in

$$\frac{V_o}{V_i} = \frac{I_1 R_1 + I_f R_f}{I_1 R_1}$$

And $I_f = I_1$. Factoring I_1 results in

$$\frac{V_o}{V_i} = \frac{R_f + R_1}{R_1} \qquad (14.26)$$

Equation (14.26) is the expression for closed-loop gain in a noninverting feedback OP AMP. From this equation, we learn that the gain is dependent only on the impedances (resistance) used in the feedback circuit and not on any amplifier parameters.

The closed-loop bandwidth for the noninverting amplifier is equal to the gain-bandwidth product divided by the closed-loop gain. Thus,

$$BW_{3dB} = \frac{GBP}{\dfrac{R_f + R_1}{R_1}} \qquad (14.27)$$

As in the inverting amplifier, R_3 is used to minimize the offset voltage error due to bias current. R_3 is the parallel combination of R_1 and R_f, as noted in Equation (14.22).

Sec. 14.5 Basic Amplifier Configurations

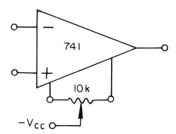

Figure 14.21 The circuit connection used to trim the offset voltage.

14.5.3 Trimming Input Offset Voltage

The input circuit of the 741 OP AMP, pictured in Figure 14.15, represents the external connection available for balancing the input stage to null (cancel) the input offset voltage. A 10 kΩ, 5- or 10-turn trimmer is connected to the negative supply, as shown in Figure 14.21.

With the input shorted to ground, the trimmer is adjusted until a sensitive voltmeter (10 mV full scale) connected across the output reads zero. This adjustment must be repeated every time the OP AMP is replaced with a new component. The 741 data sheet may be found in Appendix A.

REVIEW QUESTIONS

1. What is a differential amplifier?
2. If the two input signals to a differential amplifier are equal, what should the output ideally be? Explain why.
3. What should be the input impedance of an ideal differential amplifier? Why?
4. What is the improvement in performance caused by the current generated ($Q3$) in Fig. 14.9?
5. What is the difference gain in a DIFF AMP?
6. What is the common-mode gain in a DIFF AMP?
7. What is the common-mode rejection ratio a measure of in a DIFF AMP?
8. What should the *CMRR* be for an ideal DIFF AMP? What is it for a practical IC DIFF AMP?
9. What is the advantage of IC DIFF AMPs as compared to discrete component ones?
10. What are the differences and similarities between DIFF AMPs and OP AMPs?
11. What are the properties of an ideal OP AMP? Compare these properties with those of a practical IC OP AMP.
12. What is a virtual ground? How is it different from a conventional ground?
13. How is the input offset voltage nulled in the 741 OP AMP?
14. How is the offset voltage due to bias current corrected in the inverting amplifier?

PROBLEMS

*1. The transistor used in Ex. 14.1 is replaced by one having $h_{ie} = 500\ \Omega$, $h_{fe} = 50$, and $h_{oe} = 12.5\ \mu S$. Determine the current gain and the input impedance.

2. The DIFF AMP in Fig. 14.10 has the following circuit values: $R_C = 560\ \Omega$, $R_3 = 400\ \Omega$, $R_Z = 1.5\ k\Omega$, with $+6$ and -6 V supplies. All transistors are identical silicon NPNs with: $\beta = h_{fe} = 50$, $h_{ie} = 800\ \Omega$, and negligible h_{oe}. Determine the dc voltages and currents in the circuit, for a 4 V Zener diode.

3. In Problem 2, determine the difference and common-mode gains, as well as the CMRR.

*4. Repeat Problem 2 if the Zener diode is replaced by a 5.6 V Zener diode.

5. Determine the values of R_f, R_3, and the BW_{3dB} for the inverting amplifier of Fig. 14.19 when $R_1 = 470\ \Omega$, $GBP = 5$ MHz, and the closed-loop gain is 50.

6. Determine the values of R_f, R, and the BW_{3dB} for the noninverting amplifier of Fig. 41.20 when $R_1 = 220\ \Omega$, $R_s = 50\ \Omega$, $GBP = 1.5$ MHz, and the closed-loop gain is 10.

Chapter 15

Sinusoidal Oscillators

Oscillators are basic electronic circuits. They have no ac input, but they provide an ac output at a specific frequency. The only input needed for an oscillator is the power supply voltage to bias the active device or devices used in an oscillator. Oscillators in general are feedback amplifiers with positive or regenerative feedback.

15.1 CRITERIA FOR OSCILLATION

Consider the general oscillator circuit shown in Figure 15.1. The amplifier (not necessarily an OP AMP) is characterized by a negative voltage gain A_V, output impedance R_o, and an extremely large input impedance R_i. In Figure 15.2, we have redrawn the circuit to show more clearly the feedback network comprised of Z_1 and Z_2. The circuit is a form of voltage feedback (see Section 13.2). We can rewrite the gain (G) of the circuit as

$$G = \frac{A}{1 - \beta A} \qquad (15.1)$$

where β is the feedback factor as defined in Section 13.2. However, if the circuit is to oscillate, the gain must be infinite For the gain to be infinite, the denominator in Equation (15.1) must go to zero. Thus,

$$|1 - \beta A| = 0$$

or $\quad |\beta A| = 1 \quad$ and $\quad$ Phase angle of $(\beta A) = 0 \qquad (15.2)$

where βA is called the *loop gain* and both (or either) β and A are functions of frequency and so are complex numbers.

The condition in Equation (15.2) is called the *Barkhausen criterion*; it specifies the conditions that must be met in order for oscillations to be sustained. According to the Barkhausen criterion, the frequency of oscillation is the frequency at which the signal travels around the loop. As indicated in Figure 15.2,

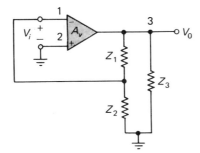

Figure 15.1 General oscillator circuit.

the signal starts at the input terminal; it must remain in phase (ensuring positive feedback), and the amplitude of the signal must not diminish in its trip around the loop. The frequency of oscillation is determined by the proper phase shift of the feedback loop; oscillation itself is ensured by sufficient loop gain. Note that a loop gain much larger than 1 will cause distortion in the signal and the output will not be sinusoidal.

Replacing the amplifier by its equivalent circuit yields the circuit shown in Figure 15.3. When the circuit of Figure 15.3 is redrawn, as shown in Figure 15.4, we can obtain the gain without feedback A:

$$A = A_v \frac{Z_L}{Z_L + R_o} \tag{15.3}$$

where we have defined Z_L as the effective load without feedback:

$$Z_L = \frac{(Z_1 + Z_2)Z_3}{Z_1 + Z_2 + Z_3} \tag{15.4}$$

Similarly, we can determine the feedback factor β from Figure 15.5:

$$\beta = \frac{Z_2}{Z_1 + Z_2} \tag{15.5}$$

Substituting Equations (15.3), (15.4), and (15.5) into the Barkhausen criterion, we find that Equation (15.2) gives us the frequency of oscillation and the amplifier gain needed. We shall consider the special case where all three impedances are purely reactive:

$$Z_1 = jX_1 \quad Z_2 = jX_2 \quad Z_3 = jX_3 \tag{15.6}$$

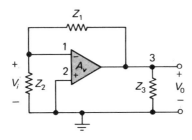

Figure 15.2 General oscillator circuit rearranged.

Sec. 15.1   Criteria for Oscillation

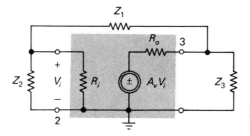

Figure 15.3 Equivalent circuit for Figure 15.2.

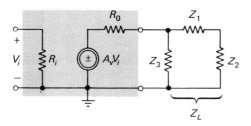

Figure 15.4 Determining the gain without feedback.

Using these relationships in the loop-gain equation, we obtain:

$$\beta A = A_V \frac{-X_2 X_3}{-X_3 X_1 - X_2 X_3 + jR_o(X_1 + X_2 + X_3)} \tag{15.7}$$

For the phase angle of βA to be zero requires that the imaginary part in the denominator of Equation (15.7) be zero. Thus,

$$X_1 + X_2 + X_3 = 0 \tag{15.8}$$

In a given circuit, Equation (15.8) yields the frequency of oscillation. If we set the magnitude of the loop gain to 1, we have

$$|A_V| = \frac{X_3}{X_2} \tag{15.9}$$

Equation (15.9) gives the critical value of the magnitude of the amplifier gain required for oscillation. Note that in reality the amplifier gain is negative.

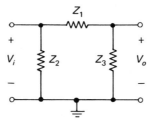

Figure 15.5 Determining the feedback factor.

15.2 HARTLEY OSCILLATORS

A Hartley oscillator using an OP AMP is illustrated in Figure 15.6. We can verify that the circuit is of the basic form shown in Figure 15.1 if we consider the amplifier to be the OP AMP together with its gain-setting resistors R_1 and R_f. The voltage gain from V_i to V_o is given by

$$A_V = -\frac{R_f}{R_1} \tag{15.10}$$

We can match the position of the inductors and capacitor of Figure 15.6 with the impedances in Figure 15.1 to get

$$X_1 = \frac{-1}{\omega C_1} \quad X_2 = \omega L_2 \quad X_3 = \omega L_3 \tag{15.11}$$

The frequency of oscillation can be determined if we substitute Equation (15.11) into Equation (15.8):

$$\frac{-1}{\omega C_1} = \omega(L_2 + L_3) \tag{15.12}$$

which yields the frequency of oscillation f_o in hertz, where $\omega = 2\pi f$. Thus,

$$f_o = \frac{1}{2\pi\sqrt{(L_2 + L_3)C_1}} \tag{15.13}$$

The minimum gain needed is evaluated from Equation (15.9):

$$|A_V| = \frac{L_3}{L_2} \tag{15.14}$$

A FET may be used for the amplifier to form the Hartley oscillator circuit shown in Figure 15.7. Operation is the same as for the OP AMP Hartley oscillator with one exception: The gain in the FET circuit is given by $g_m r_d$. In Figure 15.7,

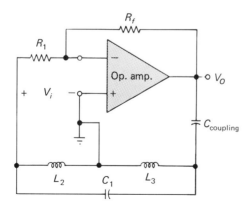

Figure 15.6 OP AMP Hartley oscillator.

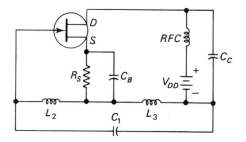

Figure 15.7 FET Hartley oscillator.

C_c is a coupling capacitor which is a short circuit at the desired frequency of oscillation. It blocks the dc drain voltage from being fed to the gate through the low dc resistance of L_2 and L_3. The radio-frequency choke (RFC) is selected to have a very high impedance at the frequency of oscillation so it blocks the ac signal and provides a dc path between the battery and the drain. FET bias is accomplished through R_s, which is bypassed by C_B at the frequency of oscillation. Equations (15.13)—for the frequency of oscillation—and (15.14)—for the gain— are still valid.

A Hartley oscillator may be constructed using a BJT. Because of the very low input impedance, however, the analysis is complicated and will not be attempted here.

The procedure for determining the frequency of oscillation for a Hartley oscillator is illustrated in Example 15.1.

Example 15.1

We wish to construct a Hartley oscillator, as shown in Fig. 15.6, with $L_3 = 0.4$ mH, $L_2 = 0.1$ mH, and $C_1 = 0.002$ μF. We want to determine the frequency of oscillation and the values of R_1 and R_f to ensure oscillation.

Solution. The frequency of oscillation is given in Eq. (15.13):

$$f_o = \frac{1}{2\pi[(0.4 + 0.1)(2 \times 10^{-12})]^{1/2}} \text{ Hz} \cong 159 \text{ kHz}$$

The minimum gain needed is given by Eq. (15.14):

$$|A_V| = \frac{0.4}{0.1} = 4$$

Therefore, if we choose R_1 to be, say, 100 kΩ, we can use R_f of 430 kΩ to give a voltage gain of 4.3, which should ensure oscillation.

15.3 COLPITTS OSCILLATORS

If the basic oscillator circuit is connected with capacitors and inductors reversed from the Hartley configuration, a Colpitts oscillator (shown in Figure 15.8) results. It should be obvious that the Hartley and Colpitts oscillators are duals. We can

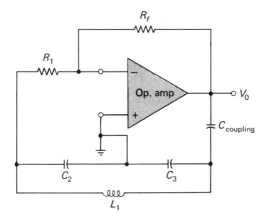

Figure 15.8 OP AMP Colpitts oscillator.

analyze the Colpitts oscillator by using the results of the general discussion in Section 15.1. Note that

$$X_1 = \omega L_1 \qquad X_2 = \frac{-1}{\omega C_2} \qquad X_3 = \frac{-1}{\omega C_3} \qquad (15.15)$$

Setting the sum of these reactances equal to zero, we can determine that the frequency of oscillation is

$$f_o = \frac{1}{2\pi\sqrt{L_1 C_S}} \qquad (15.16)$$

where C_S is the effective series combination of C_2 and C_3; then,

$$C_S \equiv \frac{C_2 C_3}{C_2 + C_3}$$

The minimum gain is determined from Equation (15.9) where $|X_2| = 1/(\omega C_2)$ and $|X_3| = 1/(\omega C_3)$.

$$A_V = \frac{C_2}{C_3} \qquad (15.17)$$

Because the gain is larger than 1, obviously C_2 must be larger than C_3.

Figure 15.9 gives an example of an FET Colpitts oscillator. You can easily

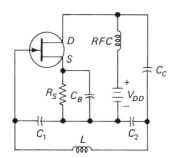

Figure 15.9 FET Colpitts oscillator.

Sec. 15.3   Colpitts Oscillators

see the basic similarity between this circuit and that of Figure 15.7. The equations determined for the OP AMP Colpitts oscillator hold for the FET version, with the exception that the FET gain is given by $g_m r_d$.

The procedure for determining the oscillation and minimum gain needed for a Colpitts oscillator is illustrated in Example 15.2.

Example 15.2

The OP AMP Colpitts oscillator of Fig. 15.8 is constructed with $L_1 = 0.1$ mH, $C_2 = 800$ pF, and $C_3 = 400$ pF. We want to determine the frequency of oscillation and the minimum gain needed.

Solution. We first calculate the equivalent capacitance:

$$C_S = \frac{(800)(400)}{800 + 400} \text{ pF} \cong 267 \text{ pF}$$

The frequency of oscillation can now be found:

$$f_o = \frac{1}{2\pi\sqrt{2.67 \times 10^{-14}}} \text{ Hz} \cong 0.97 \text{ MHz}$$

The minimum gain needed is

$$A_V = \frac{800}{400} = 2$$

We would, therefore, have $R_1 = 100$ kΩ and perhaps make R_f slightly higher than the required 200 kΩ—say, 220 kΩ—to ensure oscillation.

15.4 RC PHASE-SHIFT OSCILLATORS

Basically, both the Hartley and Colpitts oscillators operate in such a way that, at the frequency of oscillation, the feedback network provides a 180° phase shift. The amplifier provides the other 180° to ensure that the net phase shift around the loop is zero (or any multiple of 360°). We can obtain similar operation with the RC phase-shift circuit, illustrated in Figure 15.10 where a minimum of three RC sections is necessary to provide the desired 180° phase shift. (At best, a single capacitor can provide only 90°—hence the need for at least three RC sections, although the circuit may be built with four or even more.)

With three identical RC networks, as shown in Figure 15.10, the feedback network will provide 180° of phase shift at a frequency given by

$$f_o = \frac{1}{2\pi(2.45)RC} \tag{15.18}$$

At this frequency the feedback network gain is $\frac{1}{29}$. The amplifier must then have

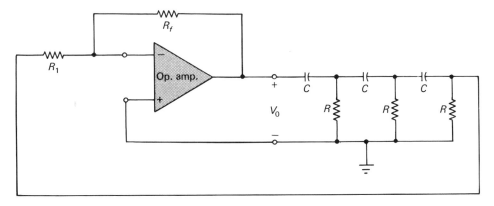

Figure 15.10 OP AMP RC phase-shift oscillator.

a gain of at least 29 to ensure oscillation. Thus, the equation

$$|A_V| = 29 = \frac{R_f}{R_1} \qquad (15.19)$$

allows us to determine the resistors needed. We may choose $R_1 = 100$ kΩ and $R_f = 3.3$ MΩ to get a gain of 33. This procedure is illustrated in Example 15.3.

Example 15.3

We want to design the circuit of Fig. 15.10 for a frequency of oscillation of 10 kHz.

Solution. We start by choosing the capacitor value $C = 0.001$ µF. We next calculate the value of R from Eq. (15.18):

$$R = \frac{1}{2\pi(2.45)(0.001 \times 10^{-6})(10 \times 10^3)} \approx 6.5 \text{ k}\Omega$$

To provide the proper gain and prevent loading down the feedback network, we can make $R_1 = 100$ kΩ and $R_f = 3.3$ MΩ.

A FET RC phase-shift oscillator is depicted in Figure 15.11. The operation is the same as the OP AMP version, with the exception of the FET gain. This gain is given by $g_m R_L$, where R_L is the total effective load resistance seen by the output of the FET; it includes r_d, R_D, and the loading effect of the feedback network. The equations developed for the frequency of oscillation and minimum gain are valid for this circuit.

Figure 15.12 shows a BJT RC phase-shift oscillator circuit. To obtain three matched RC sections, we have to break up the last R because of the heavy loading of the feedback network by the input impedance (h_{ie}) of the BJT. To preserve the frequency of oscillation, the parallel combination of R_1, R_2, and h_{ie}, all in

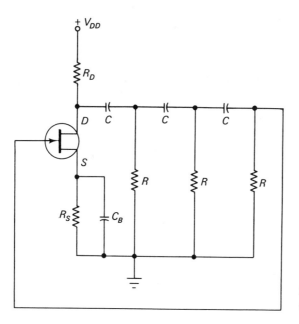

Figure 15.11 FET RC phase-shift oscillator.

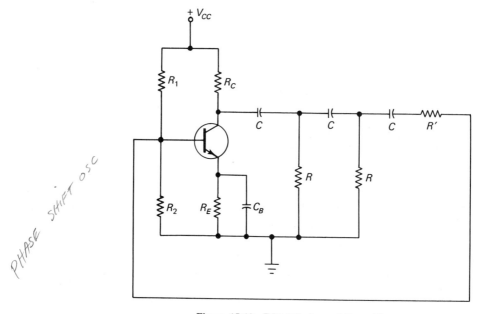

Figure 15.12 BJT RC phase-shift oscillator.

series with R', is made equal to R. Otherwise, the analysis is similar; that is, the gain of the BJT is determined in the same manner as for the FET. But we have to take into account the loading of the feedback circuit on the output of the BJT.

15.5 TUNED-OUTPUT OSCILLATORS

We can also achieve positive feedback by using a tuned transformer, illustrated in Figures 15.13 and 15.14. Capacitor C_B in each circuit is a bypass capacitor and is large enough to be a short circuit at the frequency of oscillation. Bias is provided by resistors R_1, R_2, and either R_E or R_S (depending on which circuit is used). Note that the transformer coupling is designed in such a way that it provides a 180° phase shift. The frequency is selected by the tuning of the L_1C tank circuit.

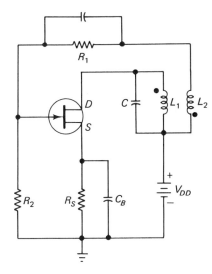

Figure 15.13 FET tuned-output oscillator.

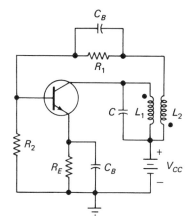

Figure 15.14 BJT tuned-output oscillator.

In effect, we have a tuned amplifier (with a maximum gain at resonance) and a transformer feedback network that provides the necessary phase shift. The frequency of oscillation is given by

$$f_o = \frac{1}{2\pi\sqrt{L_1 C}} \qquad (15.20)$$

The gain may be calculated by the methods discussed in Chapter 11 for the tuned amplifier.

15.6 TWIN-T OSCILLATORS

An OP AMP twin-T oscillator circuit is shown in Figure 15.15. The feedback network in this case is comprised of a dual (twin) T-section filter, which is sometimes referred to as a *notch* filter. This name describes its transfer impedance frequency characteristic. The impedance is very low except in the vicinity of the resonant frequency, where it is very high. At the null or resonant frequency (the twin-T circuit response resembles that of a tank circuit), the phase shift through the feedback network is 180°, so that if the amplifier gain is adjusted to compensate for the signal loss in the feedback network, oscillations will occur at

$$f_o \cong \frac{1}{2\pi RC\sqrt{8}} \cong \frac{1}{17.8RC} \qquad (15.21)$$

where $R_a = 4R_b = R$ and $C_a = \frac{1}{2}C_b = C$. The gain needed for oscillation may be determined experimentally by adjusting the ratio of R_f to R_1. Typically, the minimum gain is around 25, i.e., R_f should be at least 25 times larger than R_1.

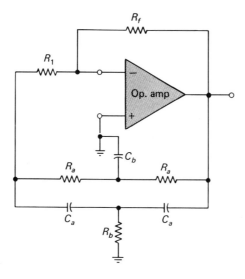

Figure 15.15 OP AMP twin-T oscillator.

15.7 WIEN-BRIDGE OSCILLATOR

An OP AMP Wien-bridge oscillator circuit is illustrated in Figure 15.16. The positive feedback network consists of resistor-capacitor combinations: R_1 in parallel with C_1 and R_2 in series with C_2. Resistors R_3 and R_4 set the gain of the OP AMP. We have redrawn the circuit in Figure 15.17 to make the analysis somewhat simpler; it is easier to recognize different parts of the circuit from Figure 15.17 than from Figure 15.16. The two circuits, nevertheless, are identical.

Oscillations occur when the phase shift through the feedback network is zero and the gain provided by the $R_3 - R_4$ combination is large enough to overcome the loss in signal in the feedback network.

The frequency of oscillation is determined from the condition that the impedance of the $R_1 - C_1$ branch be the same as the impedance of the $R_2 - C_2$ branch. Thus,

$$f_o = \frac{1}{2\pi\sqrt{R_1 R_2 C_1 C_2}} \tag{15.22}$$

At this frequency, in order to provide a loop gain of 1, the amplifier gain has to be

$$A_V \geq \frac{R_1}{R_2} + \frac{C_2}{C_1} + 1 \tag{15.23}$$

Resistors R_1 and R_2 and capacitors C_1 and C_2 are chosen, therefore, to provide the desired frequency in accordance with Equation (15.22). Resistors R_3 and R_4 are chosen to provide the gain in Equation (15.23). The gain in terms of R_3 and R_4 is

$$A_V = \frac{R_3 + R_4}{R_4} = 1 + \frac{R_3}{R_4} \tag{15.24}$$

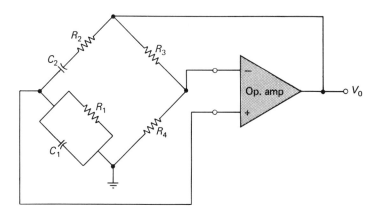

Figure 15.16 OP AMP Wien-bridge oscillator.

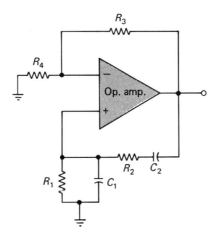

Figure 15.17 Redrawn OP AMP Wien-bridge oscillator.

Thus, we obtain the required values of R_3 and R_4 by combining Equations (15.23) and (15.24):

$$\frac{R_3}{R_4} \geq \frac{R_1}{R_2} + \frac{C_2}{C_1} \tag{15.25}$$

In the special case where $R_1 = R_2 = R$ and $C_1 = C_2 = C$, the frequency of oscillation is given by

$$f_o = \frac{1}{2\pi RC} \tag{15.26}$$

and the minimum gain for oscillation is 3. Consequently, $R_3 \geq 2R_4$ will ensure that the circuit will oscillate. (See Equation (15.24).)

15.8 AMPLITUDE-STABILIZED OSCILLATORS

In many applications the oscillator amplitude is critical; therefore, it must be precisely controlled. The basic principle in stabilizing the amplitude of oscillation involves the use of an additional feedback loop employing gain control for the device in the oscillator. We may follow this procedure with any of the oscillators discussed. For example, we can stabilize the FET phase-shift oscillator amplitude by rectifying the output signal and feeding the equivalent dc voltage to bias the FET. In this way, should the amplitude at the output increase, so does the dc level feedback, which biases the FET for a slightly lower gain. Consequently, an essentially constant output voltage is maintained.

A somewhat more involved example that uses the same principle is given in Figure 15.18. This is the Wien-bridge oscillator, with additional circuit components for amplitude stabilization. The output signal in excess of the Zener voltage of diode $D2$ is rectified by diode $D1$ and applies a negative dc voltage to the gate of FET $Q1$. This dc voltage is developed across the $R_5 - C_4$ combination. If the output voltage should increase, so does the negative bias on the gate of $Q1$. As a result, the drain-to-source resistance is increased, thus effectively reducing

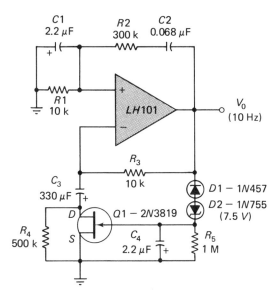

Figure 15.18 Practical amplitude-stabilized Wien-bridge oscillator. (*Courtesy of National Semiconductor Corporation*)

the gain of the OP AMP. Should the output voltage decrease, the negative bias on the gate of Q also decreases, in turn decreasing the drain-to-source resistance. As a result, the gain of the OP AMP is increased. Note that the gain of the OP AMP is set by R_3 and the parallel combination of R_4 and the drain-to-source resistance of the FET.

The frequency of oscillation is still determined by the positive feedback path provided by the $R_2 - C_2$ and $R_1 - C_1$ branches.

15.9 CRYSTAL OSCILLATORS

There are many different versions of crystal oscillators. A crystal provides a very temperature-stable tuned circuit, around which an oscillator may be built.

The frequency of oscillation is controlled by the crystal. It is stable over a long period of time and over a relatively wide temperature range.

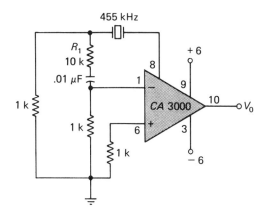

Figure 15.19 Practical DIFF AMP crystal oscillator ($f_o = 455$ kHz).

An example of a crystal oscillator using an IC DIFF AMP is shown in Figure 15.19. The crystal is connected between the inverting input and the noninverted output, supplying positive feedback so that oscillation can take place. The amount of feedback is controlled by the variable resistor R_1. It can also be adjusted for a sinusoidal output. The circuit shown is adjusted for oscillation at 455 kHz; however, the CA3000 DIFF AMP may be used in a crystal oscillator providing frequencies up to 1 MHz.

REVIEW QUESTIONS

1. What is an oscillator? What is the input to an oscillator? What is its output?
2. What are the criteria for oscillation to be sustained?
3. What determines the frequency of oscillation?
4. What comprises the feedback network in a Hartley oscillator?
5. What circuit values determine the frequency of oscillation in a Hartley oscillator?
6. What is the feedback network in a Colpitts oscillator?
7. What circuit values determine the frequency of oscillation in a Colpitts oscillator?
8. What is the feedback network in an RC phase-shift oscillator?
9. What circuit values determine the frequency of oscillation in a phase-shift oscillator?
10. What is the feedback network in a tuned output oscillator?
11. What circuit values determine the frequency of oscillation in a tuned output oscillator?
12. What is the feedback network in a twin-T oscillator?
13. What circuit values determine the frequency of oscillation in a twin-T oscillator?
14. What is the feedback network in a Wien-bridge oscillator?
15. What circuit values determine the frequency of oscillation in a Wien-bridge oscillator?
16. In what manner can the amplitude of oscillation be stabilized in an oscillator using an FET as the amplifying device?
17. In what way may an FET be used to stabilize the amplitude of oscillation in an OP AMP oscillator?
18. What properties of a crystal make it suitable for use in very stable oscillator circuits?
19. What are the basic similarities and differences between the oscillator circuits in this chapter?
20. What conditions must be satisfied for the oscillator output to be sinusoidal with a minimum of harmonic distortion?

PROBLEMS

1. The FET Hartley oscillator of Fig. 15.7 is to be designed for oscillation at 455 kHz. The FET parameters are $g_m = 6000$ µS and $r_{ds} = 100$ kΩ. Determine the circuit values.

2. Repeat Problem 1, using the OP AMP version of the Hartley oscillator shown in Fig. 15.6.

3. The OP AMP Colpitts oscillator of Fig. 15.8 is constructed using $C_2 = 5000$ pF, $C_3 = 500$ pF, and $L = 1$ mH. Determine the frequency of oscillation and the resistor values needed to ensure oscillation.

*4. Repeat Problem 3 for the FET version of the Colpitts oscillator if the FET has $g_m = 1$ mS and $r_{ds} = 50$ kΩ.

5. We wish to design the OP AMP phase-shift oscillator for a frequency of 18 kHz. Specify the circuit values. (Use all capacitors equal to 0.1 µF.)

6. Repeat Problem 5 for the FET phase-shift oscillator shown in Fig. 15.11 if the FET parameters are: $g_m = 5$ mS and $r_{ds} = 80$ kΩ. Specify the value of R_D.

7. Repeat Problem 5 for the BJT phase-shift oscillator shown in Fig. 15.12 if the BJT parameters are $h_{ie} = 2$ kΩ, $h_{fe} = 100$ with h_{re}, and h_{oe} negligible. Specify the value of R' needed.

8. The twin-T oscillator of Fig. 15.15 is connected with $C_a = 0.01$ µF, $C_b = 0.02$ µF, $R_a = 10$ kΩ, and $R_b = 2.5$ kΩ. Determine the frequency of oscillation and the approximate values of R_3 and R_4 needed for oscillation.

*9. Design the twin-T oscillator of Fig. 15.15 to provide an output at 50 kHz. Specify the approximate resistor and capacitor values needed.

10. The Wien-bridge oscillator of Fig. 15.16 is constructed with $R_1 = R_2 = 1$ kΩ, $C_1 = C_2 = 0.02$ µF. What is the frequency of oscillation and the values of the gain-setting resistors to ensure oscillation?

11. Assuming that the gain of the amplitude-stabilized Wien-bridge oscillator in Fig. 15.18 is sufficient to provide oscillation, determine the values of R_1, R_2, C_1, and C_2 to give an output at a frequency of 100 Hz.

Chapter 16

Clipping, Clamping, and Wave-Shaping Circuits

This chapter introduces a variety of relatively simple circuits used for signal processing. In many applications it is necessary to pass one portion of the signal and reject another portion. Clipping circuits perform this function. Sometimes we must clamp or latch the output at a certain level—a function accomplished by clamping circuits. The general wave-shaping circuits are used to modify the input waveshape into another waveshape or to prevent the input waveshape from being modified.

16.1 SINGLE-LEVEL CLIPPING CIRCUITS

In general, *single-level* clipping circuits are used to select the part of the input waveshape that lies either above or below a certain reference level. You may contrast the single-level clipping circuit with the *two-level* clipping circuit that is used to select the part of the input signal which lies between two levels.

16.1.1 Positive Clipping Circuits

The circuit shown in Figure 16.1 clips the input waveshape at a positive voltage determined by the reference voltage V. In order to understand the operation of this circuit, suppose that we have a sinusoidal input waveshape whose peak amplitude is larger than the reference voltage V. When the input is positive and less than V, the diode is reverse biased and therefore does not conduct. We may assume that a reverse-biased diode offers an open circuit. The output waveshape then follows the input because no current flows through R. When the input voltage exceeds the reference voltage V (in the positive direction), the diode is forward biased and conducts. In the forward direction, we can assume that the diode acts as a short circuit. During this time, the difference between the input voltage and the reference voltage appears across R, and the output is fixed at V. (Note that for low-amplitude signals, the forward drop across the diode should not be neglected where low-amplitude is $<10\times$ the diode drop.)

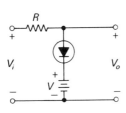
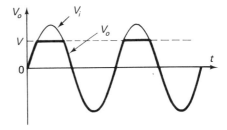

Figure 16.1 Positive clipping circuit.

Thus, the operation of the positive clipping circuit of Figure 16.1 may be summarized as follows. The output voltage follows the input voltage as long as the reference voltage is greater than the input (diode, reverse-biased); the output voltage is fixed at the reference voltage when the input voltage exceeds the reference voltage (diode, forward-biased).

An alternative for the positive clipping circuit is shown in Figure 16.2. When the input waveshape is at a level lower than the reference voltage, V, the diode is forward biased (a short circuit) and the output waveshape follows the input. When the input waveshape is of a level higher than the reference voltage, the diode is reverse biased (an open circuit) and the output voltage is equal to the reference voltage (no current through R).

16.1.2 Negative Clipping Circuits

If both the diode and the polarity of the voltage reference in Figure 16.1 are reversed, as shown in Figure 16.3, the circuit clips the input waveshape at a negative voltage.

When the input is positive and less negative than the reference voltage, V, in Figure 16.3, the diode is reverse biased and effectively open. Since there is no current through R, the output voltage is essentially the same as the input voltage. However, when the input becomes more negative than the reference voltage, V, the diode becomes forward biased and effectively a short circuit. The output is then equal to the reference voltage, as indicated in Figure 16.3.

We can also obtain a negative clipping circuit by reversing the diode and

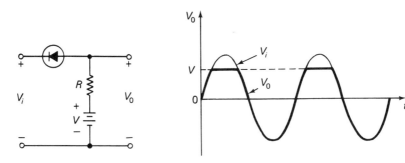

Figure 16.2 Alternate positive clipping circuit.

Sec. 16.1 Single-Level Clipping Circuits

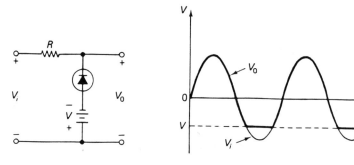

Figure 16.3 Negative clipping circuit.

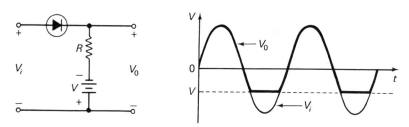

Figure 16.4 Alternate negative clipping circuit.

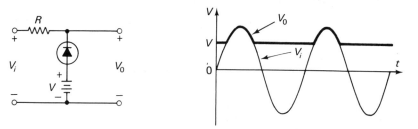

Figure 16.5 Positive comparator circuit.

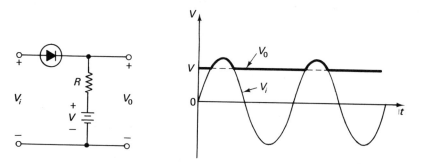

Figure 16.6 Alternate positive comparator circuit.

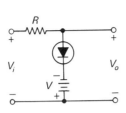
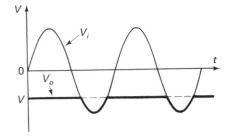

Figure 16.7 Negative comparator circuit.

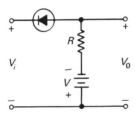
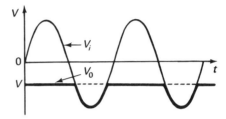

Figure 16.8 Alternate negative comparator circuit.

the polarity of the reference voltage in Figure 16.2. The resulting circuit is illustrated in Figure 16.4. When the input is either positive or less negative than the reference voltage, then the diode is forward biased and effectively a short circuit. The output voltage then follows the input voltage. However, when the input becomes more negative than the reference, the diode is reverse biased and effectively an open circuit. Therefore, we set the output voltage at the negative reference voltage, as shown in Figure 16.4.

Consider the four clipping circuits when a load is placed across the output terminals. It should be obvious that the circuits in Figures 16.1 and 16.3 will cause a *voltage* of the waveshape shown to exist across the load. At the same time, the circuits of Figures 16.2 and 16.4 will cause a *current* of the given waveshape to flow through the load.

Figures 16.5, 16.6, 16.7, and 16.8 show the clipping circuits discussed above with the diode reversed. The resulting output waveshapes are shown. The verification of these waveshapes is left as an exercise.

16.2 TWO-LEVEL CLIPPING CIRCUITS

In certain applications we have to clip the input waveshape at two preselected levels. The circuits shown in Figures 16.9, 16.10, 16.11, and 16.12 perform this operation.

Consider the circuit in Figure 16.9. If we assume that V_2 is greater than V_1, we can see that when the input is positive and less than V_1, diode $D1$ is forward

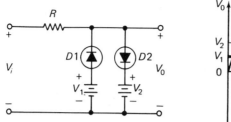 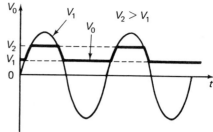

Figure 16.9 Circuit for clipping at two positive voltages.

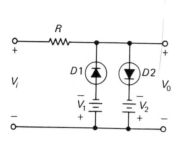 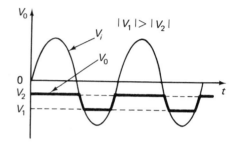

Figure 16.10 Circuit for clipping at two negative voltages.

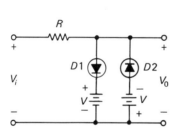 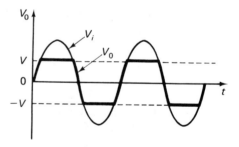

Figure 16.11 Circuit for clipping at one positive and one negative voltage.

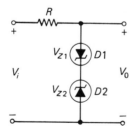 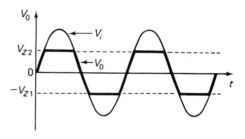

Figure 16.12 Clipping circuit using Zener diodes.

302 Clipping, Clamping, and Wave-Shaping Circuits Chap. 16

biased, whereas diode $D2$ is reverse biased. The output voltage is then fixed at V_1. When the input is more positive than V_1 but less positive than V_2, $D1$ and $D2$ are both reverse biased. The output then follows the input. When the input exceeds both V_1 and V_2, $D1$ is still reverse biased and $D2$ is forward biased. The output voltage becomes fixed at V_2. When the input voltage goes negative, $D2$ is reverse biased with $D1$ forward biased and conducting, so that the output is at V_1. The circuit in Figure 16.9 is used when we want to clip the input voltage between two positive reference voltages V_1 and V_2.

If both $D1$ and $D2$ as well as the polarity of both reference voltages V_1 and V_2 in Figure 16.9 are reversed, the circuit shown in Figure 16.10 results. This circuit clips the input voltage between two predetermined negative reference levels and operates quite similarly to the circuit of Figure 16.9.

When the input waveshape needs to be clipped between one positive and one negative level, we use the circuit of Figure 16.11. (Although the reference voltages are shown to be the same, this need not be the case.) If the input voltage is less negative than the reference voltage and at the same time less positive than the other reference voltage, both $D1$ and $D2$ are reverse biased, and the output voltage follows the input. When the input becomes more positive than the $D1$ reference voltage, $D1$ conducts and the output is forced to the $D1$ reference voltage (V). When the input is more negative than the $D2$ reference voltage, $D2$ is forward biased (with $D1$ reverse biased), and the output is forced to the negative reference voltage of $D2$, as indicated.

An alternate way of achieving clipping at a negative and positive voltage is illustrated in Figure 16.12. Here we use two Zener diodes. They need not be identical or have the same breakdown voltages. When the input is positive, $D1$ is forward biased, whereas $D2$ is reverse biased. When the input exceeds the breakdown voltage of $D2$ (this voltage is in the positive direction), the output is fixed at V_{Z2}. When the input is negative, $D2$ is forward biased, and $D1$ is reverse biased. When the input is more negative than the breakdown voltage of $D1$, $D1$ goes into breakdown and limits the output voltage to V_{Z1}. Note that when we deal with Zener diodes that have breakdown voltages lower than, say, 10 V, we cannot neglect the forward diode drop. The clipping levels then will be $V_Z + 0.7$ V.

16.3 CLAMPING CIRCUITS

One operation that we must carry out frequently is establishing the specific maximum or a minimum for the waveshape. These extremes specify the voltage to which the output is said to be *clamped*. The circuit accomplishing this function is called a *clamping circuit*.

16.3.1 DC Restorer Circuits

A circuit that clamps the output voltage to ground is often called a *dc restorer circuit*. It is shown in Figure 16.13. With a sinusoidal input, circuit operation is as follows. As the input starts on its positive upswing, the diode is forward biased

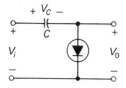

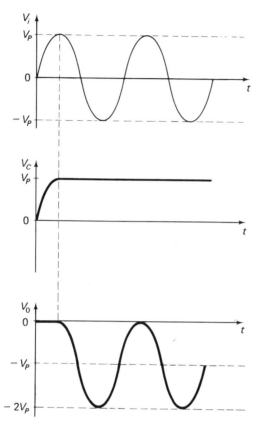

Figure 16.13 Negative dc restorer (clamping) circuit and waveshapes.

and conducts. The output voltage is essentially zero, and the diode current charges the capacitor (which is assumed to be uncharged initially). The capacitor voltage follows the input voltage until the positive peak is reached. At this time the voltage on the capacitor is V_p. When the input falls below V_p, the voltage across the capacitor exceeds the input voltage and the diode is reverse biased. Because of the reverse-biased diode, no more current flows and the voltage across the capacitor (V_p) is maintained. Thus, the output now follows the input but starts at ground when the input is at V_p. From then on, the diode is always reverse biased (assuming ideal operation), the capacitor maintains its voltage (V_p), and the output is a sine wave with a negative dc level equal to the peak value of the input. The circuit thus has restored a dc level to the output. In other words, it has clamped the output at zero, not allowing it to go above zero.

If the diode in Figure 16.13 is reversed, the circuit (shown in Figure 16.14) restores a positive dc level to the output. Let us consider the circuit in Figure 16.14, with the capacitor uncharged initially and a sinusoidal input as before. The input is positive, so the diode is reverse biased and the output follows the input as shown. There is no clamping action to this point. Clamping action begins when the input starts on its negative downswing. During this time the diode is forward biased and allows the capacitor to charge up. Note that the capacitor charges to $-V_p$. Once the input becomes less negative than V_p, the diode is again reverse

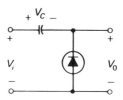

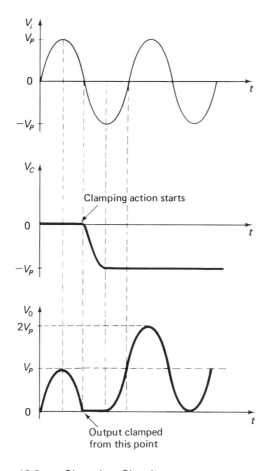

Figure 16.14 Positive dc restorer (clamping) circuit and waveshapes.

Sec. 16.3 Clamping Circuits

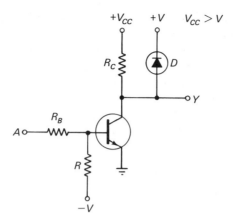

Figure 16.15 Diode-clamped gate.

biased and the output is clamped at ground and follows the input. The output in this case has a positive dc level equal to the value of the input wave peak.

16.3.2 Latching Circuits

In many applications we have to modify the performance of a circuit by causing the output to be clamped to a voltage other than the specific supply voltage being used for that circuit. An example is depicted in Figure 16.15, where the diode (called a *latching* diode) and its reference supply, V, are added to the simple transistor inverter circuit.

The addition of the latching diode does not change the basic circuit operation; the circuit is still an inverter. With a high input at A, the transistor saturates, point Y goes essentially to zero, and the diode is reverse biased. Thus, with a high input, the diode and its reference supply have no effect on the performance of the inverter. When the input to the inverter is low, the transistor is cut off and draws no current. We have redrawn the circuit in Figure 16.16 to show the transistor disconnected. Because V_{CC} is larger than V, the diode is now forward biased and conducts. The current I is essentially given by

$$I = \frac{V_{CC} - (V - 0.7)}{R_C} \tag{16.1}$$

where a forward diode drop of 0.7 V has been assumed. The output is now at $V - 0.7$ and not at V_{CC}, as would be the case were the latching circuit removed. We add the latching circuit in order to improve the fan-out of this type of gate.

To observe this improvement, let us consider the inverter circuit without

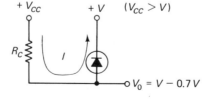

Figure 16.16 Equivalent circuit for Figure 16.15 when the transistor is off with the output clamped to approximately V.

the latching network. The logical 1 level at the output is V_{CC}. However, as soon as another gate is added to the output, it draws current and the logical 1 output voltage falls below V_{CC}. If enough gates are tied to the output, the voltage would fall below the specified minimum for a logical 1 level, and the gate would cease to function.

When the circuit of Figure 16.15 is used, additional gates may be connected to the output without decreasing the logical 1 output voltage. This operation is illustrated in Example 16.1.

Example 16.1

Determine the logical 1 output voltage $V_0(1)$ in the circuit of Fig. 16.17 with R_x of: (1) 22 kΩ and (2) 10 kΩ. Assume the forward diode drop to be 0.7 V.

Solution. Without R_x connected and with the transistor cut off, $V_0(1) = 5.7 - 0.7 = 5$ V. This figure will be the output voltage so long as I_x is smaller than I. We see that $I_x = I - I_D$. From Eq. (16.1):

$$I \cong \frac{10 - (5.7 - 0.7)}{2.2} \text{ mA} \cong 2.27 \text{ mA}$$

For $R_x = 22$ kΩ:

$$I_x = \frac{V_o(1)}{R_x} \cong \frac{5}{22} \text{ mA} \cong 0.23 \text{ mA}$$

Thus, I_D is approximately 2.0 mA. For $R_x = 10$ kΩ:

$$I_x \cong \frac{5}{10} \cong 0.5 \text{ mA}$$

and I_D is approximately 1.8 mA. Note that the diode must be conducting for the output voltage to remain at 5 V. Thus, $I_D \geq 0$. As a result, we have a maximum I_x equal to I. For this worst-case condition, the value of R_x is

$$R_{x(\min)} \cong \frac{5}{2.27} \text{ k}\Omega \cong 2.2 \text{ k}\Omega$$

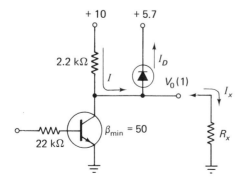

Figure 16.17 Latched inverter (see Example 16.1).

Sec. 16.3 Clamping Circuits

If R_x represents the net effect of connecting additional gates to the output and all the gates are simulated by an input resistance of R_B (22 kΩ), then we see that a maximum of 10 gates may be connected to the output of the gate in Figure 16.17 and the logical 1 level of 5 V still be preserved. With the next gate added (the 11th), the output voltage would drop below 5 V.

The performance of the latched inverter of the preceding example can be compared to an inverter without the latching circuit. The logical 1 output voltage for the inverter without the latching diode would vary from 10 V for no loading to 5 V with 10 identical gates connected to the output.

The diode in the latching circuit may be reversed and referenced to a voltage that is higher than V_{CC} if we wish to latch the output to a higher voltage.

The inverter example of a latching circuit shows only one of the many uses for diodes as latches.

16.4 WAVE-SHAPING CIRCUITS

In this section we consider some of the more common wave-shaping circuits using RC combinations. In many applications it is necessary (1) to form a "spike" at the output with a square-wave input or (2) to detect the average or peak value of the input waveshape. We can perform both of these functions with RC networks.

16.4.1 Spike-forming Circuits (Differentiators)

Consider the circuit shown in Figure 16.18. If the input is a square wave as shown, the output may contain both positive and negative going spikes. To see how these spikes are obtained and what conditions must be satisfied, we can analyze the circuit. With the input at zero and no initial charge on the capacitor, the output is also at zero. When the input switches to $+V$, the capacitor voltage cannot change instantaneously, so the change is transmitted to the output. If the RC time constant is very short in comparison to the duration of the input pulse, τ (typically $\frac{1}{5}$ of τ or shorter), the capacitor quickly charges to the input voltage V and the output falls to zero.

When the input switches to zero, the capacitor voltage is still $+V$, so the output switches from zero to $-V$, as shown. The capacitor now quickly discharges to zero through the resistor, and the output voltage goes to zero, as shown. When the input pulse is repeated, the operation just described is also repeated. Note that before the next pulse arrives, the capacitor is completely discharged. Consequently, the initial assumption of no charge on the capacitor is valid.

In order to achieve spikes at the output with a square wave at the input of the RC circuit illustrated in Figure 16.18, the RC time constant must be much smaller than the pulse duration; that is, RC must be less than $\frac{1}{5}\tau$.

In some applications we wish to generate only positive or only negative going spikes. In such cases, the RC circuit is coupled to a diode clipper circuit, as shown in Figure 16.19. The operation of the two circuits (the RC and clipping

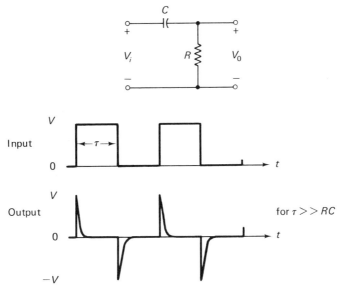

Figure 16.18 RC spike-forming (differentiator) circuit and waveshapes.

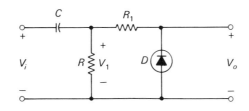

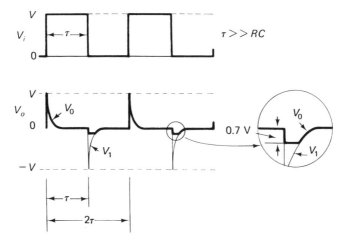

Figure 16.19 Circuit for forming positive-going spikes.

Sec. 16.4   Wave-Shaping Circuits

circuits) is the same as it was when they were used individually. With the diode connected as indicated, the negative going spikes are clipped and the output contains only positive going spikes. If the diode is reversed, the output will contain only negative spikes. Note that the output has a positive going spike at every point where the input voltage makes a transition from zero to V.

If the voltage levels are low (say, less than 10 V) as they would be in most digital circuits, the output waveshape contains a small part of the negative spike, as shown in the detail in Figure 16.19. This waveshape results because the forward drop across the diode is not zero but is 0.7 V for a silicon diode. This effect becomes very pronounced when the input pulse level is just a few volts. We can eliminate the undesired part of the waveshape if the diode clipper is replaced with a precision rectifier.

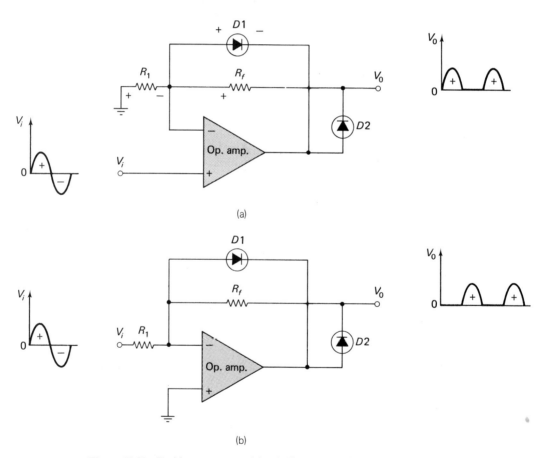

Figure 16.20 Positive-output precision half-wave rectifier using an OP AMP: (a) noninverting mode and (b) inverting mode.

Precision rectifier circuits are shown in Figures 16.20 and 16.21. These circuits use an OP AMP to effectively reduce the diode forward drop. The two circuits in Figure 16.20 give a positive output; the two in Figure 16.21 give a negative output.

Consider first the circuit in Figure 16.20(a). The input is applied to the non-inverting input-terminal of the OP AMP. When the input is positive, $D1$ is reverse biased and off; $D2$ is forward biased and conducting. The gain of the OP AMP in this case is given by $(R_1 + R_f)/R_1$; the output follows the input with the given gain. When the input is negative, $D1$ is forward biased and conducting; $D2$ is reverse biased and off. The gain, with $D1$ effectively shorting the inverting input to the output, is essentially zero. Thus, the output is zero when the input is negative.

In the circuit of Figure 16.20(b), the input is positive. Then $D1$ is forward biased and conducting; $D2$ is reverse biased and off. With $D1$ effectively shorting the inverting input to the output, the gain is essentially zero and the output is at

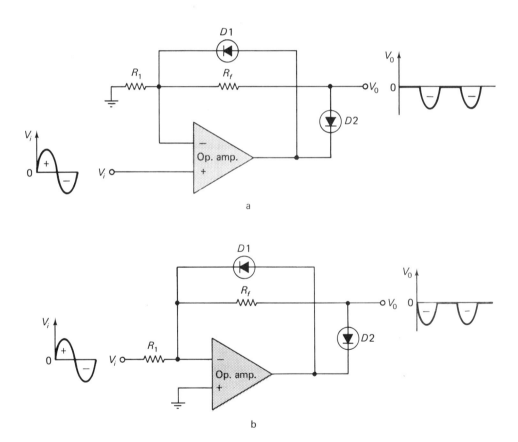

Figure 16.21 Negative-output precision half-wave rectifier using an OP AMP: (a) non-inverting mode and (b) inverting mode.

zero. When the input is negative, $D1$ is off and $D2$ is on. As a result, the gain is given by $-R_f/R_1$. The output then is positive and equal to $-R_f/R_1$, with V_i negative.

Operation of the two precision rectifiers with negative outputs, illustrated in Figures 16.21(a) and 16.21(b), may be examined in the same manner. Note that the only difference between the circuits of Figures 16.20 and 16.21 is that both $D1$ and $D2$ are reversed, thus giving a positive output in one case and a negative one in the other.

The main advantage of these precision rectifier circuits is their close approximation to an ideal diode. Many times we have to rectify signals whose amplitude is lower than a volt, and it becomes necessary to use a precision rectifier in place of a diode.

As an example, suppose that the input pulse has an amplitude V of less than 1 V. The precision rectifier can be used together with the RC wave-shaping network, as shown in Figure 16.22, to give an output which contains only a negative going spike. (For a positive going spike at the output, one of the two precision rectifiers shown in Figure 16.20 can be used with the RC network.) When $\tau \gg RC$, voltage V_1 contains the indicated positive and negative spikes. The precision rectifier then eliminates the positive position of V_1 and gives just negative going spikes. The amplitude is dependent on the gain set by R_f and R_1.

16.4.2 Peak and Average Detecting Circuits

In some cases, we must detect either the average or peak value of a time-varying signal. If the signal is sufficiently larger than the forward diode voltage, the simple diode detectors shown in Figure 16.23 may be used.

Consider the circuit in Figure 16.23(a). In the beginning there is no charge on the capacitor. When the input is positive, the diode is forward biased and conducts, charging the capacitor to a value somewhat below the input voltage. If the input voltage falls below the capacitor voltage (or becomes negative), the diode turns off. The output, therefore, is proportional to the average value of the input signal.

In Figure 16.23(b), the resistor is placed in parallel with the capacitor, and the circuit acts as a peak detector. Again the capacitor is uncharged at the start. When the input goes positive, the diode conducts and the capacitor charges up quickly through the low diode forward resistance. The output voltage then follows the input. When the input falls below the voltage stored across the capacitor, the diode turns off. The capacitor may discharge somewhat through the parallel resistance (which should be high). In essence, the capacitor voltage, and therefore the ouput, follows the peak value of the input voltage.

The circuits of Figure 16.24 are used when the amplitude of the input signal is very small. The operation of the precision detector circuits in Figure 16.24 is identical to that discussed for the diode circuits. The only exception is that the diodes are replaced with precision rectifiers.

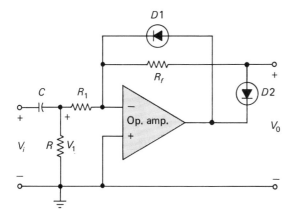

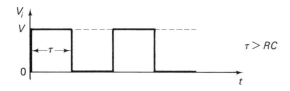

$\tau > RC$

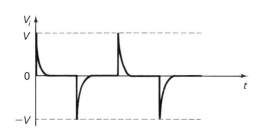

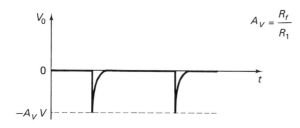

$A_V = \dfrac{R_f}{R_1}$

Figure 16.22 Improved spike-forming circuit for generating negative-going spikes using a precision half-wave rectifier; and wave-shapes.

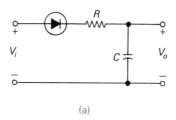

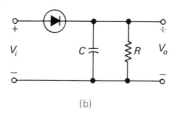

Figure 16.23 Diode detectors: (a) average and (b) peak.

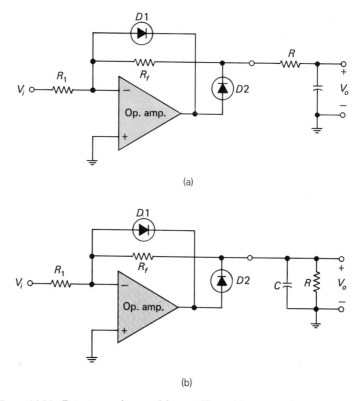

Figure 16.24 Detectors using precision rectifiers: (a) average detector and (b) peak detector.

REVIEW QUESTIONS

1. What is a clipping circuit? What function does it perform?
2. What is the output of a single-level positive clipping circuit? Discuss in terms of the reference voltage.
3. What is the output of a single-level negative clipping circuit? Discuss in terms of the reference voltage.
4. What is the output of either the positive or negative single-level clipping circuit if the reference voltage is zero?
5. What is the operation of the two-level clipping circuit in Fig. 16.9? Discuss in terms of diode conduction and nonconduction.
6. What is the difference in the operation of the two-level clipping circuits shown in Figs. 16.9 and 16.10?
7. What is the operation of the two-level clipping circuit in Fig. 16.11? What is the output if the two reference voltages are not the same?
8. What is the difference in operation between the two-level clipping circuits shown in Figs. 16.11 and 16.12?
9. What is a clamping circuit? What function does it perform?
10. Why is the name of "dc restorer" given to the circuits in Figs. 16.13 and 16.14?
11. What is the role of the capacitor in the dc restorer circuits of Figs. 16.13 and 16.14?
12. What is a latching circuit? What function does it perform?
13. Explain the function of the latching diode used in the transistor inverter circuit of Fig. 16.15.
14. What function does the spike-forming circuit in Fig. 16.18 perform?
15. What is meant by a *precision* rectifier? How is it different from a regular diode rectifier?
16. Under what conditions must a precision rectifier be used (that is, when is a simple diode rectifier insufficient)?
17. Explain the operation of the peak (an average) detecting circuit.

PROBLEMS

1. The input to the clipping circuit of Fig. 16.11 is a sinusoid with peak amplitude of 10 V. With $V = 2$ and $R = 1$ kΩ determine and sketch the diode current and output voltage.
2. The same input signal as in Problem 1 is applied to the circuit in Fig. 16.2. If the circuit values are also the same as in Problem 1 determine and sketch the diode current and output voltage.
3. A square wave with a positive peak of 5 V and a negative peak of -10 V is applied to the circuit of Fig. 16.3. The circuit values are: $V = 5$ V and $R = 100\Omega$. Determine and sketch the diode current and the output voltage.

*4. A 115 V-rms signal is applied to the circuit in Fig. 16.9. Determine the values of V_1 and V_2 to give an output centered around 60 volts, with a peak-to-peak amplitude of 30 V.

5. Determine the value of R needed in Problem 4 if the maximum diode forward current of 50 mA is not to be exceeded.

6. We wish to obtain a symmetrical wave with a 5 V peak-to-peak amplitude from the circuit in Fig. 16.11. The input is a 10 V peak-amplitude sine wave. Silicon diodes ($V_{on} = 0.6$ V) are used with $R = 2$ kΩ. Determine the reference voltages needed.

7. In Problem 6 determine the two diode currents and sketch their waveshapes.

8. Silicon Zener diodes with $V_Z = 5.6$ V and $V_{f(on)} = 0.5$ V are used in the circuit shown in Fig. 16.12. The input signal is a 20 V peak-amplitude sine wave. Sketch the output waveshape.

*9. If $R = 50\Omega$ in Problem 8, what is the peak power dissipated in the Zener diode?

10. Sketch the output waveshape if a 115 V-rms sine wave is applied to the circuit in Fig. 16.13.

11. Repeat Problem 10 for the circuit in Fig. 16.14.

12. Determine the logic voltages in the inverter circuit of Fig. 16.15 if $R_B = 10$ kΩ, $R = 50$ kΩ, $R_C = 200\Omega$, $V_{CC} = 20$ V, $V = 10$ V, and $\beta > 100$. Both the diode and the transistor are silicon.

*13. How many inverter circuits identical to that in Problem 12 could be tied to the collector of the inverter in Problem 12 without destroying the logic levels.

14. A square wave amplitude of 0.5 V is applied to the input of the circuit shown in Fig. 16.22. If the on and off times of the square wave both equal 1 millisecond, determine the circuit values needed to provide a negative going spike with an amplitude of 5 V at the output.

Part III

Systems

In these days of high technology, the importance of large-scale electronic systems is obvious. Therefore, we shall describe representative electronic systems in this section as a logical extension of the circuits covered in Part II. From rectifiers and feedback amplifiers, the natural next step to discuss is regulated power supplies. In addition, power control systems using many of the relatively simple circuits developed in Part II are covered. Similarly, from our study of differential and operational amplifiers, we go on to analog systems.

Chapter 17

Regulated Power Supplies

The dc power supply is one of the most important and necessary subsystems in electronics. Whether we are dealing with communication, instrumentation, computers, or any electronic system, small or extremely large, a source of dc power furnished by a power supply is needed. The basic function of a power supply is to convert the readily available 60 Hz 115 V-rms ac into a specific dc voltage. The power supply contains several circuits: (1) the transformer, which, depending on the need, either steps up or steps down the available line voltage; (2) the rectifier circuit, which converts the alternating current into unidirectional or pulsating direct current; (3) a filter circuit, which reduces or minimizes the ripple; and (4) a regulator circuit, which maintains the dc level at the output constant with varying loads. All the components of a power supply except the regulator circuit were discussed in Chapter 9. In this chapter, we will study various discrete and integrated circuit regulators.

17.1 REGULATORS

A regulator is any circuit that maintains a rated output voltage under all conditions: either no load (open circuit)—supplying no output current—or full load—supplying an output current (see Figure 17.1). No circuit provides perfect regulation; that is, no circuit maintains the output voltage at V_{OC} while supplying any current to the load. A practical regulator may have characteristics like those indicated in Figure 17.2, where the output voltage under the load, V_L, is somewhat lower than the no-load output voltage, V_{OC}.

As indicated in Equation (17.1), the amount of regulation provided by a circuit is measured as the ratio between the difference in the output voltage with and without a load and the output voltage under load conditions. Obviously, for a perfect regulator, this ratio is zero since $V_{OC} - V_L = 0$. So the smaller the

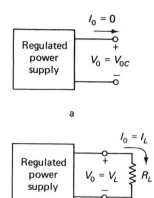

Figure 17.1 Voltage regulation: (a) no load (open circuit) and (b) full load.

ratio, the better the regulation will be for any given circuit. The amount of regulation in percentage is given by

$$\% \text{ regulation} = \frac{V_{OC} - V_L}{V_L} \times 100 \quad (17.1)$$

Note that this ratio is defined with respect to a specific load condition or rated current.

Figure 17.1(b) tells us that $V_L = I_L R_L$, so that Equation (17.1) may be rewritten:

$$\% \text{ regulation} = \frac{V_{OC} - V_L}{I_L R_L} \times 100 \quad (17.2)$$

The output resistance of the regulator as pictured in Figure 17.3 is the ratio of the difference between the open circuit voltage and the voltage under load divided by the amount of current drawn. Thus,

$$R_o = \frac{V_{OC} - V_L}{I_L} \quad (17.3)$$

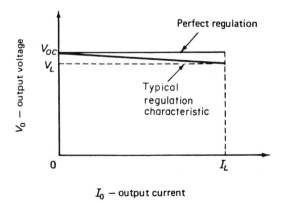

Figure 17.2 Voltage characteristics of a power supply.

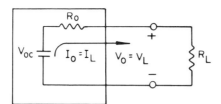

Figure 17.3 Equivalent circuit of a regulated power supply with the output resistance (R_o) of the regulator shown.

By substituting Equation (17.3) into Equation (17.2), the percentage regulation then may be rewritten:

$$\% \text{ regulation} = \frac{R_o}{R_L} \times 100 \qquad (17.4)$$

For a given load, therefore, regulation improves (i.e., percentage regulation becomes lower) as the output resistance of the regulator decreases. One of the primary characteristics of a good regulator is low-output resistance. A perfect regulator has zero output resistance.

17.1.1 Zener Regulator Circuits

The simplest form of voltage regulator uses a Zener diode, as shown in Figure 17.4. The raw or unregulated direct current, labeled by V, is applied to the series current limiting resistor R, and the regulated output is taken across the Zener diode. Note that the unregulated dc voltage at the input reverse biases the Zener diode and must be *larger* than the Zener voltage of the diode.

The output voltage of this regulator is essentially equal to the Zener voltage. It does, however, change somewhat when a load is connected because of the nonzero resistance of the diode. We can see this change if we replace the Zener diode by its equivalent circuit consisting of a voltage V_Z and equivalent resistance R_Z, as indicated in Figure 17.5. The output resistance of this regulator is the parallel combination of R and R_Z. Typically, the Zener resistance is 2 to 100 Ω. Therefore, the Zener regulator provides good regulation as long as the load resistance is sufficiently higher than R_Z.

In order for the Zener regulator to be cost effective, it is usually limited to low power (<1 W) applications.

The procedure for determining the regulation and voltage for a Zener regulator circuit is illustrated in Example 17.1.

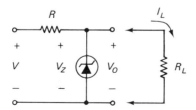

Figure 17.4 Basic Zener regulator.

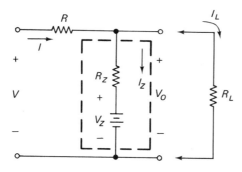

Figure 17.5 Equivalent circuit for Zener regulator.

Example 17.1

The Zener regulator circuit illustrated in Fig. 17.3 is constructed to provide a nominal regulated dc voltage of 10 V, with $V = 20$ V, $R = 100$ Ω, using a 1N5240 (10 V Zener diode). The load is to be 200 Ω. We want to determine the percentage regulation and the actual load voltage under full load.

Solution. The 1N5240 diode has a maximum resistance of 17 Ω at a test current of 20 mA (from spec sheet in Appendix A). We may assume a typical resistance of, say, 10 Ω. The output resistance of the regulator is then

$$R_o = \frac{RR_Z}{R + R_Z} = \frac{(10)(100)}{10 + 100} = 9.1 \text{ Ω}$$

The open-circuit voltage is determined from the equivalent circuit in Fig. 17.4 with $I_L = 0$, or $I = I_Z = (20 - 10)/(100 + 10) \cong 90$ mA. Thus,

$$V_{OC} = V_Z + I_Z R_Z \cong 10 + (0.091)(10) \cong 10.91 \text{ V}$$

The equivalent circuit for the regulator, therefore, is an open-circuit voltage of 10.9 V and a resistance of 9.1 Ω as shown in Fig. 17.6. With the load connected, the output voltage is given by

$$V_L = \frac{R_L V_{OC}}{R_L + R_o} = \frac{(200)(10.9)}{200 + 9} \cong 10.44 \text{ V}$$

The percentage regulation using Eq. (17.4) is

$$\% \text{ regulation} = \frac{R_o}{R_L} \times 100 = \frac{(9.1)(100)}{200} \cong 4.6$$

The regulation characteristics for this circuit are also given in Fig. 17.6.

In the design of Zener regulators, we must make certain of two facts. First, the power rating of the Zener should not be exceeded. Secondly, under the worst-case load (lowest R_L) the Zener must still draw some minimum current. In the previous example, the maximum current for the circuit occurs when the Zener draws zero current and its voltage is barely 10 V. Under these conditions the load

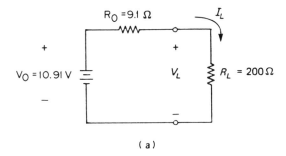

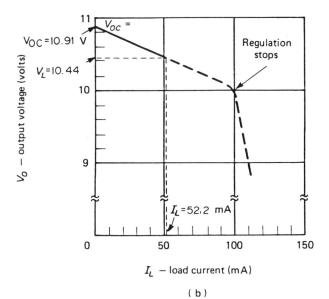

Figure 17.6 (a) Zener regulator equivalent circuit and (b) regulation characteristics.

is being supplied with the absolute maximum current and the Zener is about to stop regulating. Thus, if the load in the example were 100 Ω or less, the Zener would not regulate; the load voltage would rapidly decrease from 10 V as the load was decreased in size. This decrease is shown in Figure 17.6(b). Furthermore, the circuit could not safely be operated without a load equal to or greater than 220 Ω in order to keep the dissipation in the Zener at or below its rated maximum of 500 mW.

17.1.2 Basic Series Regulator

Regulation over a wider range of loads is possible with the *series regulator* circuit, illustrated in Fig. 17.7. The output voltage of this circuit is approximately the Zener voltage minus the base-emitter voltage of the transistor ($V_o = V_Z - 0.6$).

The unregulated dc input voltage must exceed the desired output voltage by at least 1 V. The transistor ceases to provide regulation once it saturates. Its

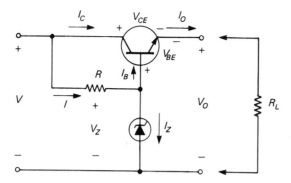

Figure 17.7 Basic series regulator constructed from discrete components.

collector-emitter voltage is the difference between the unregulated input and regulated output voltages.

The series regulator operates in the following way. The current supplied to the load is essentially the same as the collector current of the transistor. So long as the transistor is operating in its active region, the collector current is essentially βI_B. Therefore, when the load resistance changes, the transistor base current changes. As a result, the collector current also changes, thus providing the proper output current in order to maintain the output voltage constant. The actual amount of change in the output voltage will be very small; it is governed by the amount of change in V_{BE} to cause the desired change in I_B. This procedure is illustrated in Example 17.2.

Example 17.2

Consider the series regulator depicted in Fig. 17.7, with a silicon transistor $\beta = 100$, $V_{BE} = 0.6$ V, a 1N5232 Zener diode ($V_Z = 5.6$ V, $R_Z < 11\,\Omega$ at 20 mA), $V = 10$ V, and $R = 220\,\Omega$. We want to determine the circuit conditions with R_L of 100 Ω and R_L of 50 Ω.

Solution. The output voltage with 100 Ω load is

$$V_o = 5.6 - 0.6 \cong 5.0 \text{ V}$$

The output current is essentially the same as the collector current and must be

$$I_C = \frac{V_o}{R_L} \cong \frac{5}{100} = 50 \text{ mA}$$

The current through R is given by

$$I = \frac{V - V_Z}{R} \cong \frac{10 - 5.6}{220} = 20 \text{ mA}$$

For the collector current calculated, the base current must be

$$I_B = \frac{I_C}{\beta} = \frac{50}{100} \text{ mA} \cong 0.5 \text{ mA}$$

Thus, the Zener current is

$$I_Z = I - I_B = 20 - 0.5 = 19.5 \text{ mA}$$

With a 50 Ω load, the circuit values become $I_C \cong 100$ mA, $I_B \cong 1$ mA, and $I_Z \cong 19$ mA. For this collector current, a typical value of h_{ie} may be roughly 500 Ω. With the increase in base current, the approximate increase in V_{BE} may be about 0.25 V. Thus, the output voltage decreases by the same amount; it becomes approximately 4.75 volts. ($V_o = 5.6 - 0.85 = 4.75$ V).

17.1.3 Shunt Regulators

The shunt regulator circuit is shown in Figure 17.8. The regulated output voltage in this circuit is the sum of the Zener and base-emitter voltages. To understand the operation of this regulator, suppose that we decrease the load resistance, thus attempting to decrease the output voltage. A decrease in the output voltage must be reflected as a decrease in the base-emitter voltage because the Zener voltage is constant. A decrease in base-emitter voltage, therefore, causes both the base and collector currents to decrease. (The decrease in I_C is the larger and more important of the two.) If we assume that the input current I is essentially fixed, the current I' is also fixed. (Typically I_B is negligibly small when compared to either I or I'.) Because I_C has decreased, the output current I_o must increase if I and I' are constant. Thus, the cycle of cause and effect is complete. A tendency in the output voltage to decrease as a result of additional loading is eventually followed by an increase in the output current that tends to keep the output voltage constant.

The limit in the regulation for this circuit occurs when the load tries to draw all the current supplied by the unregulated supply (I). As a result, the transistor and Zener become cut off.

17.1.4 Series Regulator with Transistor Feedback

We can improve the performance of the basic series regulator of Figure 17.7 if the output voltage is sensed and the series transistor is forced to adjust to the load. This procedure is the basic concept of feedback. One form of a circuit that uses feedback to regulate the output voltage is depicted in Figure 17.9.

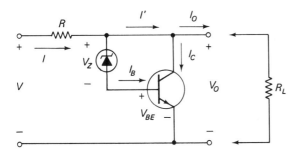

Figure 17.8 Basic shunt regulator.

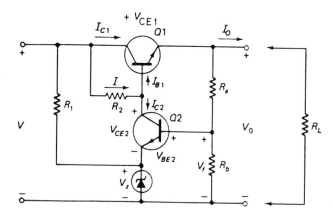

Figure 17.9 Series regulator with transistor feedback.

The output voltage in the circuit of Figure 17.9 is always given by

$$V_o = V - V_{CE1} \tag{17.5}$$

Regulation is achieved by forcing V_{CE1} to decrease by the same amount as the unregulated input, thus maintaining the desired constant output.

Suppose that the output voltage has decreased as a result of either a decrease in the unregulated input voltage (which might be caused by line voltage decrease) or an increase in the current drawn by the load. This decrease in the output voltage is sensed across resistor R_b in terms of a decrease in the feedback voltage V_f. We can see that this voltage is given by

$$V_f = V_{BE2} + V_Z \tag{17.6}$$

Because the Zener voltage is fixed, any decrease in the feedback voltage is reflected as a decrease in V_{BE2}. Consequently, I_{C2} decreases. I_{B1} is given by

$$I_{B1} = I - I_{C2} \tag{17.7}$$

Therefore, the decrease in I_{C2} causes I_{B1} to increase, which means that I_{C1} also increases (with V_{CE1} decreasing). In this way, it supplies additional current to the load to cancel out the decrease in output voltage. This procedure is illustrated in Example 17.3.

Example 17.3

To illustrate the operation of the regulator circuit of Fig. 17.9, assume that the unregulated input is 30 V, the load voltage is 15 V, with $R_L = 30\ \Omega$, and the series transistor has a β of 20. Its output characteristics are shown in Fig. 17.10. We want to determine the increase in base current needed to maintain the specified output voltage when the load is 15 Ω. (Assume that the input voltage remains constant.)

Solution. With the 30 Ω load, the load line labeled case 1 on Fig. 17.10 applies. The conditions from the load line for transistor 1 are: $I_C \cong I_o \cong 0.5$ A, $I_B = 25$ mA, and $V_{CE} = 15$ V. (The current through R_a is negligible when

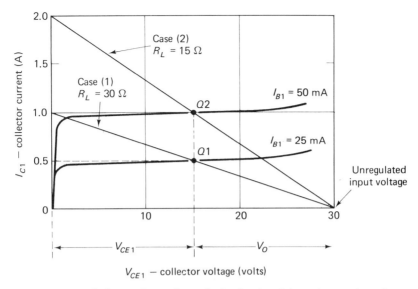

Figure 17.10 Series regulator: change in the Q-point of the series transistor due to increased load current requirement.

compared to either I_C or I_o.) When the load is decreased to 15 Ω, load line labeled case 2 applies. If the output voltage is to remain at 15 V, then the collector current must increase to 1 A, as indicated by the new operating point Q2. The increase in base current needed to maintain the output voltage is 25 mA. In this case, V_{CE1} remains essentially fixed.

In the preceding example we assumed that the unregulated input voltage remains fixed to illustrate how the circuit can adjust for increased current demands and maintain the regulated output voltage. In Example 17.4 we will consider the case when the unregulated input voltage decreases because of a decrease in the ac line voltage.

Example 17.4

Consider the same situation as in Ex. 17.3, except that the load is constant at 30 Ω and that the input voltage drops from 30 to 25 V as a result of a decrease in ac line voltage. We want to determine the needed change in the operating point of the series transistor to maintain the output voltage.

Solution. With the input voltage at 30 V, we have the same load line and operating point as in the previous example. This result is shown in Fig. 17.11, labeled case 1. When the input drops to 25 V, the load line labeled case 2 applies. The slope is the same because the load is still 30 Ω. The base and collector currents are essentially unchanged, but there is a change in V_{CE1}, as shown, from 15 to 10 V. Thus, even though the input has decreased by 5 V, the output voltage remains essentially unchanged at 15 V.

Sec. 17.1 Regulators

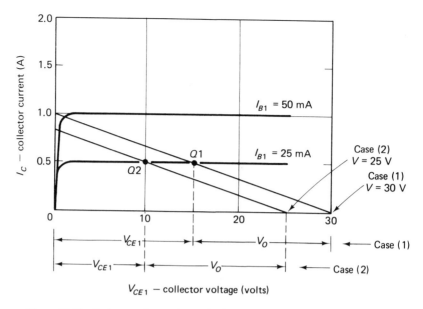

Figure 17.11 Series regulator: change in the Q-point of the series transistor due to a decrease in the unregulated input voltage. Notice that V_o remains constant at 15 V in each case.

Let us now consider in Example 17.5 the *worst case*, when the load demands additional current at the same time that the unregulated input voltage decreases. Note that this is usually the case; that is, when additonal current is drawn from the unregulated dc supply, its terminal voltage decreases.

Example 17.5

With a 30 Ω load and a 30 V input, the output is at 15 V. When a 15 Ω load is connected, the additional current requirement forces the input voltage to fall to 25 V. We want to determine the change in the Q-point of the series transistor.

Solution. Under the original conditions, the load line is the same as case 1 in the two previous examples. When the load is changed to 15 Ω, both the starting point and the slope of the load line shift. The starting point moves from 30 V to 25 V; the slope now has to correspond to a 15 Ω resistor instead of a 30 Ω resistor. This change is indicated in Fig. 17.12, with the new conditions labeled by case 2. Note that the new operating point, $Q2$, gives: $V_{CE1} \cong 10$ V, $I_{C1} \cong 1$ A, and $I_{B1} \cong 50$ mA. The series transistor can adjust to maintain the output voltage at 15 V even when an increase in output current is required at the same time that the input voltage decreases.

We can modify the series regulator circuit shown in Figure 17.9 by replacing the series-pass transistor $Q1$ with a Darlington pair if the change in the collector

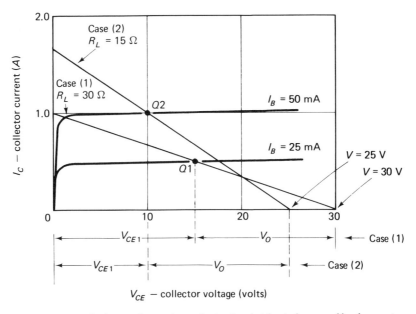

Figure 17.12 Series regulator: change in the Q-point due to increased load current requirement as well as a decrease in unregulated input voltage. Notice that V_o remains constant at 15 V in each case.

current of $Q2$ is not sufficient to produce the desired change in the output current. The operation of the circuit will be essentially unchanged. However, the same change in the collector current of $Q2$ causes a larger change in I_{C1} and thus in I_o. This procedure is illustrated in Example 17.6.

Example 17.6

The series regulator circuit as shown in Fig. 17.13 is constructed with the series transistor having a β of 40. The regulated output is to be 10 V at a maximum current of 1 A ($R_L \geq 10\ \Omega$). The unregulated input voltage is between 25 and 35 V. Determine the parameters for the transistors and Zener diode.

Solution. The worst-case conditions for the circuit occur when the input voltage is a maximum (35 V) and the output current is also a maximum (1 A). Assuming the output voltage to be approximately 10 V under these worst-case conditions, we calculate that $V_{CE1} = 25$ V. Thus, the series-pass transistor $Q1$ is going to have to withstand 25 V and 1 A. The minimum power handling capacity of this transistor should be 30 W.

The base of $Q1$ is at approximately 10 V, so the current I is

$$I = \frac{35 - 10}{330} = 76\ \text{mA}$$

I_{B1} is $I_{C1}/\beta \cong 1A/40 \cong 25$ mA; since $I_{C2} = I - I_{B1}$, I_{C2} must be $76 - 25 \cong$

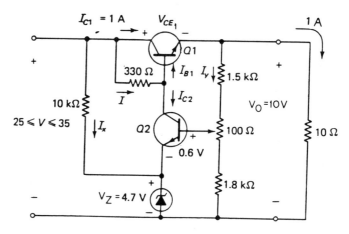

Figure 17.13 Series regulator example.

51 mA. The net current through the Zener is the sum of I_{C2} and the current through the 10 kΩ resistor, I_x. $I_x = (35 - 4.7)/10 \cong 3$ mA. Thus, the maximum current in the Zener is approximately $51 + 3 \cong 54$ mA.

The Zener must be able to dissipate (54 mA) (4.7 V) = 260 mW of power. A safe choice would be a 500 mW Zener. Transistor $Q2$ has a worst-case collector-emitter voltage of approximately $V_o - V_Z = 6$ V. The maximum collector current is slightly in excess of 50 mA, so that a 500 mW transistor would be acceptable for $Q2$.

17.1.5 Series Regulator with DIFF AMP Feedback

Figure 17-14 illustrates an improved series regulator circuit using a basic DIFF AMP in the feedback loop. The DIFF AMP consists of transistors $Q3$ and $Q4$, together with their bias resistors and reference Zener $ZD2$. The circuit values are

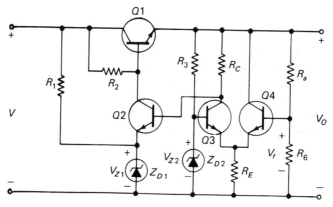

Figure 17.14 Series regulator with DIFF AMP feedback.

adjusted as follows. Resistors R_a and R_6 are chosen to provide a reference voltage V_f that is the same as the Zener voltage of $ZD2$ when the output is at the desired value. The output of the DIFF AMP (at the collector of $Q3$) should be at about 0.6 V above V_{Z1} to insure that $Q2$ is conducting. Resistors R_1 and R_3 are chosen to limit the respective Zener currents to safe limits.

Although the DIFF AMP provides additional sensitivity to changes in the output voltage, its operation is the same as that described in the previous section for series regulators with transistor feedback.

17.1.6 Series Regulator with OP AMP Feedback

An OP AMP can be used to give the highest sensitivity in the feedback loop. An example of a series regulator using an OP AMP in the feedback loop is depicted in Figure 17.15. You can understand the circuit operation by considering that V_f and V_Z are approximately equal when the output voltage is at its desired value. The large open-loop gain of the OP AMP causes even the slightest difference between the two to be significant enough to change the Q-point of the series-pass transistor.

Consider the action when the output voltage tries to decrease: V_f becomes slightly smaller than V_Z; the output of the OP AMP becomes more positive and causes a larger base current in the series transistor, which, in turn, causes the collector current to increase. The output voltage is, therefore, maintained at its original level.

Should the output voltage try to increase (because of a lower output current demand), the reference voltage momentarily exceeds the Zener voltage; the output of the OP AMP is driven in the negative direction, thus reducing the base current drive for the transistor. This action, in turn, decreases the collector current, and the output voltage is again adjusted to its original value.

The circuit of Figure 17.15 may be used for a variety of output voltages. However, the output current is limited because the OP AMP cannot provide a base current in excess of, typically, 10 to 20 mA. If additional output current is needed, the series transistor ($\beta \approx 40$) may be replaced with a power Darlington circuit ($\beta \approx 400$), where the OP AMP output current is amplified by a factor of 400 instead of 40.

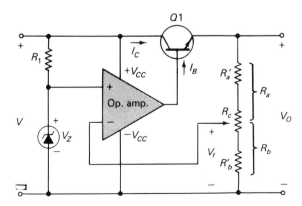

Figure 17.15 Series regulator with OP AMP feedback.

The minimum supply voltage for which the OP AMP will operate also limits the use of the circuit. The OP AMP takes its supply from the unregulated input voltage; therefore, the minimum value of the input voltage should be sufficient to bias the OP AMP. For a 741 OP AMP, this minimum supply voltage is typically ±5 volts. The absoute minimum unregulated voltage is approximately 10 volts. If a potentiometer (R_c) is used, the output voltage is adjustable. Resistor values (R'_a, R'_b, and R_c) are chosen so that the arm of the potentiometer (pot) can be adjusted to a voltage at or near V_Z.

17.2 CURRENT LIMITING CIRCUITS

All power supplies need some form of protection from overcurrent conditions caused by a component failure in the circuit that is being supplied or by accidental short circuits.

17.2.1 Diode Overcurrent Protection

We can modify the series regulator circuit as indicated in Figure 17.16 to include overcurrent protection. As long as the load current is below the desired limit, the regulator behaves in the way already described. Diodes $D1$ and $D2$ will be essentially off (not conducting) and $I_D = 0$. When the load tries to draw an excessive current, the voltage drop across R_{limit} becomes sufficient to forward bias both diodes, which now conduct and limit the series-transistor emitter current. With both diodes conducting, the maximum voltage drop across the limiting resistor is $V_D - V_{BE1}$, where V_D is the sum of the diode voltage drops. With silicon diodes and transistors, $V_{BE1} = 0.7$ V and $V_D = 2(0.7)$ V. In this case, the emitter current

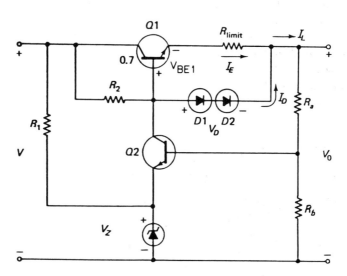

Figure 17.16 Current limiting in a series regulator using diodes $D1$ and $D2$.

(I_E) is limited to a value of approximately $0.7\ V/R_{\text{limit}}$. For example, suppose that we need to limit the output current to 1 A. Using silicon devices, we would need a limiting resistor of approximately 0.7 Ω with a power rating of 1 W.

If, for some reason, a short-circuit condition existed at the output, the current would be limited to $0.7\ V/R_{\text{limit}} + I_D$, where I_D is determined by R_2 and is approximately V/R_2.

17.2.2 Transistor Overcurrent Protection

The same principle is used to limit the output current in the circuit shown in Figure 17.17. Here the current is limited by a transistor. While the output current is below the desired limit, circuit operation is unaffected by the presence of $Q3$, because its base-emitter voltage is not high enough to cause it to conduct. However, when the output current increases to the point that the voltage drop across R_{limit} nears 0.7 V, $Q3$ begins to turn on and draws collector current. This collector current of $Q3$ is supplied through R_2, which also supplies the base current for the series-pass transistor $Q1$. Thus, when an excessive current is demanded by the load, $Q3$ begins to conduct. It diverts the additional current that would otherwise become an increase in the base current of $Q1$. The load current is effectively limited to approximately $0.7\ V/R_{\text{limit}}$ when $Q3$ is a silicon transistor.

Under short-circuit conditions, the output current will not exceed $0.7\ V/R_{\text{limit}} + V/R_2$.

Either the diode or transistor current-limiting scheme can be used with any of the various discrete device series regulator circuits discussed in the previous sections.

Since most Zener diodes exhibit noise, they are usually bypassed by placing a ceramic capacitor, typically 0.01 to 0.1 μF, in parallel with the Zener diode.

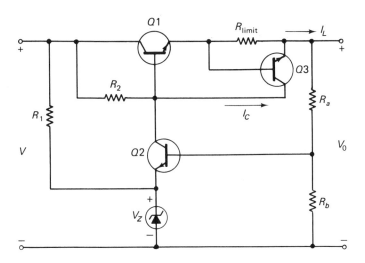

Figure 17.17 Current limiting in a series regulator using a transistor ($Q3$).

17.3 IC REGULATORS

Low-cost fabrication techniques have made many commercial integrated circuit (IC) regulators available. These devices range from fairly simple, fixed-voltage types to high-quality, precision regulators. Today, low-cost IC regulators are available that have improved performance over regulators made from discrete components. These modern IC regulators may be used by themselves or in more powerful regulators as the control circuit in conjunction with discrete transistors. In studying the series regulator of Figure 17.15, you saw the application of an IC OP AMP as an error amplifier. This application does not have the sophistication of the integrated circuit regulator.

IC regulators have a multitude of features built into them. In discrete component form, implementing these features would require a lot of extra space and would significantly increase the cost of the regulator. Among these features are current limiting (either variable or fixed), self-protection against overtemperature, remote control, remote shutdown, operation over a wide range of input voltages, and foldback current limiting. Figure 17.18 pictures the block diagram of a popular low-power IC voltage regulator. IC regulators of this type provide a very sophisticated device to use in designing voltage regulators.

The 723* is a precision voltage regulator that can accommodate input voltages between 9.5 and 40 V and can provide output (or regulated) voltages between 2 and 37 V. It can supply an output current up to 150 mA without an external series-pass transistor (limited to a programmable value). The output current may be increased by adding an external power transistor. We can employ this IC

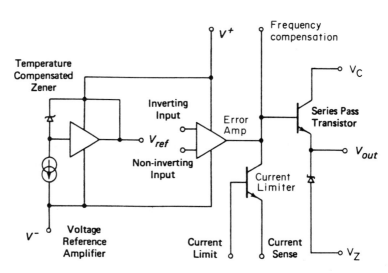

Figure 17.18 The 723 regulator block diagram.

---

* The 723 is a generic number. The 723 is also known as the μA723, LM723, and MC1723. The MC1723 data sheet is in Appendix A.

regulator as a negative as well as a positive voltage regulator. As noted in Figure 17.18, 723 contains a temperature-compensated reference amplifier, an error amplifier, a series-pass transistor, and a current limiter with remote shutdown access.

Typical applications of the 723 regulator are illustrated in Figures 17.19, 17.20, and 17.21. Note the graph of output voltage versus current. It illustrates the principle of *foldback* current limiting.

The output current is limited to a certain value under load conditions but is forced to an even lower value when the resistance of the load is too low or a short circuit condition exists. Observe how useful foldback current limiting is. When the output is short-circuited as a worst case, the series transistor must dissipate the highest power because the voltage across it is the highest possible. To keep dissipation down, the current is "folded back" or decreased from the normal limit.

17.3.1 IC Regulator Terminology

You do not need to be completely familiar with the circuit inside the IC regulator in order to use it. However, it is necessary to understand the terminology used in the specification sheets.

> *Input voltage range:* The upper and lower limit of the input voltage (unregulated or *raw* dc) that may safely be applied.
>
> *Output voltage range:* The range of possible regulated output voltages obtainable from the regulator. (Note: The regulated dc output voltage is always lower than the unregulated or raw dc input voltage. In order

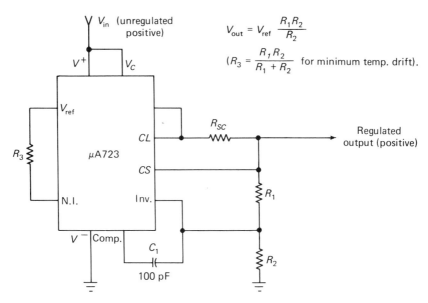

Figure 17.19 The μA723 as a positive voltage regulator; $7 \text{ V} < V_{out} < 37 \text{ V}$. *(Courtesy of Signetics Corp.)*

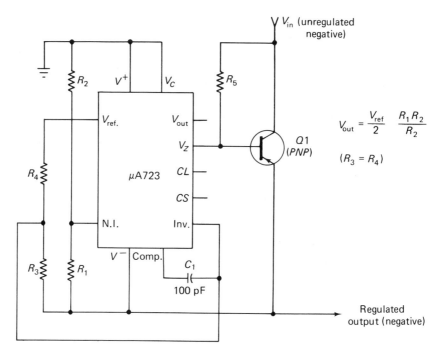

Figure 17.20 The μA723 as a negative voltage regulator. (*Courtesy of Signetics Corp.*)

for the regulator to work, some voltage must be dropped across the series transistor or transistors. See dropout voltage.)

Line regulation: The percentage of change in regulated output voltage for a specified change in unregulated input voltage.

Load regulation: The percentage of change in regulated output voltage for a change in load current from zero to the specified or rated maximum load current.

Ripple rejection: The amount of decrease (specified as a percentage or in dB) in the ac component from the input to the output. It is the ratio of rms input ripple voltage to rms output ripple voltage.

Temperature stability: An indication of the change in the output voltage for a change in operating temperature.

Quiescent current: The amount of current drawn off by the regulator when no load is connected. (Called standby current for dual tracking regulators.)

Output noise voltage: The rms value of the ac voltage at the output under load with no ripple at the input.

Dropout voltage: The voltage difference between input and output at which the regulator stops regulating. Dependent upon junction temperature and load current.

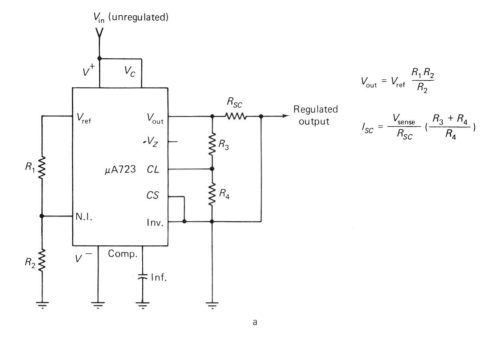

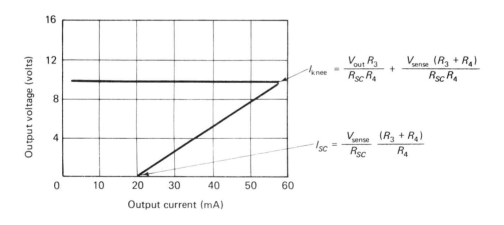

Figure 17.21 The μA723 as a positive voltage regulator with foldback current limiting: (a) circuit and (b) output VI characteristics. (*Courtesy of Signetics Corp.*)

17.3.2 Three-Terminal IC Power Regulators

The LM117/LM317 series of 3-terminal adjustable regulators is representative of the latest advancement in 3-terminal IC power regulator technology. In Figure 17.22 you see the LM117/LM317 being used as (a) a 5 V regulator, (b) an adjustable (1.2 V − 25 V) regulator, and (c) an adjustable constant current (10 mA to 100 mA) battery charger.

Sec. 17.3   IC Regulators   337

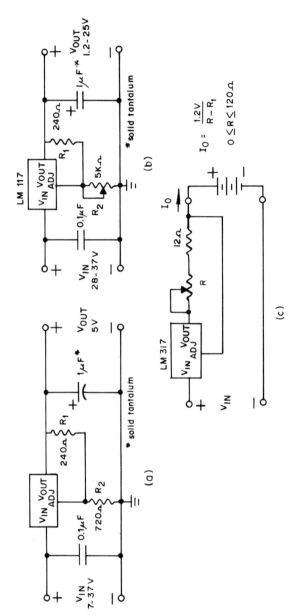

Figure 17.22 Typical applications of the LM117/LM317 adjustable regulator: (a) fixed 5 V output; (b) adjustable 1.2-25 V output; (c) adjustable 10-100 mA constant current battery charger. (Note: $V_{out} = 1.25\,(1 + R2/R1)$.

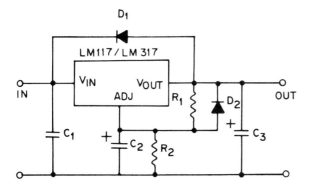

Figure 17.23 LM117/LM317 regulator circuit with diodes added to shunt the discharge currents of $C2$ and $C3$ around the IC.

The LM117/LM317 IC regulator has excellent line (0.01%/V) and load (0.1%) regulation and ripple rejection (80 dB) as well as a full overload protection. The regulator is virtually fail-safe; that is, the output current is automatically limited to a safe value. An internal thermal shutdown circuit is provided to sense when internal dissipation becomes too great and to turn OFF the output to prevent burnout.

Using the LM117/LM317 requires only two resistors to set the output voltage between 1.2 and 37 V. With most applications, no capacitors are needed. However, an input capacitor of 0.1 μF (ceramic disc) or 1 μF (solid tantalum) is added when the filter capacitor is more than a few inches (>4 inches) from the regulator. An optional 1 μF (solid tantalum) output capacitor will ensure trouble-free transient response. To optimize the ripple rejection, a 10 μF (sold tantalum) capacitor is added from the adjustment terminal of the IC to the ground.

Diodes are added to the regulator circuit (Figure 17.23) to protect the IC from the discharge of capacitors C_2 and C_3. D_1 is added when the output voltage is greater than 25 V and/or C_3 is greater than 25 μF. D_2 is added to protect the IC from C_2 when the output terminals are shorted to ground.

The LM117/LM317 has become the workhorse of IC regulators. This has resulted from its wide range of voltages and its 1.5 A current capability as well as its outstanding regulation properties. Because of its adaptability, the LM117/LM317 has eliminated the need for many of the earlier fixed IC regulators.

17.4 COMPLETE POWER SUPPLY

Separately, we have discussed the operation of all the components in a regulated dc power supply. In this section, we will examine the interfacing of the components to form a complete system.

In a dc power supply, the desired output voltage (or range of output voltages) and the maximum current needed are known. Typically, the input is derived from the single phase 117 V-rms, 60 Hz power line.

The unregulated voltage available from the filter is always larger than the final regulated output voltage. The transformer, rectifier, and filter should be chosen so that they can safely handle the maximum anticipated current and be able to supply a dc voltage *under full load* that is sufficiently higher than the desired regulated output voltage within the rated input voltage range for the regulator. It is important to specify "under load" because the dc (or average) level at the output of the filter decreases under load from its no-load value. This decrease is especially true of capacitive filters.

As an example of a minimum number of components in a well-regulated supply, examine the circuit of Figure 17.24. There are only five components in the complete supply: (1) a transformer, (2) a full-wave bridge rectifier assembly, (3) a capacitor ($C1$) for ripple filtering, (4) a 3-terminal IC voltage regulator mounted on a heat sink, and (5) the external resistors to set the regulator to 5 V. The procedure for determining the specifications of these components is illustrated in Example 17.7.

Example 17.7

We want to determine the specifications of each of the five components of the power supply pictured in Fig. 17.24 and specify a heat sink for a maximum ambient air temperature of 40°C.

Solution. From the information in Tables 8.1 and 8.2, as well as the equations in Chapter 8, we will determine an appropriate value for C_1. To begin the solution, we will assume an rms ripple voltage (v_{or}) of 1 mV at the output of the LM317 regulator. From the LM117/LM317 data sheet in Appendix A, we learn that the worst-case ripple rejection ratio is ≈ -55 dB at 5 V, 1 A, and 120 Hz when $C_{adj} = 0$. The rms ripple (v_r) at the filter, C_1, is determined:

$$N_{dB} = 20 \log \left(\frac{v_{or}}{v_r}\right)$$

$$v_r = v_{or}/10^{N_{dB}/20} = 1 \times 10^{-3}/10^{-55/20}$$

$$v_r = 562 \text{ mV}$$

The ripple voltage across C_1 is 562 mV. Assuming a dc voltage of 8 V at the input of the regulator, then the ripple factor (r) at C_1 may be determined from Equation (8.5).

$$r = v_r/V_{dc} = 562 \text{ mV}/8 = 0.07$$

C_1 is determined from Table 8.1.

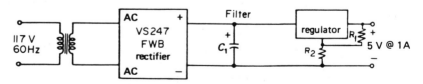

Figure 17.24 A 5-V, 1-A regulated power supply.

$$r = \frac{2400}{R_L C} \quad \text{where} \quad R_L = 8\text{V}/1\text{A} = 8\ \Omega$$

$$C = \frac{2400}{8(0.07)} = 4300\ \mu\text{F}$$

Select 5000 μF at 10 V.

Determine the transformer specifications from Table 8.2 and Equation (8.21).

$$V_{\text{trans}} = \frac{0.707(V_{\text{load}} + V_{\text{rect}} + v_r)}{0.81}$$

where $V_{\text{load}} = 8$ V; $V_{\text{rect}} = 2.4$ V; $v_r = 0.562$ V

$$V_{\text{trans}} = \frac{0.707(8 + 2.4 + 0.562)}{0.81} = 10\text{ V}$$

From Table 8.2, $I_{\text{trans}} = 1.8 \times I_{dC}$. Thus,

$$I_{\text{trans}} = 1.8 \times 1 = 1.8\text{ A}$$

Select a 115 V 60 Hz power transformer with a secondary rated at 10 V, 2 A.

The rectifier is selected with a diode current of twice the load current or 2×1 A $= 2$ A. The peak-reverse voltage rating of the bridge should be at least $2V_m$. Determine V_m from Table 8.1.

$$V_{dC} = V_m - \frac{4200 I_{dC}}{C}$$

$$V_m = V_{dC} + \frac{4200 I_{dC}}{C} = 8 + \frac{4200 \times 1}{5000}$$

$$V_m = 8 + 0.84 = 8.8\text{ V}$$

$$2V_m = 18\text{ V}$$

Select a 2 A epoxy bridge rated at a minimum of 18 V. Locate the data sheet for 2 A epoxy bridges in Appendix A.

The value of R_1 and R_2 for the regulator are determined from the expression $V_o = 1.25(1 + R_2/R_1)$, assume $R_1 = 240\ \Omega$. Thus,

$$5 = 1.25 + 1.25 R_2/240$$

$$0.00521 R_2 = 3.75$$

$$R_2 = 3.75/0.00521 = 720\ \Omega$$

The heat sink is determined from the thermal specifications of the LM317 data sheet in Appendix A.

Select a T package (TO-220). The junction to case thermal impedance (θ_{JC}) is 4°C/W and the maximum junction temperature (T_J) is 125°C. The maximum ambient temperature (T_A) was specified as 40°C.

The regulator power dissipation (P_D) is determined from the following expression:

$$P_D = (V_{IN} - V_{OUT})I_o = (8 - 5)1 = 3 \text{ W}$$

The thermal impedance of the heat sink is determined by solving the following equation:

$$\Delta T = \theta P_D$$

$$\theta = \Delta T / P_D$$

$$\theta = (125 - 40)/3$$

$$\theta = 28°C/W$$

$$\theta_{CA} = \theta - \theta_{JC} = 28 - 4 = 24°C/W$$

where $\Delta T = T_J - T_A$ (°C)
$\theta = \theta_{JC} + \theta_{CA}$ (°C/W)
P_D = device dissipation (W)

From the information in Appendix A, select an Aham Tor No. 191 20°C/W heat sink for the TO-220 case style of the LM317T regulator.

Figure 17.25 depicts an adjustable 0 to 22 V power supply using an LM338 adjustable 3-terminal, 5 A power regulator. Besides outstanding load (0.1%) and line regulation (0.005%/V), the LM138/338 series has a unique time-dependent current limiting feature that can accommodate transient loads up to 12 A (peak) for short periods of time. Besides this feature, the regulator has thermal overload protection as well as safe area protection. Safe area protection significantly reduces the maximum output current for large differences in $V_{IN} - V_{OUT}$ and for elevated regulator temperatures.

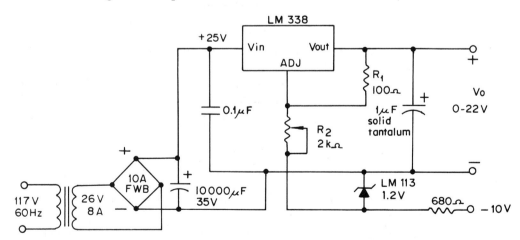

Figure 17.25 Regulated adjustable power supply 0-22 V. (*Regulator circuit courtesy of National Semiconductor Corp.*)

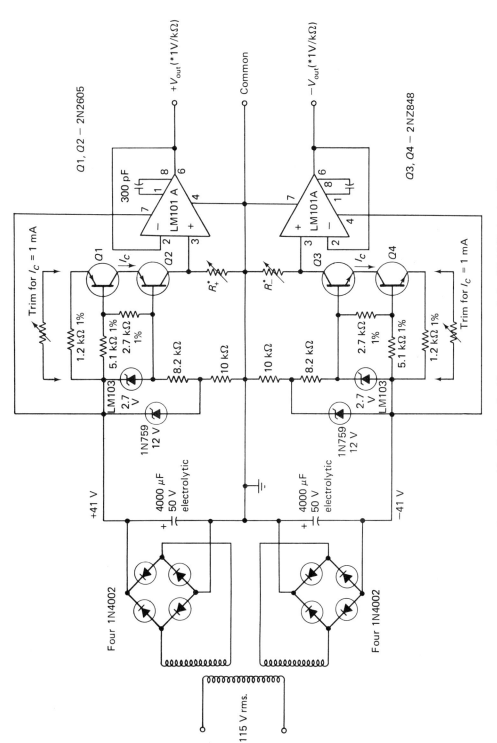

Figure 17.26 Programmable split (+ and −) low power supply for OP AMP circuits. (*Regulator circuits courtesy of National Semiconductor Corp.*)

Figure 17.26 shows a dual (positive and negative) low-current supply suitable for OP AMP or other IC circuits. The transformer windings should produce $+$ and -41 V at the output of the filter, as indicated. Rectification is performed by means of two bridge circuits, with a capacitive filter in each side. Two OP AMPs, each connected in a voltage-follower configuration, provide regulation. Because the inverting input of each OP AMP is connected to the output, whatever voltage is applied to the noninverting input appears at the output. The input reference voltage is developed across variable resistors R_+ and R_-, which allow the output voltage to be varied at the rate of 1 V for each 1000 Ω of resistance. We achieve this variation by using constant current generators comprised of two Zener diodes and two transistors in each half of the supply. In each case, the 1.2 kΩ resistor should be trimmed for exactly 1 mA collector current through the two transistors. In this way, a constant current of 1 mA is provided through the programming resistors R_+ and R_-.

REVIEW QUESTIONS

1. What are the components of a regulated power supply?
2. What is a regulator?
3. What distinguishes an unregulated power supply from a regulated one?
4. For best regulation, what should be the output resistance of a voltage supply?
5. How may a Zener diode together with a resistor be used to provide voltage regulation?
6. In the basic series regulator shown in Fig. 17.7, how is voltage regulation obtained?
7. How is regulation provided in the shunt regulator illustrated in Fig. 17.8?
8. How does the addition of transistor feedback improve the regulation in a series regulator?
9. How does the use of DIFF AMP feedback improve the performance of a series regulator?
10. How does the use of an OP AMP in the feedback of a series regulator improve its performance?
11. What are the advantages of the different feedback schemes used with a series regulator? The disadvantages?
12. Make a comparison of the different regulator circuits in terms of (a) performance, (b) circuit complexity (cost), and (c) operation.
13. What is meant by current limiting in a voltage supply?
14. How does the simple diode circuit of Fig. 17.16 provide current limiting?
15. How does the addition of $Q3$ in Fig. 17.17 provide current limiting?
16. What is meant by foldback current limiting?
17. What is the line regulation in a power supply?
18. What is load regulation in a power supply?

19. What is meant by the ripple rejection of a regulator?
20. What is the function of each part of a complete power supply?

PROBLEMS

1. In Fig. 17.1 the power supply voltage is 25 V with no load. When a 100 Ω load is connected, the voltage is 24 V. What is the percentage regulation and what is the output resistance of the supply?

*2. The simple Zener diode regulator shown in Fig. 17.4 is connected using a 12 V Zener diode with $R = 100$ Ω and $V = 15$ V. Determine (a) the power rating of the Zener necessary for operation without a load: (b) the highest current that can be supplied with the diode still regulating, assuming a Zener resistance of 10 Ω and I_{Zmin} of 5 mA.

3. Determine the equivalent resistance and voltage for the power supply in Problem 2. Make a plot of its regulation curve (see Fig. 17.6).

4. The series regulator in Fig. 17.7 is constructed with: $R = 180$ Ω, a TIP29 transistor and a 1N5231 Zener diode. Make a plot of the output voltage as a function of output current when the load is varied from 1 kΩ to 10 Ω. Assume the unregulated input voltage to be 10 V with a source resistance of 25 Ω.

*5. Circuit components for the shunt regulator in Fig. 17.8 are: $R = 50$ Ω, $V_Z = 15$ V, and $R_Z = 10$ Ω; a silicon transistor with β of 50 is used. For an input voltage of 20 V, determine the no-load power dissipation in the transistor. Also determine the output voltage with no load and again with a 250 Ω load.

6. In the series regulator circuit in Fig. 17.9, circuit values are: $V = 40$ V, $R_1 = 3.3$ kΩ, $R_2 = 220$ Ω, with both silicon transistors having a β of 50. The desired output voltage is to be nominally 20 V. If we use a 6.8 V Zener together with $R_b = 6.8$ kΩ, determine the circuit conditions (voltages and currents) as well as the value of R_a needed.

7. When an output current of 100 mA is drawn, the unregulated input voltage drops to 35 V in Problem 6. Determine the new circuit conditions and the output voltage.

8. The OP AMP in the regulator circuit in Fig. 17.15 is capable of supplying a maximum of 20 mA output current. The circuit values are $R_1 = 5.6$ kΩ, $V = 35$ V, $V_Z = 6.8$ V, and the desired output voltage is 15 V. Determine the values of R_a and R_b needed. (This pair of resistors should not draw more than 5 mA.)

*9. If the supply in Problem 8 is to be able to provide the rated output voltage into a 5 Ω load, what must be the β of $Q1$? Also determine the worst-case power dissipation in $Q1$. (Assume that at the highest output current V falls to 20 V.)

Chapter 18

Power Control Systems

In many industrial and consumer applications, the net amount of power delivered must be controlled. These systems range from the simple light dimmer to very sophisticated lighting-control installations or motor-speed controls.

18.1 PRINCIPLES OF POWER CONTROL

The basic control element is a *thyristor*. Depending on the specific application, this device may be an SCR, triac, SCS, or any other member of the thyristor family. Most often, the source of power is the line, either 115 V-rms or 230 V-rms, at 60 Hz (or sometimes at 50 Hz). The most common form of power control is *phase* control. In this mode of operation, the thyristor is held in an off condition, in which it blocks all current flow in the circuit except for a very small leakage current. It stays off for a portion of the positive (and/or negative) half-cycle; then it is *triggered* or *fired* into conduction at a time in the half-cycle determined by the control circuitry.

A single SCR in series with the load, as shown in Figure 18.1, can be used to control the amount of power delivered to the load. The control signal keeps the SCR in its OFF state for the first part of the positive half-cycle of the ac input and then fires it at the desired point to allow current to flow. When the SCR is conducting, almost all the applied ac voltage appears across the load (with the exception of approximately 1 V, which is across the SCR). The current in the circuit is only a function of the applied voltage and the load. Thus, the SCR controls only the voltage. Once the control circuit has fired the SCR by applying a sufficiently high gate signal, it loses control, but the SCR continues to conduct so long as its anode is positive with respect to the cathode. When the positive half cycle is terminated and the input voltage starts to go negative, the SCR reverts to its off or nonconducting state. (Actually, the SCR turns off when the anode currrent falls below the holding current; see Section 6.3.1. This decrease occurs while the input is still slightly positive.) The SCR cannot be fired again until the

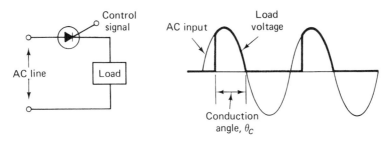

Figure 18.1 A single SCR in a power control circuit and the resulting waveform.

input becomes positive; thus, a single SCR can control only one half-cycle. The amount of power delivered to the load is proportional to the length of time during which the SCR is conducting. This duration is called the *conduction angle*; it can be varied from almost **50% of the available line input power** (SCR on half the time; conduction angle = 180°) to almost 0% (SCR off all the time; conduction angle = 0°).

The circuit shown in Figure 18.2 uses two SCRs connected in inverse parallel (anode of one to the cathode of the other) and can control both the negative as well as the positive half-cycle of the ac input. During the positive half-cycle, the anode of *SCR2* is negative and *SCR2* is off, no matter what the control signal. *SCR1* is fired into conduction during the positive half-cycle in the same manner as in the single SCR circuit of Figure 18.1. When the input goes from positive to negative, *SCR1* turns off and *SCR2* can be turned on by an appropriate signal to its gate. Thus, *SCR1* controls the positive half-cycle, whereas *SCR2* controls the negative half-cycle to achieve full-wave power control.

The two SCRs in the circuit of Figure 18.2 can be replaced with a single triac as shown in Figure 18.3. Circuit operation is unchanged because the triac is nothing more than a bilateral SCR; it can conduct in either direction (assuming that the proper gating signal is applied). In a full-wave control system, the power delivered to the load can be varied from **100% of the available line power** (conduction angle 180°) to almost 0% (conduction angle 0°).

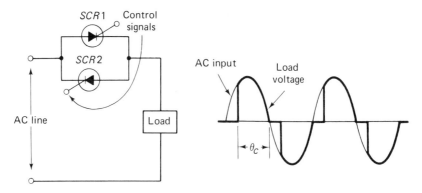

Figure 18.2 Two SCRs in a power control circuit and the resulting load voltage waveform.

Sec. 18.1   Principles of Power Control

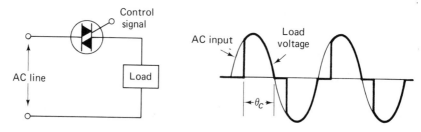

Figure 18.3 Triac power control circuit and the resulting load voltage waveform.

An alternate means of full-wave power control is shown in Figure 18.4. The ac input is full-wave rectified by the bridge consisting of diodes $D1$ through $D4$. It is applied to the load through an SCR. We can vary the amount of power delivered to the load from 0% and to essentially 100% of the available input power by controlling the conduction angle through the gating signal to the SCR. Similar operation can be achieved with the circuit depicted in Figure 18.5, where diodes $D1$ and $D2$ have been replaced with SCRs. During the positive half-cycle of the input, power is applied to the load only when $SCR1$ is gated and load current flows through $SCR1$ and $D3$. During the negative half-cycle of the input, power is applied to the load only when $SCR2$ is gated and load current flows through $SCR2$ and $D4$. Note that the output is a controlled full-wave rectified waveshape.

Figure 18.6 indicates the percentage of the available input power that is

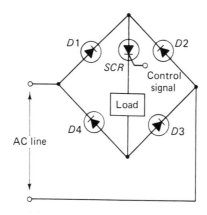

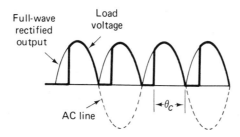

Figure 18.4 Full-wave bridge with one SCR used for power control.

348   Power Control Systems   Chap. 18

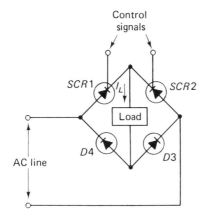

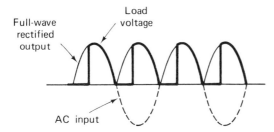

Figure 18.5 Full-wave bridge with two SCRs used for power control.

delivered to the load in a half-wave power control circuit as a function of the conduction angle. The maximum power delivered to the load is 50% for a conduction angle of 180°. However, approximately 45% is delivered at a conduction angle of 150°. It is not practical, therefore, to try to achieve conduction angles in excess of 150°. Similarly, only about 2% of the available input power is delivered

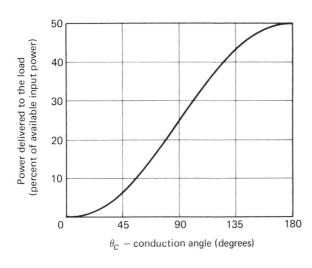

Figure 18.6 Power delivered to the load as a function of the conduction angle in a half-wave circuit.

Sec. 18.1 Principles of Power Control

to the load at a conduction angle of 30°, so it becomes pointless to try to achieve conduction angles of less than 30°.

The full-wave power control characteristics as a function of the conduction angle are given in Figure 18-7. In full-wave power control, conduction angles between 30° and 150° provide almost the entire range (from about 3% to 97%) of power delivered to the load. Note that the full-wave control circuits can deliver 100% of the available input power to the load, whereas the half-wave control circuits can deliver a maximum of 50%.

A simple means of firing an SCR is depicted in Figure 18.8. During the positive half-cycle, the capacitor charges up through the adjustable resistor R. When the capacitor voltage reaches the gate firing potential for the SCR, the SCR is turned on. It conducts until the input voltage goes to zero. When the SCR fires, its gate current discharges the capacitor, so that the cycle can be repeated when the input becomes positive again. The rate at which C_1 charges determines how quickly its voltage becomes high enough to fire the SCR; therefore, it controls the conduction angle. The charging rate is a function of both R_1 and C_1. By adjusting R_1, the conduction angle may be controlled. For conduction angles below 90°, the voltage across the capacitor is affected by the decreasing input voltage. It does not increase at the same rate above 90° that it did below 90°. This problem can be minimized by the addition of a series resistor and a parallel capacitor between $C1$ and the SCR gate as shown in Figure 18.9. Such a circuit offers superior performance for conduction angles below 90°. It also can be used to control the firing of an SCR to give conduction angles between essentially 0° and 170°.

We can further refine the SCR by inserting a trigger diode in series with the SCR (or triac) gate, as indicated in Figure 18.9. This circuit operates as follows. During the positive half-cycle capacitor $C1$ charges through $R1$; the voltage across $C1$ causes $C2$ to charge through $R2$. The SCR remains off as long as the voltage across $C2$ is below the breakover voltage of the trigger diode. When the voltage across $C2$ reaches the breakover voltage of the trigger diode, the diode fires (i.e.,

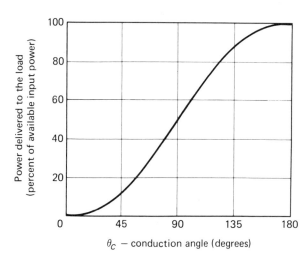

Figure 18.7 Power delivered to the load as a function of the conduction angle in a full-wave circuit.

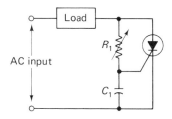

Figure 18.8 Simple means of controlling the conduction angle of an SCR.

conducts), and supplies a fast-rising, high-current gate pulse that in turn fires the SCR. The two capacitors are discharged through the trigger diode and gate of the SCR. The point at which the voltage across $C2$ reaches the firing potential for the trigger diode controls the conduction angle. It can be adjusted by the variable resistor $R1$: the smaller $R1$, the faster the charging of the capacitors and the sooner the firing of both the trigger diode and SCR. Thus, a decreasing $R1$ will increase the conduction angle. A unilateral (Shockley or four-layer) diode or a silicon unilateral switch (SUS), which fires in one direction only, may be used if the control element is an SCR (providing half-wave control). A bilateral diode [diac, trigger diode or silicon bilateral switch (SBS)] together with a triac can be used if full-wave control is desired.

The UJT (see Section 6.1) is another device that we frequently use to fire an SCR. A typical UJT control circuit is shown in Figure 18.10. The UJT is connected as a relaxation oscillator. The ac input is full-wave rectified by the bridge consisting of $D1$-$D4$. This rectified signal is then clipped by the Zener diode $D5$ to give the waveshape shown in Figure 18.11. When the UJT circuit has a voltage applied, capacitor C_T charges exponentially through R_T. When the capacitor voltage reaches the UJT peakpoint voltage V_p, the UJT fires and the emitter-base-1 resistance becomes very low. Capacitor C_T is then discharged through this low emitter-base-1 resistance and $R3$. The discharge current causes a voltage spike across $R3$, which is used as a control signal to fire an SCR. The capacitor and base-1 waveshapes are depicted in Figure 18.12. The timing of the charging and discharging of C_T can be controlled by R_T. Increasing R_T increases the charging time of C_T and thus increases the length of time between control signal pulses. The UJT relaxation oscillator is synchronized to the line because at the end of every half-cycle of the line input, the input to the relaxation oscillator (shown in Figure 18.11) goes to zero. Thus, we can make certain that the timing capacitor is discharged at the beginning of the next half-cycle.

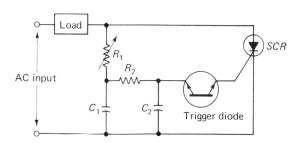

Figure 18.9 Improved circuit for firing an SCR.

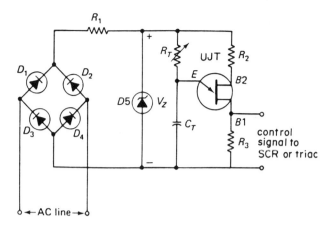

Figure 18.10 UJT line-synchronized relaxation oscillator, which provides the signal for firing an SCR or Triac.

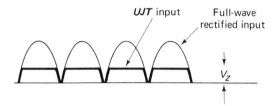

Figure 18.11 Input waveshapes for UJT circuit of Figure 18.10.

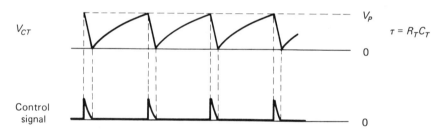

Figure 18.12 UJT relaxation oscillator waveshapes.

18.2 TRIAC LIGHT-INTENSITY CONTROL

We can use the triggering circuit shown in Figure 18.9 in a simple light-intensity control (light dimmer) circuit, as shown in Figure 18.13. On each half-cycle (both positive and negative) capacitor $C2$ charges through the phase-shift network of $R1$, $C1$, and $R2$, until it reaches the firing potential of diode $D1$ (about 20 V). Once the trigger diode has fired, its voltage drops and the triac is turned on. For the rest of the half-cycle, the timing network is effectively shorted through the trigger diode and the triac gate. No further pulsing can occur until the next half-cycle. During the positive half-cycle, the discharging current flows into the triac gate; on the negative half-cycle, the triac gate supplies the discharge current. Because the triac and the trigger diode are symmetrical, the conduction angle is

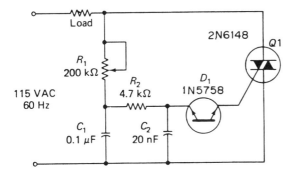

Figure 18.13 Triac 800-watt light dimmer. (*Courtesy of Motorola, Inc.*)

the same during both the negative and positive half-cycles. It is controlled by the setting of $R1$. Increasing $R1$ has the effect of reducing the conduction angle and thus reducing the amount of power to the load—and thus dimming the light.

18.3 SCS ALARM CIRCUIT

There are many applications in which we want an audible or a visible indication of an event. In such cases, the SCS alarm circuit (illustrated in Figure 18.14) may be used. We have incorporated it here because it can indicate certain conditions that may be useful in power control applications. The circuit may be expanded to accommodate as many inputs as desired.

The figure gives one possible form for an input. It does not matter whether the sensor resistor R_s is sensitive to temperature or to light or to radiation so long as it is normally in its high-resistance state. Then when it experiences a decrease in its resistance, the condition to turn on the alarm occurs. Resistor R_1 is set to a value that allows the input to the SCS to be below the triggering level with the sensor resistance high. When the sensor resistance decreases because of a change in its temperature, light, or radiation, a positive pulse activates the gate of the

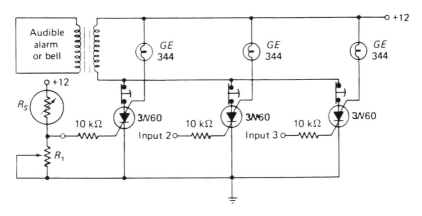

Figure 18.14 Silicon controlled switch (SCS) alarm circuit.

SCS, causing the SCS to conduct. The audible alarm is triggered and at the same time gives a visual indication by lighting the appropriate lamp. Any of the inputs thus triggered will cause an audible alarm. But only a specific input will cause the lamp to light. The circuit is reset by opening the anode circuit.

18.4 SCR UNIVERSAL MOTOR SPEED AND DIRECTION CONTROL

In a series-wound dc motor, the speed is governed by the voltage and current in the armature, whereas direction is determined by the direction of current through the field winding. A circuit for controlling both the direction and speed of such a motor is depicted in Figure 18.15. The ac input is full-wave rectified by the bridge comprised of diodes $D1$-$D4$. As the voltage V_x increases, capacitor C_1 charges through R_1, R_2, and the primary resistance of either $T1$ or $T2$, depending on the position of switch $S1$. Let us assume that for the time being $S1$ is directing current through $T1$. The Zener diode $D5$ is off until the capacitor charges up to and a little beyond the Zener voltage (about 51 V). At this point, the Zener diode conducts, and a positive voltage is developed across R_3. This voltage eventually fires $SCR5$, which now discharges the capacitor through the winding of transformer $T1$. The discharge current causes a positive pulse to appear at the sec-

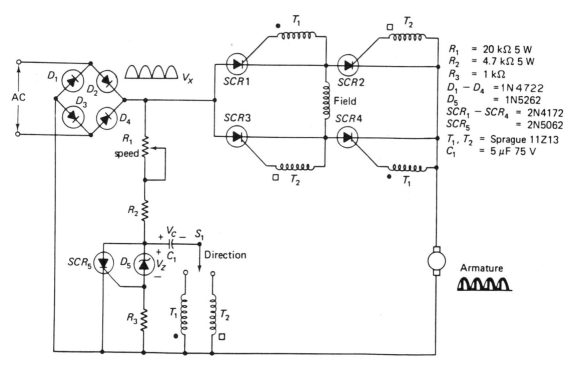

Figure 18.15 Direction and speed control for a series-wound universal motor. (*Courtesy of Motorola, Inc.*)

ondary of transformer $T1$; the pulse, in turn, causes $SCR1$ and $SCR4$ to conduct. As a result, power is applied to the motor and a current flows downward through the field winding. The speed of the motor is determined by the conduction angle of $SCR1$ and $SCR4$. This conduction angle can be varied by R_1. Increasing R_1 decreases the conduction angle and also decreases the speed of the motor.

The direction of the motor can be reversed by connecting $S1$ to the primary winding of $T2$. Thus, when $SCR5$ fires, it discharges the capacitor through the primary of $T2$, causing a positive pulse on the gates of $SCR2$ and $SCR3$. These two SCRs are then fired and power is once again applied to the motor. However, in this case, the current through the armature is in the upward direction, causing the motor to turn in a direction opposite to that when $S1$ is energizing $T1$. The timing is still determined by R_1, R_2, C_1, and the Zener diode.

We can use the same circuit to control the speed and direction of a shunt-wound motor if the field and armature windings are reversed. In a shunt motor the speed is governed by the power applied and the direction is determined by the direction of the current through the armature.

18.5 12-VOLT BATTERY CHARGER

The 12-volt battery charger circuit shown in Figure 18.16, which is capable of supplying 8 A, uses a programmable UJT (PUT) in a relaxation oscillator circuit. The circuit uses the programmable peak-point voltage characteristic of the PUT. When power is applied, the battery to be charged supplies current to charge capacitor $C1$ through $R1$. When the capacitor voltage reaches the peak-point voltage of the PUT, the PUT fires and, in turn, causes a positive voltage pulse

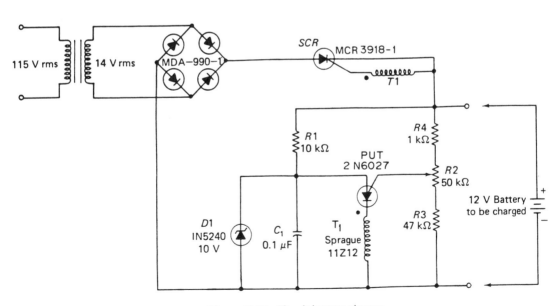

Figure 18.16 12-volt battery charger.

in the secondary of transformer T1. This voltage pulse fires the SCR and charging begins. So long as the battery voltage is low, the PUT relaxation oscillator supplies the firing pulses to maintain charging current through the battery. As the battery charges up, its voltage increases. As the battery voltage increases, the peak-point voltage of the PUT also increases, so that the capacitor must charge to a slightly higher voltage before the PUT (and, in turn, the SCR) can fire. The maximum capacitor voltage (in this case, 10 V) is set by the Zener diode, and the capacitor cannot charge above this voltage. Thus, when the battery is charged and its voltage is above a certain limit (set by $R2$), the PUT relaxation oscillator ceases to function. The capacitor cannot reach the PUT peak-point voltage; therefore, the SCR firing pulses are no longer supplied. The charger turns itself off once the battery is charged.

Note that the charger will not function unless the battery is connected properly. In order to start, the relaxation oscillator takes its voltage and current from the battery; if the polarity of the battery is inadvertently reversed, the oscillator will not start. The charged voltage of the battery is variable (with $R2$) between the lower limit set by the Zener diode (10 V) and the voltage available from the bridge rectifier (14 V).

18.6 ELECTRONIC CROWBAR

An *electronic crowbar* is used to shut down electronic equipment when an overvoltage condition develops due to regulator failure or other cause. A short circuit is placed across the supply line, which causes the supply fuse or circuit breaker to open (*blow*).

The circuit pictured in Figure 18.17 uses a silicon bilateral switch (SBS) to trigger the triac. Since both the SBS and the triac are bilateral, the crowbar circuit may be used to protect both ac and dc supplies. Note that this circuit does not respond to short, random power line transients.

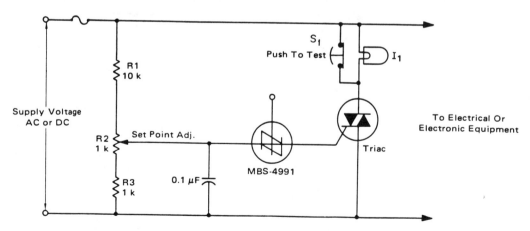

Figure 18.17 An electronic crowbar circuit using a silicon bilateral switch (SBS) to trigger the triac. (*Courtesy of Motorola, Inc.*)

The specified resistors R_1, R_2, and R_3 may be changed to adjust the 60- to 120-V dc (42- to 82-V ac) range to a new range. When adjusting the voltage range, a triac must be selected with a voltage rating greater than the highest voltage set by R_2. The lamp, I_1, is selected with a voltage rating equal to the supply voltage.

The crowbar trip point is set by adjusting R_2 with S_1 open (pushed to test). The *firing* of the triac is indicated by the lamp lighting. In normal operation, the test lamp is by-passed by the normally closed switch. When the triac fires, the supply line is shorted through the low ($\approx 0\ \Omega$) ON state impedance of the triac.

In the previous sections, we have examined a small but representative sample of the power control circuits. If you wish information beyond this introduction, we recommend you consult the manufacturers' application notes and manuals that are available from most power device manufacturers.

REVIEW QUESTIONS

1. What is power control?
2. What is the most common method of power control?
3. What is phase control?
4. What is the conduction angle? What determines it?
5. In a half-wave control circuit, what is the maximum percentage of the available power that can be applied to the load?
6. What is the relationship between the conduction angle and the percentage of the available power that reaches the load?
7. In the half-wave control circuit of Fig. 18.1, what determines the conduction angle?
8. In the half-wave control circuit of Fig. 18.1, once the SCR is fired, how is it turned off?
9. Explain how two SCRs, as shown in Fig. 18.2, may be used for full-wave control.
10. Directly below the load voltage for Fig. 18.2 sketch the waveshapes for the two control signals.
11. How is a triac used for full-wave control? (Refer to Fig. 18.3.)
12. Explain the operation of the bridge control circuits in Figs. 18.4 and 18.5.
13. What are the advantages and disadvantages of the two bridge control circuits in Figs. 18.4 and 18.5?
14. Why is the firing circuit of Fig. 18.9 superior to that of Fig. 18.8?
15. How does varying R_1 in Fig. 18.9 control the conduction angle for the SCR?
16. What function does the trigger diode in Fig. 18.9 perform?
17. Explain the operation of the UJT relaxation oscillator.
18. How is the UJT relaxation oscillator in Fig. 18.10 synchronized to the line frequency?

19. The light dimmer circuit of Fig. 18.13 provides for full-wave control. How is this control accomplished and what determines the conduction angle?
20. How should the motor speed and direction circuit shown in Fig. 18.15 be modified to provide control for a shunt-wound motor?
21. How is the battery charger in Fig. 18.16 turned off once the battery has been fully charged?
22. What determines when the battery charger (Fig. 18.16) turns off?
23. If the battery is connected with the polarity reversed, the battery charger in Fig. 18.16 will not operate. Why?

Chapter 19

Analog Systems

In this chapter we introduce the functions performed in an analog computer. We shall also show how programs can be specifically implemented to solve certain problems. Analog circuits have a wide range of applications outside analog computers. Most are extremely easy to implement, requiring only an OP AMP and a few discrete components.

19.1 PRINCIPLES OF ANALOG COMPUTATION

Some experts maintain that analog computers have become obsolete since the advent of the high-speed digital computer. Although the use of analog computers has definitely declined with the increased availability of sophisticated programming subroutines for high-speed digital computers, the analog computer can still be useful, mainly as a simulator. The analog computer is able to simulate the physical world in real time. Because both analog and digital computers have applications for which each is best suited, they will continue to exist side by side. This is evidenced by the number of hybrid computers using both digital and analog signal processing.

In an analog system, the input and output waveshapes are linearly related; that is, the output waveshape is proportional to or, at least a linear function of, the input waveshape.

Let us discuss the chief building block in an analog computer: the OP AMP. The basic *inverting configuration*, using generalized impedances, is shown in Figure 19.1. As described in Chapter 14, the output voltage for this configuration is given by:

$$V_o = -\frac{Z_f}{Z_1} V_1 \qquad (19.1)$$

The function performed by the circuit is determined by the specific elements used for the feedback and input impedances. To minimize the error in the output voltage

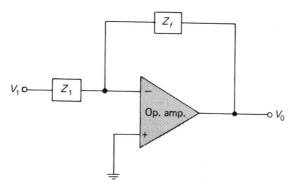

Figure 19.1 General negative feedback circuit using the inverting input.

caused by input offset current, the noninverting input should be returned to ground through a resistance equal to the parallel combination of Z_f and Z_1 (if these are resistive).

Figure 19.2 depicts the basic noninverting connection, which has the input signal applied to the noninverting terminal. In this case, the output voltage is given by

$$V_o = \frac{Z_f + Z_1}{Z_1} V_1 \qquad (19.2)$$

These two general circuits may be used to implement a variety of analog functions.

19.1.1 Sign Changer (Inverter)

Among the simplest yet most important analog functions is sign changing. Although we call it inversion, it should not be confused with the function of the digital inverter. The analog inverter is implemented by making the two impedances in Figure 19.1 equal; precision resistors are commonly used, as shown in Figure 19.3. The output voltage then is the negative of the input voltage.

19.1.2 Scaler

In many applications we have to scale or change the amplitude of a signal. This process may involve either increasing or decreasing the amplitude. For the scaler circuit shown in Figure 19.4, the *scale factor K* is negative, given by $-R_f/R_1$, and may be either greater than or less than 1.

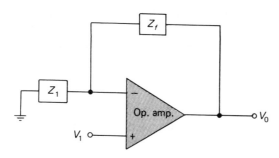

Figure 19.2 General negative feedback circuit using the noninverting input.

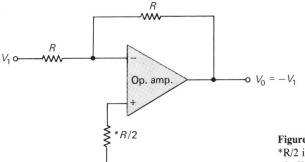

Figure 19.3 Sign changer (inverter). *R/2 is included to minimize the effect of offset current.

The noninverting scaler circuit is illustrated in Figure 19.5. Note that this circuit can only scale up; that is, the output voltage is always larger than the input voltage.

Precision resistors are used in the scaler circuits to improve the accuracy of the scale factor K. A variable scale factor can be achieved by using a calibrated (usually 5 or 10 turn) potentiometer for either R_1 or R_f.

19.1.3 Summing Amplifier (Adder)

Figure 19.6 shows the OP AMP connected as a summing amplifier. The output voltage is proportional to the sum of the input voltages. Although three inputs are shown, any number may be used by connecting additional input resistors. To see how the addition is performed, remember that the input of the OP AMP is a virtual ground and draws no input current. The current through R_1 is labeled I_1; through R_2, I_2; and through R_3, I_3. These currents are given by

$$I_1 = \frac{V_1}{R_1} \quad I_2 = \frac{V_2}{R_2} \quad I_3 = \frac{V_3}{R_3} \qquad (19.3)$$

Because the input of the OP AMP draws no current, the current through the

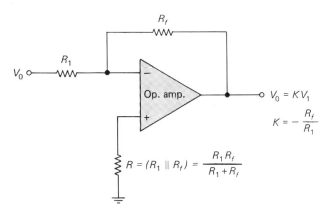

Figure 19.4 Inverting scaler.

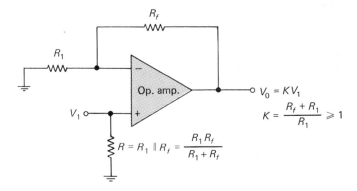

Figure 19.5 Noninverting scaler.

feedback resistor, labeled I_f, is the sum of I_1, I_2, and I_3. Thus,

$$I_f = I_1 + I_2 + I_3 = -\frac{V_o}{R_f} \tag{19.4}$$

Making use of Equation (19.3), we obtain

$$V_o = -(K_1 V_1 + K_2 V_2 + K_3 V_3) \tag{19.5}$$

where we have defined scale factors:

$$K_1 = \frac{R_f}{R_1} \quad K_2 = \frac{R_f}{R_2} \quad K_3 = \frac{R_f}{R_3} \tag{19.6}$$

Note that the output voltage given in Equation (19.5) is proportional to the negative of the sum of the input voltages. If $R_1 = R_2 = R_3 = R_f$, it is obvious that $V_o = -(V_1 + V_2 + V_3)$.

Inverting summing amplifier. The circuit of Figure 19.6 offers flexibility by allowing us to use a different scale factor for each of the voltages to be added. The following example, although not very practical, illustrates the use of different scale factors.

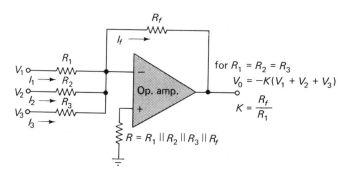

Figure 19.6 Inverting summing amplifier (adder).

Suppose that the three input voltages represent the number of coins; that is, the number of millivolts for V_1 corresponds to the number of quarters, the number of millivolts for V_2 corresponds to the number of dimes, and the number of millivolts for V_3 corresponds to the number of nickels. We can have the output voltage correspond to the number of total coins (regardless of worth) by making all the resistors in Figure 19.6 equal. The total number of coins would then be given by the number of millivolts at the output. (Note that this number would be negative.) If, on the other hand, we wanted to know the net amount of money represented by the coins, we might choose $K_1 = 25$, $K_2 = 10$, and $K_3 = 5$. Thus, the output would give us an amount (in cents) totaling the worth of all the coins represented by the number of millivolts for V_o. If we wanted an output where each millivolt represents a dollar, we would use scale factors $K_1 = 0.25$, $K_2 = 0.10$, and $K_3 = 0.05$.

Noninverting summing amplifier. The noninverting summing amplifier circuit is shown in Figure 19.7. With all the input resistors equal, the scale factor K is set by the values of R_1 and R_f, as given by Equation (19.2). The output voltage is

$$V_o = K(V_1 + V_2 + V_3) \tag{19.7}$$

In this case, the output voltage is proportional to the sum of the input voltages and is not negative (as happened with the circuit of Fig. 19.6).

19.1.4 Difference Amplifier (Subtractor)

The circuit whose output voltage is proportional to the difference of the two input voltages is depicted in Figure 19.8. The scale factor K is once again set by R_1

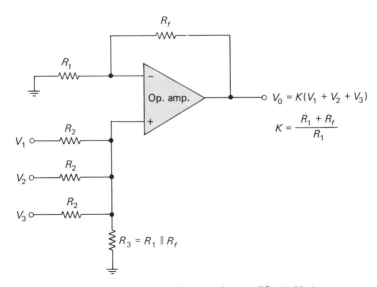

Figure 19.7 Noninverting summing amplifier (adder).

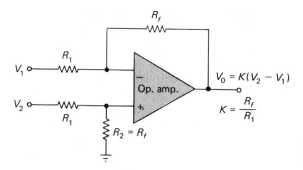

Figure 19.8 Difference amplifier (subtractor).

and R_f, and the output voltage is given by

$$V_o = K(V_2 - V_1) \tag{19.8}$$

The circuit of Figure 19.8 is the basic differential configuration for the OP AMP and is often used in instrumentation applications. For example, we may use it to replace a microammeter in a bridge circuit, as shown in Figure 19.9. Any imbalance in the bridge circuit is reflected as a voltage between points A and B and is amplified by the difference amplifier. When the bridge is balanced, $R = R_b R_x / R_a$, and the output of the difference amplifier goes to zero. As the bridge is brought closer to balance, R_f should be increased to provide higher gain and therefore better sensitivity.

19.1.5 Integrator

The basic OP AMP integrator circuit is illustrated in Figure 19.10. In this case, the feedback impedance is a capacitor whose impedance in operational form is

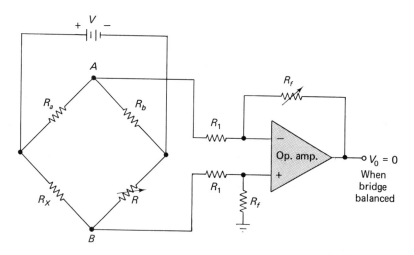

Figure 19.9 Difference amplifier used to indicate balance in the bridge.

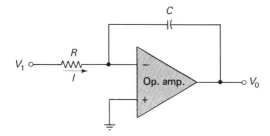

Figure 19.10 Integrator circuit.

$1/sC$. The output voltage is given by

$$V_o = -\frac{1}{RCs} V_1 \qquad (19.9)$$

where both V_1 and V_o are the functions of s, the complex *frequency domain* variable. Network theory tells us that dividing by s in the frequency domain corresponds to integrating in the time domain. Thus, the output voltage is proportional to the input voltage divided by s in the frequency domain, and so the circuit performs integration on the input voltage. To fully understand this point, assume that the input voltage is zero and then abruptly rises to a voltage V, as shown in Figure 19.11. While the input is at zero, the output is also at zero. We assume there is no initial charge on the capacitor. When the input becomes V, the voltage across the capacitor cannot change instantaneously, so that the output voltage builds up as the capacitor charges up. The current through R is the current charging the capacitor. The input side of R is at V; the other side is connected to the inverting input of the OP AMP, which is essentially at ground (i.e., virtual ground). Therefore, the current through R is constant throughout the charging period and given by V/R. For a capacitor, the rate at which the terminal voltage changes is directly proportional to the charging current. Thus, if the charging current is constant, the rate of change in the capacitor terminal voltage is also

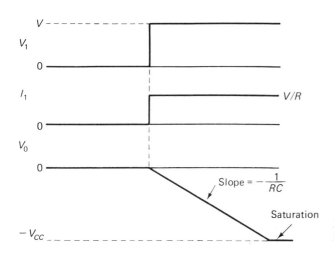

Figure 19.11 Waveshapes for the integrator circuit.

Sec. 19.1 Principles of Analog Computation

constant. The capacitor is charging to provide a *linear* increase in output voltage inversely proportional to the circuit time constant RC.

There is another way in which we can approach the operation of the integrator circuit of Figure 19.10. One end of the capacitor is permanently held at ground by the inverting input of the OP AMP. If it were not for the capacitor, once the input is applied the output would try to reach a voltage that is essentially the input voltage multiplied by the difference gain of the amplifier. Obviously, this voltage is extremely large. For example, a 1 V input with a typical difference gain of 200,000 would mean that the output is trying to reach $-200,000$ V. A practical OP AMP is limited to operating between its negative and positive supply voltages. However, as far as the capacitor is concerned, it is trying to charge up to this extremely large voltage. The fact that it can never reach it is unimportant. The important fact is that during the first fractions of time constant, the charging of a capacitor is linear. Thus, the OP AMP has the effect of providing a linear charging characteristic.

When we integrate mathematically, we are finding (between specified limits) the area under the curve of the function we are integrating. The circuit shown in Figure 19.10 can be used to integrate a function of time over a specified interval. The output voltage provides a plot of the area under the curve of the input waveshape in Figure 19.11. (Note: The output is actually the *negative* of the integral of the input.) When the output voltage reaches the supply voltage V_{CC}, the OP AMP is said to be saturated, and the output voltage cannot increase further. At this point, the circuit ceases to function as an integrator.

Discharging the integrator capacitor. If the input voltage is a succession of voltage pulses, the output waveshape becomes a linear triangle wave. If we wish to integrate a repetitive function, the capacitor must be discharged to zero at the end of each cycle. This discharge is accomplished by placing some form of a transistor (usually an FET) switch in parallel with the capacitor, as shown in Figure 19.12. The switch is normally open during the integration interval. At the end of the integration, the gate of the switch is pulsed, causing the transistor to saturate and discharge the capacitor through its very low resistance. A typical set of waveshapes is given in Figure 19.13. The circuit operates as an integrator

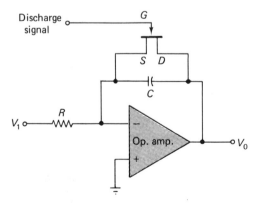

Figure 19.12 The FET provides a discharge path in the integrator circuit.

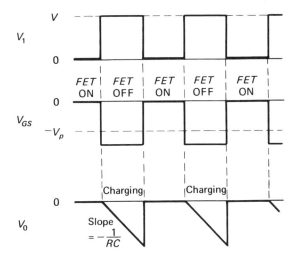

Figure 19.13 Waveshapes for the circuit in Figure 19.12.

so long as the FET is off; that is, its gate-source junction is biased beyond pinchoff. The capacitor is discharged when the gate-source junction is biased above pinchoff.

In certain cases, it is necessary to start the integration with certain initial values, which are introduced as an initial charge (supplied by a dc power source) on the capacitor.

19.1.6 Differentiator

A useful but less common circuit is the *differentiator* circuit, illustrated in Figure 19.14. The resistor and capacitor of the integrator circuit has been interchanged. The input impedance is now $1/Cs$, so that the output voltage is given by

$$V_o = -RCsV_1 \tag{19.10}$$

where both V_1 and V_o are functions of the complex frequency domain variable s. Multiplication by s in the frequency domain corresponds to differentiation in the time domain. Consequently, the output voltage (as a function of time) is the negative of the differentiated input voltage.

Figure 19.15 shows that the output waveshape for a simple case of a ramp input voltage is a step function. Circuit operation is as follows. While no input is

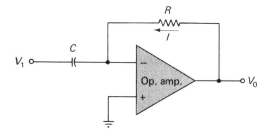

Figure 19.14 Differentiator circuit.

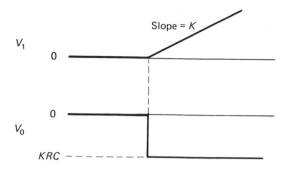

Figure 19.15 Differentiator waveshapes.

applied, the capacitor is uncharged and the output is zero. When the input begins to increase at a linear rate (i.e., slope = constant), the current through the capacitor must be constant. It is supplied through R, one end of which is at virtual ground (the inverting input of the OP AMP) and the other end at the output. Because the current through R is constant, the voltage drop across R must also be constant. This voltage drop is equal to the output voltage.

If repetitive waveshapes are to be differentiated, the capacitor must be discharged at the end of each cycle. Discharge may be accomplished with the use of an FET switch using a procedure similar to that shown in Figure 19.12 for the integrator; the source of the FET would be connected to the end of the capacitor at the inverting input.

19.2 ASSORTED ANALOG CIRCUITS

In this section we shall take up a few assorted analog circuits that do not rightfully belong under the classification of analog computer circuits but that are useful enough to warrant our attention.

19.2.1 Voltage-to-Current Converter

In many cases a voltage signal needs to be converted into a proportional current signal. We can perform this conversion by using the circuit illustrated in Figure 19.16. The load is placed in the feedback slot. This converter is especially useful when it is necessary to have both ends of the load isolated from ground.

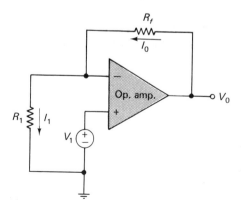

Figure 19.16 Voltage-to-current converter.

The circuit is nothing more than a noninverting amplifier. Because the OP AMP draws no input current, I_o and I_1 must be equal. Moreover, the differential input voltage for the OP AMP must be zero, so $V_1 = I_1 R_1 = I_o R_1$. Thus,

$$I_o = \frac{V_1}{R_1} \qquad (19.11)$$

The output current is directly proportional to the input voltage. Note that for a given V_1 the output current is independent of the load resistance R_f; therefore, the magnitude of the output current can be adjusted by changing R_1.

19.2.2 Current-to-Voltage Converter

The circuit shown in Figure 19.17 provides an output voltage that is proportional to the input current. It is useful in applications where the source provides an extremely low voltage under load conditions, for example, to amplify the output of a photodiode.

Because of the virtual ground at the input of the OP AMP, V_1 is zero, and the source resistance R_1 draws no current. All of the input current flows through R_f, causing an output voltage. Note that the output voltage is not a function of the source resistance R_1; and, for a given I_1, the output voltage may be varied by simply changing R_f.

We can use this circuit to amplify very low currents. However, for extremely low currents, the OP AMP input bias current causes an error; that is, the actual circuit would have a part of the input current flowing in R_1 and into the input of the OP AMP. Both of these currents are negligible for input currents sufficiently larger than the OP AMP input bias current of 80 nA for the bipolar 741 and 50 pA for the JFET input OP AMP (LF351).

19.2.3 Voltage Follower (Buffer)

In some instances it is necessary to amplify a voltage from a source that has a very high source resistance. The simple circuit shown in Figure 19.18 may be used in such cases.

Because the output is tied to the inverting input, which is *virtually* tied to

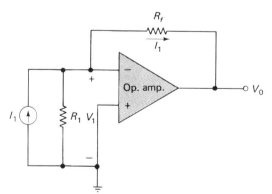

Figure 19.17 Current-to-voltage converter.

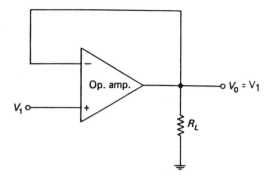

Figure 19.18 Voltage follower used to buffer or transform impedance levels.

the noninverting input, the output voltage follows (or is equal to) the input voltage. OP AMPs are specifically designed for such applications. They have a very high input impedance and an extremely low input current drain. An example is the LF351 JFET input OP AMP, which has a typical input bias current of 50 pA (50 × 10^{-12} A) and a typical input resistance of 1 TΩ (10^{12} Ω). Therefore, the noninverting input of the LF351 offers a very high input impedance to the source, while the output offers a low impedance (≤50 Ω) to the load. You may find the complete specifications of the LF351 in Appendix A.

19.3 APPLICATION OF ANALOG SYSTEMS

Analog computation is useful in problems where we need to evaluate (1) the effect of varying one or more parameters and (2) the solution to specific values. For example, let us again determine the net worth of a given number of quarters, dimes, and nickels. The circuit for solving this problem is given in Figure 19.19. Each coin to be counted corresponds to (i.e., is analogous to), say, 1 millivolt at one of the input potentiometers. The number of quarters is represented by the

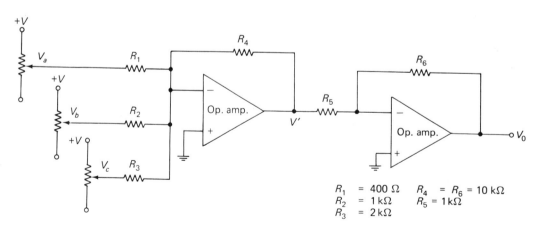

R_1 = 400 Ω R_4 = R_6 = 10 kΩ
R_2 = 1 kΩ R_5 = 1 kΩ
R_3 = 2 kΩ

Figure 19.19 Example for computing the worth of a given number of coins.

number of millivolts at the input to $R_1(V_a)$. The number of dimes is represented by the number of millivolts at the input to $R_2(V_b)$. The number of nickels is represented by the number of millivolts applied to the input of $R_3(V_c)$. For the resistor values shown, voltage V' is given by

$$V' = -(25V_a + 10V_b + 5V_c) \quad (19.12)$$

where $K_a = 10\text{ k}/0.4\text{ k} = 25$, $K_b = 10\text{ k}/1\text{ k} = 10$, and $K_c = 10\text{ k}/2\text{ k} = 5$.

The final output voltage is

$$V_o = 250V_a + 100V_b + 50V_c \quad (19.13)$$

where the scale factor for the second amplifier is $10\text{ k}/1\text{ k} = 10$.

Assume that the input voltages are set to represent 16 quarters, 21 dimes, and 7 nickels; that is, $V_a = 16\text{ mV}$, $V_b = 21\text{ mV}$, and $V_c = 7\text{ mV}$. The output voltage for this case would be

$$V_o = (250)(16) + (100)(21) + (50)(7) = 6450\text{ mV} = 6.45\text{ V}$$

The answer in volts corresponds to the sum of the coins in dollars, so that 16 quarters, 21 dimes, and 7 nickels equal $6.45.

You can see that the real power of this method is the ease of changing the numbers of the coins involved by simply resetting the input voltages. Although we are not suggesting that this problem is best solved by analog computer methods, it is a simple problem that easily illustrates these methods.

As another example, consider the solution of simultaneous equations. Assume that two items costing X and Y dollars are purchased. The first time, three of one item and four of the other are purchased for a net cost of $0.95. The second time, one of each item is purchased for a net cost of $0.30. This is expressed in equation form as

$$3X + 4Y = 0.95 \quad (19.14)$$
$$X + Y = 0.30$$

Let us use the analogy that each dollar is equivalent to a volt. We can rearrange the equations, solving one for X and the other for Y:

$$Y = 0.2375 - 0.75X \quad (19.15)$$
$$X = 0.30 - Y$$

These two equations are implemented as indicated in Figure 19.20. The output voltage of OP AMP 1 ($OA1$) is analogous to X and the output voltage of $OA2$ is analogous to Y. Verification of the answers $X = 0.25$ and $Y = 0.05$ is left as an exercise.

As a more practical example, suppose that we want to examine the response (current waveshape) of the series RLC circuit shown in Figure 19.21, for (1) different excitations (input voltages), (2) different initial conditions, and (3) different circuit parameters. One obvious way would be to connect the circuit in the laboratory and perform the tests on the actual circuit. The testing is easily done

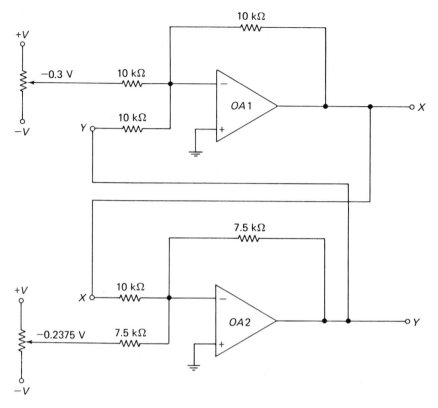

Figure 19.20 Solution of simultaneous equations.

with simple circuits, but the problems involved with performing experiments on actual complicated circuits should be obvious. However, with an analog computer we can easily simulate the operation of this simple circuit as well as more complicated circuits.

The voltage applied to the circuit of Figure 19.21 is given by

$$V = IR + \int \frac{1}{C} I \, dt + L \frac{dI}{dt} \tag{19.16}$$

To simplify matters, we can use a variable other than the current, namely, the charge (Q) in the circuit. Remember that current is the rate of flow of charge or

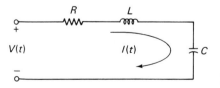

Figure 19.21 Series RLC circuit.

that the charge is the integral of the current. Thus,

$$V = R\frac{dQ}{dt} + \frac{Q}{C} + L\frac{d^2Q}{dt^2} \tag{19.17}$$

To implement this second order differential equation on an analog computer, we first solve for the highest derivative of the variable Q.

$$-\frac{d^2Q}{dt^2} = -\frac{V}{L} + \frac{Q}{LC} + \frac{R}{L}\frac{dQ}{dt} \tag{19.18}$$

The implementation of this equation on an analog computer is shown in Figure 19.22. The inputs to the inverting or summing amplifier $OA1$ are the functions on the righthand side of Equation (19.18). The output of $OA1$ then must be equal to the lefthand side of Equation (19.18), namely, $-d^2Q/dt^2$. This signal is integrated by $OA2$ to provide an output dQ/dt. This signal is again integrated by $OA3$ to provide $-Q$ and also multiplied by a -1 in inverter $OA4$.

We can now obtain the circuit response for any excitation voltage V, which

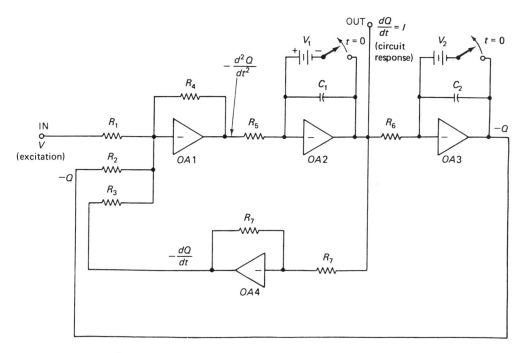

Analogies $\frac{R_4}{R_1} = \frac{1}{L}$; $\frac{R_4}{R_2} = \frac{1}{LC}$; $\frac{R_4}{R_3} = \frac{R}{L}$; $\frac{1}{R_5C_1} = \frac{1}{R_6C_2} = 1$;

$V_1 = I_0$ (current in L at $t = 0$) ; $V_2 = Q_0$ (charge on C at $t = 0$)

Figure 19.22 Analog computer simulation for a series RLC circuit.

is derived from a function generator. It may be obtained for any given set of initial conditions—provided by voltages V_1 and V_2—and for any set of circuit values—provided by the scaling resistors R_1, R_2, R_3, and R_4. The current is the circuit response. Its analog is obtained at the output of $OA2$ (because $I = dQ/dt$). The analog may be displayed on an oscilloscope or recorded on an $x - y$ plotter or strip-chart recorder.

The three examples just discussed are by no means the only problems that can be handled on an analog computer. They simply illustrate the basic operation of analog systems, which are used to simulate very complex problems. If you are interested in analog computer programming we invite you to consult one of the many books on the subject.

REVIEW QUESTIONS

1. What is the basic building block in an analog system?
2. What must be the conditions in the circuit of Fig. 19.3 for the circuit to change the sign of the input signal?
3. What is a scaler?
4. What are the uses of a scaler? Illustrate.
5. What is an adder? What are the requirements if the output is to be truly equal to the sum of the inputs?
6. What is a subtractor?
7. What is an integrator?
8. If the input to an integrator is a dc voltage, what is the output?
9. What are the limitations of a practical OP AMP integrator?
10. What is a differentiator?
11. If the input to a differentiator is a voltage increasing linearly with time, what is the output?
12. How is a voltage signal converted into a proportional current signal?
13. Why is it not enough to simply connect a resistor in series with the voltage signal to produce a current signal?
14. How is a current signal converted into a proportional voltage signal?
15. Why is it not enough to simply connect a resistor in parallel with the current signal to convert to a proportional voltage signal?
16. What is a buffer?
17. What are some applications for a voltage follower?
18. Give some concrete examples of the application of analog systems.
19. For what type of problems is the analog computer best suited?
20. If the OP AMPs used in the example of Fig. 19.19 have (output) saturation voltages of $+20$ V and -20 V, what is the highest count that the system can accomplish? Why?

Appendix A

Manufacturers' Data Sheets

This appendix contains a compilation of manufacturers' data sheets for many of the devices discussed in the text. These data sheets were selected as typical of the types of devices readily available rather than as superior in performance characteristics. The authors wish to convey their gratitude to the manufacturers who made these data sheets available.

Device	Description	Courtesy of
1N4001—1N4007	Rectifier	Motorola
VS048X—VS648X	Epoxy bridge rectifier	Varo Semiconductor
1N5221—1N5272	Zener diode	Motorola
1N5139—1N5148	Varactor diode	Motorola
MBD501—MBD701	Schottky barrier diode	Motorola
TIP29	Silicon power transistor	Texas Instruments
2N2218—2N2222	*NPN* silicon transistor	Motorola
2N5457—2N5459	*N*-channel JFET Type A	Motorola
2N3796—2N3797	*N*-channel MOSFET Type B	Motorola
2N4351	*N*-channel MOSFET Type C	Motorola
1N5758—1N5762	Silicon bilateral trigger	Motorola
MBS4991	Bidirectional switch (SBS)	Motorola
2N6027	Programmable unijunction	Motorola
2N5060—2N5064	Silicon controlled rectifier	Motorola
2N6068—2N6075	Triac	Motorola
MFOD2405	Integrated detector/preamp	Motorola
4N29—4N33	Optocoupler/isolator	Motorola
LM741	Operational amplifier	National Semiconductor
LF351	JFET input OP AMP	National Semiconductor
MC1723	Voltage regulator	Motorola
LM117/LM317	Adjustable positive voltage regulator	Motorola
191	Heat sink	Aham/Thor

1N4001 thru 1N4007

LEAD MOUNTED SILICON RECTIFIERS

50-1000 VOLTS
DIFFUSED JUNCTION

Designers Data Sheet

"SURMETIC"▲ RECTIFIERS

... subminiature size, axial lead mounted rectifiers for general-purpose low-power applications.

Designers Data for "Worst Case" Conditions
The Designers▲ Data Sheets permit the design of most circuits entirely from the information presented. Limit curves — representing boundaries on device characteristics — are given to facilitate "worst case" design.

*MAXIMUM RATINGS

Rating	Symbol	1N4001	1N4002	1N4003	1N4004	1N4005	1N4006	1N4007	Unit
Peak Repetitive Reverse Voltage Working Peak Reverse Voltage DC Blocking Voltage	V_{RRM} V_{RWM} V_R	50	100	200	400	600	800	1000	Volts
Non-Repetitive Peak Reverse Voltage (halfwave, single phase, 60 Hz)	V_{RSM}	60	120	240	480	720	1000	1200	Volts
RMS Reverse Voltage	$V_{R(RMS)}$	35	70	140	280	420	560	700	Volts
Average Rectified Forward Current (single phase, resistive load, 60 Hz, see Figure 8, $T_A = 75°C$)	I_O	1.0							Amp
Non-Repetitive Peak Surge Current (surge applied at rated load conditions, see Figure 2)	I_{FSM}	30 (for 1 cycle)							Amp
Operating and Storage Junction Temperature Range	T_J, T_{stg}	−65 to +175							°C

*ELECTRICAL CHARACTERISTICS

Characteristic and Conditions	Symbol	Typ	Max	Unit
Maximum Instantaneous Forward Voltage Drop ($i_F = 1.0$ Amp, $T_J = 25°C$) Figure 1	v_F	0.93	1.1	Volts
Maximum Full-Cycle Average Forward Voltage Drop ($I_O = 1.0$ Amp, $T_L = 75°C$, 1 inch leads)	$V_{F(AV)}$	—	0.8	Volts
Maximum Reverse Current (rated dc voltage) $T_J = 25°C$ $T_J = 100°C$	I_R	0.05 1.0	10 50	µA
Maximum Full-Cycle Average Reverse Current ($I_O = 1.0$ Amp, $T_L = 75°C$, 1 inch leads)	$I_{R(AV)}$	—	30	µA

* Indicates JEDEC Registered Data.

MECHANICAL CHARACTERISTICS

CASE: Transfer Molded Plastic
MAXIMUM LEAD TEMPERATURE FOR SOLDERING PURPOSES: 350°C, 3/8" from case for 10 seconds at 5 lbs. tension
FINISH: All external surfaces are corrosion-resistant, leads are readily solderable
POLARITY: Cathode indicated by color band
WEIGHT: 0.40 Grams (approximately)

▲Trademark of Motorola Inc.

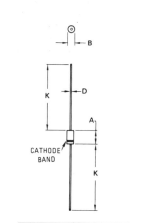

	MILLIMETERS		INCHES	
DIM	MIN	MAX	MIN	MAX
A	5.97	6.60	0.235	0.260
B	2.79	3.05	0.110	0.120
D	0.76	0.86	0.030	0.034
K	27.94	—	1.100	—

CASE 59-04
Does Not Conform to DO-41 Outline.

© MOTOROLA INC. 1982

DS 6015 R3

VARO SEMICONDUCTOR, INC., P.O. BOX 40676
1000 NORTH SHILOH, GARLAND, TEXAS 75040
(214) 271-8511 TWX 910-860-5178

DLS 044

EBR 2 Amp Fast Recovery Time Epoxy Bridge Rectifiers
January 1980

200 Nanosecond Maximum Reserve Recovery
50V, 100V, 200V, 400V, and 600V V_{RRM} Ratings
35 Amps Peak One Half Cycle Surge Current

MAXIMUM RATINGS (At T_A=25°C unless otherwise noted)	SYMBOL	VS048X	VS148X	VS248X	VS448X	VS648X	UNITS
DC Blocking Voltage, Working Peak Reverse Voltage, Peak Repetitive Reverse Voltage	V_{RM} V_{RWM} V_{RRM}	50	100	200	400	600	Volts
RMS Reverse Voltage	$V_{R(RMS)}$	35	70	140	280	420	Volts
Peak Surge Current, ½ Cycle at 60 Hz, (Non-Rep) and T_A = 45°C (Fig. 2)	I_{FSM}	35					Amps
Peak Surge Current, 1 sec. at 60 Hz and T_A - 45°C (Fig. 2)	I_{FRM}	6					Amps
DC Forward Current at T_A = 45°C, (Fig. 1)	I_O	2					Amps
Junction Operating and Storage Temperature Range	T_J, T_{STG}	−50 to +135					°C
Maximum Soldering Temperature & Time		10 Seconds at 265°C					

ELECTRICAL CHARACTERISTICS (At T_A=25°C unless otherwise noted)	SYMBOL		UNITS
Maximum Instantaneous Forward Voltage Drop (Per Diode) at 2 Amps (Fig. 3)	V_{FM}	1.5	Volts/Leg
Maximum Reverse Recovery Time, I_F = 1 Amp, I_R = 2 Amps (Fig. 6)	t_{rr}	200	nsec
Maximum Reverse Current at Rated V_{RM} at T_J = 40°C (Fig. 4)	I_{RM}	10	μA
Maximum Reverse Current at Rated V_{RM} at T_J = 135°C (Fig.4)	I_{RM}	4	mA
Insulation Strength From Circuit to Case (min.)		2000	Volts DC

Part Nos. VS048X, VS148X, VS248X, VS448X and VS648X have been recognized under the Component Program of Underwriters Laboratories, Inc.

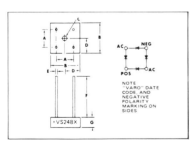

LT	INCHES	MILLIMETERS
A	.411 − .441	10.44 − 11.20
B	.590 − .610	14.99 − 15.49
C	.137 − .167 Dia.	3.48 − 4.24 Dia.
D	.295 − .305	7.49 − 7.75
E	.037 − .043 Dia.	.94 − 1.09 Dia.
F	1.0 Min.	25.4 Min.
G	.195 − .205	4.95 − 5.21

App. A   Manufacturers' Data Sheets

MOTOROLA SEMICONDUCTORS
P.O. BOX 20912 • PHOENIX, ARIZONA 85036

Designers Data Sheet

1N5221 thru 1N5272

500 MILLIWATT HERMETICALLY SEALED GLASS SILICON ZENER DIODES

- Complete Voltage Range — 2.4 to 110 Volts**
- DO-35 Package — Smaller than Conventional DO-7 Package
- Double Slug Type Construction
- Metallurgically Bonded Construction
- Nitride Passivated Die

Designer's Data for "Worst Case" Conditions

The Designer's Data sheets permit the design of most circuits entirely from the information presented. Limit curves — representing boundaries on device characteristics — are given to facilitate "worst case" design.

GLASS ZENER DIODES
500 MILLIWATTS
2.4–110 VOLTS

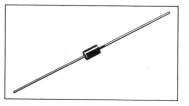

*MAXIMUM RATINGS

Rating	Symbol	Value	Unit
DC Power Dissipation @ $T_L \leq 75°C$ Lead Length = 3/8" Derate above $T_L = 75°C$	P_D	500 4.0	mW mW/°C
Operating and Storage Junction Temperature Range	T_J, T_{stg}	−65 to +200	°C

*Indicates JEDEC Registered Data
**See 1N5273 thru 1N5281 for devices > 110 volts.

MECHANICAL CHARACTERISTICS

CASE: Double slug type, hermetically sealed glass

MAXIMUM LEAD TEMPERATURE FOR SOLDERING PURPOSES: 230°C, 1/16" from case for 10 seconds

FINISH: All external surfaces are corrosion resistant with readily solderable leads

POLARITY: Cathode indicated by color band. When operated in zener mode, cathode will be positive with respect to anode

MOUNTING POSITION: Any

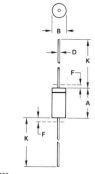

NOTES:
1. PACKAGE CONTOUR OPTIONAL WITHIN A AND B. HEAT SLUGS, IF ANY, SHALL BE INCLUDED WITHIN THIS CYLINDER, BUT NOT SUBJECT TO THE MINIMUM LIMIT OF B.
2. LEAD DIAMETER NOT CONTROLLED IN ZONE F TO ALLOW FOR FLASH, LEAD FINISH BUILDUP AND MINOR IRREGU-LARITIES OTHER THAN HEAT SLUGS.
3. POLARITY DENOTED BY CATHODE BAND.
4. DIMENSIONING AND TOLERANCING PER ANSI Y14.5, 1973.

DIM	MILLIMETERS		INCHES	
	MIN	MAX	MIN	MAX
A	3.05	5.08	0.120	0.200
B	1.52	2.29	0.060	0.090
D	0.46	0.56	0.018	0.022
F	—	1.27	—	0.050
K	25.40	38.10	1.000	1.500

All JEDEC dimensions and notes apply.
CASE 299-02
DO-204AH
(DO-35)

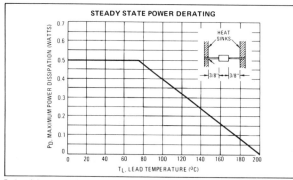

STEADY STATE POWER DERATING

Designer's is a trademark of Motorola Inc.

© MOTOROLA INC., 1982

DS 7051R1

ELECTRICAL CHARACTERISTICS

(T_A = 25°C unless otherwise noted. Based on dc measurements at thermal equilibrium; lead length = 3/8"; thermal resistance of heat sink = 30°C/W) V_F = 1.1 max @ I_F = 200 mA for all types.

JEDEC Type No. (Note 1)	Nominal Zener Voltage V_Z @ I_{ZT} Volts (Note 2)	Test Current I_{ZT} mA	Max Zener Impedance A and B Suffix only		Max Reverse Leakage Current				Max Zener Voltage Temperature Coeff. (A and B Suffix only) θ_{VZ} (%/°C) (Note 3)
					A and B Suffix only			Non-Suffix	
			Z_{ZT} @ I_{ZT} Ohms	Z_{ZK} @ I_{ZK} = 0.25 mA Ohms	I_R µA	@	V_R Volts	I_R @ V_R Used for Suffix A µA	
							A	B	
1N5221	2.4	20	30	1200	100		0.95	1.0	200	−0.085
1N5222	2.5	20	30	1250	100		0.95	1.0	200	−0.085
1N5223	2.7	20	30	1300	75		0.95	1.0	150	−0.080
1N5224	2.8	20	30	1400	75		0.95	1.0	150	−0.080
1N5225	3.0	20	29	1600	50		0.95	1.0	100	−0.075
1N5226	3.3	20	28	1600	25		0.95	1.0	100	−0.070
1N5227	3.6	20	24	1700	15		0.95	1.0	100	−0.065
1N5228	3.9	20	23	1900	10		0.95	1.0	75	−0.060
1N5229	4.3	20	22	2000	5.0		0.95	1.0	50	±0.055
1N5230	4.7	20	19	1900	5.0		1.9	2.0	50	±0.030
1N5231	5.1	20	17	1600	5.0		1.9	2.0	50	±0.030
1N5232	5.6	20	11	1600	5.0		2.9	3.0	50	+0.038
1N5233	6.0	20	7.0	1600	5.0		3.3	3.5	50	+0.038
1N5234	6.2	20	7.0	1000	5.0		3.8	4.0	50	+0.045
1N5235	6.8	20	5.0	750	3.0		4.8	5.0	30	+0.050
1N5236	7.5	20	6.0	500	3.0		5.7	6.0	30	+0.058
1N5237	8.2	20	8.0	500	3.0		6.2	6.5	30	+0.062
1N5238	8.7	20	8.0	600	3.0		6.2	6.5	30	+0.065
1N5239	9.1	20	10	600	3.0		6.7	7.0	30	+0.068
1N5240	10	20	17	600	3.0		7.6	8.0	30	+0.075
1N5241	11	20	22	600	2.0		8.0	8.4	30	+0.076
1N5242	12	20	30	600	1.0		8.7	9.1	10	+0.077
1N5243	13	9.5	13	600	0.5		9.4	9.9	10	+0.079
1N5244	14	9.0	15	600	0.1		9.5	10	10	+0.082
1N5245	15	8.5	16	600	0.1		10.5	11	10	+0.082
1N5246	16	7.8	17	600	0.1		11.4	12	10	+0.083
1N5247	17	7.4	19	600	0.1		12.4	13	10	+0.084
1N5248	18	7.0	21	600	0.1		13.3	14	10	+0.085
1N5249	19	6.6	23	600	0.1		13.3	14	10	+0.086
1N5250	20	6.2	25	600	0.1		14.3	15	10	+0.086
1N5251	22	5.6	29	600	0.1		16.2	17	10	+0.087
1N5252	24	5.2	33	600	0.1		17.1	18	10	+0.088
1N5253	25	5.0	35	600	0.1		18.1	19	10	+0.089
1N5254	27	4.6	41	600	0.1		20	21	10	+0.090
1N5255	28	4.5	44	600	0.1		20	21	10	+0.091
1N5256	30	4.2	49	600	0.1		22	23	10	+0.091
1N5257	33	3.8	58	700	0.1		24	25	10	+0.092
1N5258	36	3.4	70	700	0.1		26	27	10	+0.093
1N5259	39	3.2	80	800	0.1		29	30	10	+0.094
1N5260	43	3.0	93	900	0.1		31	33	10	+0.095
1N5261	47	2.7	105	1000	0.1		34	36	10	+0.095
1N5262	51	2.5	125	1100	0.1		37	39	10	+0.096
1N5263	56	2.2	150	1300	0.1		41	43	10	+0.096
1N5264	60	2.1	170	1400	0.1		44	46	10	+0.097
1N5265	62	2.0	185	1400	0.1		45	47	10	+0.097
1N5266	68	1.8	230	1600	0.1		49	52	10	+0.097
1N5267	75	1.7	270	1700	0.1		53	56	10	+0.098
1N5268	82	1.5	330	2000	0.1		59	62	10	+0.098
1N5269	87	1.4	370	2200	0.1		65	68	10	+0.099
1N5270	91	1.4	400	2300	0.1		66	69	10	+0.099
1N5271	100	1.3	500	2600	0.1		72	76	10	+0.110
1N5272	110	1.1	750	3000	0.1		80	84	10	+0.110

NOTE 1. **Tolerance** — The JEDEC type numbers shown indicate a tolerance of ±10% with guaranteed limits on only V_Z, I_R and V_F as shown in the electrical characteristics table. Units with guaranteed limits on all six parameters are indicated by suffix "A" for ±10% tolerance and suffix "B" for ±5.0% units.

†For more information on special selections contact your nearest Motorola representative.

NOTE 2. **Special Selections†** Available Include:
1. Nominal zener voltages between those shown.
2. Two or more units for series connection with specified tolerance on total voltage. Series matched sets make zener voltages in excess of 200 volts possible as well as providing lower temperature coefficients, lower dynamic impedance and greater power handling ability.
3. Nominal voltages at non-standard test currents.

 MOTOROLA Semiconductor Products Inc.

1N5139,A thru 1N5148,A (SILICON)

6.8-47 pF EPICAP VOLTAGE-VARIABLE CAPACITANCE DIODES

CASE 51
(DO-7)
Polarity band on cathode end

Silicon voltage-variable capacitance diodes, designed for electronic tuning and harmonic-generation applications, and providing solid-state reliability to replace mechanical tuning methods.

MAXIMUM RATINGS (T_C = 25°C unless otherwise noted)

Rating	Symbol	Value	Unit
Reverse Voltage	V_R	60	Vdc
Forward Current	I_F	250	mAdc
RF Power Input †	P_{in} †	5.0	Watts
Device Dissipation @ T_A = 25°C Derate above 25°C	P_D	400 2.67	mW mW/°C
Device Dissipation @ T_C = 25°C Derate above 25°C	P_C	2.0 13.3	Watts mW/°C
Junction Temperature	T_J	+175	°C
Storage Temperature Range	T_{stg}	-65 to +200	°C

†The RF power input rating assumes that an adequate heat sink is provided.

ELECTRICAL CHARACTERISTICS (T_A = 25°C unless otherwise noted)

Characteristic – All Types	Test Conditions	Symbol	Min	Typ	Max	Unit
Reverse Breakdown Voltage	I_R = 10 μAdc	B_{VR}	60	70	—	Vdc
Reverse Voltage Leakage Current	V_R = 55 Vdc, T_A = 25°C V_R = 55 Vdc, T_A = 150°C	I_R	— —	— —	0.02 20	μAdc
Series Inductance	f = 250 MHz, L ≈ 1/16"	L_S	—	5.0	—	nH
Case Capacitance	f = 1 MHz, L ≈ 1/16"	C_C	—	0.25	—	pF
Diode Capacitance Temperature Coefficient	V_R = 4 Vdc, f = 1 MHz	TC_C	—	200	300	ppm/°C

Device	C_T, Diode Capacitance V_R = 4 Vdc, f = 1 MHz pF			Q, Figure of Merit V_R = 4 Vdc, f = 50 MHz	α V_R = 4 Vdc, f = 1 MHz		TR, Tuning Ratio C_4/C_{60} f = 1 MHz	
	Min	Typ	Max	Min	Min	Typ	Min	Typ
1N5139	6.1	6.8	7.5	350	0.37	0.40	2.7	2.9
1N5139A	6.5	6.8	7.1	350	0.37	0.40	2.7	2.9
1N5140	9.0	10.0	11.0	300	0.38	0.41	2.8	3.0
1N5140A	9.5	10.0	10.5	300	0.38	0.41	2.8	3.0
1N5141	10.8	12.0	13.2	300	0.38	0.41	2.8	3.0
1N5141A	11.4	12.0	12.6	300	0.38	0.41	2.8	3.0
1N5142	13.5	15.0	16.5	250	0.38	0.41	2.8	3.0
1N5142A	14.3	15.0	15.7	250	0.38	0.41	2.8	3.0
1N5143	16.2	18.0	19.8	250	0.38	0.41	2.8	3.0
1N5143A	17.1	18.0	18.9	250	0.38	0.41	2.8	3.0
1N5144	19.8	22.0	24.2	200	0.43	0.45	3.2	3.4
1N5144A	20.9	22.0	23.1	200	0.43	0.45	3.2	3.4
1N5145	24.3	27.0	29.7	200	0.43	0.45	3.2	3.4
1N5145A	25.7	27.0	28.3	200	0.43	0.45	3.2	3.4
1N5146	29.7	33.0	36.3	200	0.43	0.45	3.2	3.4
1N5146A	31.4	33.0	34.6	200	0.43	0.45	3.2	3.4
1N5147	36.1	39.0	42.9	200	0.43	0.45	3.2	3.4
1N5147A	37.1	39.0	40.9	200	0.43	0.45	3.2	3.4
1N5148	42.3	47.0	51.7	200	0.43	0.45	3.2	3.4
1N5148A	44.7	47.0	49.3	200	0.43	0.45	3.2	3.4

MOTOROLA Semiconductor Products Inc.

MBD501
MBD701

SILICON HOT-CARRIER DIODE
(SCHOTTKY BARRIER DIODE)

HIGH-VOLTAGE SILICON HOT-CARRIER DETECTOR AND SWITCHING DIODES
50-70 VOLTS

... designed primarily for high-efficiency UHF and VHF detector applications. Readily adaptable to many other fast switching RF and digital applications. Supplied in an inexpensive plastic package for low-cost, high-volume consumer and industrial/commercial requirements.

- The Schottky Barrier Construction Provides Ultra-Stable Characteristics By Eliminating the "Cat-Whisker" or "S-Bend" Contact
- Extremely Low Minority Carrier Lifetime — 100 ps (Max)
- Very Low Capacitance — 1.0 pF
- High Reverse Voltage — to 70 Volts
- Low Reverse Leakage — 200 nA (Max)

CASE 182-02
TO-92

STYLE 1:
PIN 1. ANODE
2. CATHODE

DIM	MILLIMETERS MIN	MILLIMETERS MAX	INCHES MIN	INCHES MAX
A	4.32	5.33	0.170	0.210
B	4.45	5.21	0.175	0.205
C	3.18	4.19	0.125	0.165
D	0.356	0.533	0.014	0.021
F	0.407	0.482	0.016	0.019
G	1.27 BSC		0.050 BSC	
H	–	1.27	–	0.050
J	2.54 BSC		0.100 BSC	
K	12.70	–	0.500	–
L	6.35	–	0.250	–
N	2.03	2.66	0.080	0.105
P	2.93	–	0.115	–
R	3.43	–	0.135	–

MAXIMUM RATING (T_J = 125°C unless otherwise noted)

Rating	Symbol	Value	Unit
Reverse Voltage	V_R		Volts
MBD501		50	
MBD701		70	
Forward Power Dissipation @ T_A = 25°C	P_F	500	mW
Derate Above 25°C		5.0	mW/°C
Operating Junction Temperature Range	T_J	–55 to +125	°C
Storage Temperature Range	T_{stg}	–65 to +150	°C

ELECTRICAL CHARACTERISTICS (T_A = 25°C unless otherwise noted)

Characteristic	Symbol	Min	Typ	Max	Unit
Reverse Breakdown Voltage (I_R = 10 μAdc)	$V_{(BR)R}$				Volts
MBD501		50	–	–	
MBD701		70	–	–	
Total Capacitance, Figure 1 (V_R = 20 Volts, f = 1.0 MHz)	C_T	–	0.5	1.0	pF
Minority Carrier Lifetime, Figure 2 (I_F = 5.0 mA, Krakauer Method)	τ	–	15	100	ps
Reverse Leakage, Figure 3	I_R				nAdc
(V_R = 25 V) MBD501		–	7.0	200	
(V_R = 35 V) MBD701		–	9.0	200	
Forward Voltage, Figure 4 (I_F = 10 mAdc)	V_F	–	1.0	1.2	Vdc
Series Inductance (f = 250 MHz, Lead Length ≈ 1/16")	L_S	–	6.0	–	nH
Case Capacitance (f = 1.0 MHz, Lead Length ≈ 1/16")	C_C	–	0.18	–	pF

DS 2514

App. A   Manufacturers' Data Sheets

TYPES TIP29, TIP29A, TIP29B, TIP29C
N-P-N SINGLE-DIFFUSED MESA SILICON POWER TRANSISTORS

FOR POWER-AMPLIFIER AND HIGH-SPEED-SWITCHING APPLICATIONS
DESIGNED FOR COMPLEMENTARY USE WITH TIP30, TIP30A, TIP30B, TIP30C

- 30 W at 25°C Case Temperature
- 1 A Rated Collector Current
- Min f_T of 3 MHz at 10 V, 200 mA

mechanical data

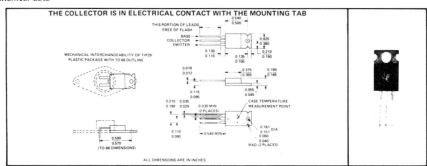

absolute maximum ratings at 25°C case temperature (unless otherwise noted)

	TIP29	TIP29A	TIP29B	TIP29C
Collector-Base Voltage	40 V	60 V	80 V	100 V
Collector-Emitter Voltage (See Note 1)	40 V	60 V	80 V	100 V
Emitter-Base Voltage	←――― 5 V ―――→			
Continuous Collector Current	←――― 1 A ―――→			
Peak Collector Current (See Note 2)	←――― 3 A ―――→			
Continuous Base Current	←――― 0.4 A ―――→			
Safe Operating Region at (or below) 25°C Case Temperature	←――― See Figure 5 ―――→			
Continuous Device Dissipation at (or below) 25°C Case Temperature (See Note 3)	←――― 30 W ―――→			
Continuous Device Dissipation at (or below) 25°C Free-Air Temperature (See Note 4)	←――― 2 W ―――→			
Unclamped Inductive Load Energy (See Note 5)	←――― 32 mJ ―――→			
Operating Collector Junction Temperature Range	←――― −65°C to 150°C ―――→			
Storage Temperature Range	←――― −65°C to 150°C ―――→			
Lead Temperature 1/8 Inch from Case for 10 Seconds	←――― 260°C ―――→			

NOTES: 1. This value applies when the base-emitter diode is open-circuited.
2. This value applies for $t_W \leq 0.3$ ms, duty cycle $\leq 10\%$.
3. Derate linearly to 150°C case temperature at the rate of 0.24 W/°C.
4. Derate linearly to 150°C free-air temperature at the rate of 16 mW/°C.
5. This rating is based on the capability of the transistor to operate safely in the circuit of Figure 2. L = 20 mH, R_{BB1} = 100 Ω, V_{BB2} = 0 V, R_S = 0.1 Ω, V_{CC} = 10 V. Energy ≈ $I_C^2 L/2$.

TEXAS INSTRUMENTS
INCORPORATED
POST OFFICE BOX 5012 • DALLAS, TEXAS 75222

TYPES TIP29, TIP29A, TIP29B, TIP29C
N-P-N SINGLE-DIFFUSED MESA SILICON POWER TRANSISTORS

electrical characteristics at 25°C case temperature

PARAMETER		TEST CONDITIONS		TIP29 MIN	TIP29 MAX	TIP29A MIN	TIP29A MAX	TIP29B MIN	TIP29B MAX	TIP29C MIN	TIP29C MAX	UNIT
$V_{(BR)CEO}$	Collector-Emitter Breakdown Voltage	$I_C = 30$ mA, See Note 6	$I_B = 0$,	40		60		80		100		V
I_{CEO}	Collector Cutoff Current	$V_{CE} = 30$ V,	$I_B = 0$		0.3		0.3					mA
		$V_{CE} = 60$ V,	$I_B = 0$						0.3		0.3	
I_{CES}	Collector Cutoff Current	$V_{CE} = 40$ V,	$V_{BE} = 0$		0.2							mA
		$V_{CE} = 60$ V,	$V_{BE} = 0$				0.2					
		$V_{CE} = 80$ V,	$V_{BE} = 0$						0.2			
		$V_{CE} = 100$ V,	$V_{BE} = 0$								0.2	
I_{EBO}	Emitter Cutoff Current	$V_{EB} = 5$ V,	$I_C = 0$		1		1		1		1	mA
h_{FE}	Static Forward Current Transfer Ratio	$V_{CE} = 4$ V, See Notes 6 and 7	$I_C = 0.2$ A,	40		40		40		40		
		$V_{CE} = 4$ V, See Notes 6 and 7	$I_C = 1$ A,	15	75	15	75	15	75	15	75	
V_{BE}	Base-Emitter Voltage	$V_{CE} = 4$ V, See Notes 6 and 7	$I_C = 1$ A,		1.3		1.3		1.3		1.3	V
$V_{CE(sat)}$	Collector-Emitter Saturation Voltage	$I_B = 125$ mA, See Notes 6 and 7	$I_C = 1$ A,		0.7		0.7		0.7		0.7	V
h_{fe}	Small-Signal Common-Emitter Forward Current Transfer Ratio	$V_{CE} = 10$ V, $f = 1$ kHz	$I_C = 0.2$ A,	20		20		20		20		
$\|h_{fe}\|$	Small-Signal Common-Emitter Forward Current Transfer Ratio	$V_{CE} = 10$ V, $f = 1$ MHz	$I_C = 0.2$ A,	3		3		3		3		

NOTES: 6. These parameters must be measured using pulse techniques. $t_w = 300$ μs, duty cycle ≤ 2%.
7. These parameters are measured with voltage-sensing contacts separate from the current-carrying contacts.

thermal characteristics

PARAMETER		MAX	UNIT
$R_{\theta JC}$	Junction-to-Case Thermal Resistance	4.17	°C/W
$R_{\theta JA}$	Junction-to-Free-Air Thermal Resistance	62.5	

switching characteristics at 25°C case temperature

PARAMETER		TEST CONDITIONS†			TYP	UNIT
t_{on}	Turn-On Time	$I_C = 1$ A,	$I_{B(1)} = 100$ mA,	$I_{B(2)} = -100$ mA,	0.5	μs
t_{off}	Turn-Off Time	$V_{BE(off)} = -4.3$ V,	$R_L = 30$ Ω,	See Figure 1	2	

†Voltage and current values shown are nominal; exact values vary slightly with transistor parameters.

App. A   Manufacturers' Data Sheets

2N2218,A/2N2219,A
2N2221,A/2N2222,A
2N5581/82

JAN, JTX, JTXV AVAILABLE

2N2218,A
2N2219,A
CASE 79-02
TO-39 (TO-205AD)

2N2221,A
2N2222,A
CASE 22-03
TO-18 (TO-206AA)

2N5581
2N5582
CASE 26-03
TO-46 (TO-206AB)

GENERAL PURPOSE TRANSISTOR

NPN SILICON

MAXIMUM RATINGS

Rating	Symbol	2N2218 2N2219 2N2221 2N2222	2N2218A 2N2219A 2N2221A 2N2222A	2N5581 2N5582	Unit
Collector-Emitter Voltage	V_{CEO}	30	40	40	Vdc
Collector-Base Voltage	V_{CBO}	60	75	75	Vdc
Emitter-Base Voltage	V_{EBO}	5.0	6.0	6.0	Vdc
Collector Current — Continuous	I_C	800	800	800	mAdc

Rating	Symbol	2N2218,A 2N2219,A	2N2221,A 2N2222,A	2N5581 2N5582	Unit
Total Device Dissipation @ T_A = 25°C Derate above 25°C	P_D	0.8 4.57	0.4 2.28	0.6 3.33	Watt mW/°C
Total Device Dissipation @ T_C = 25°C Derate above 25°C	P_D	3.0 17.1	1.2 6.85	2.0 11.43	Watts mW/°C
Operating and Storage Junction Temperature Range	T_J, T_{stg}	−65 to +200			°C

ELECTRICAL CHARACTERISTICS (T_A = 25°C unless otherwise noted.)

Characteristic		Symbol	Min	Max	Unit
OFF CHARACTERISTICS					
Collector-Emitter Breakdown Voltage (I_C = 10 mAdc, I_B = 0) Non-A Suffix		$V_{(BR)CEO}$	30	—	Vdc
A-Suffix, 2N5581, 2N5582			40	—	
Collector-Base Breakdown Voltage (I_C = 10 μAdc, I_E = 0) Non-A Suffix		$V_{(BR)CBO}$	60	—	Vdc
A-Suffix, 2N5581, 2N5582			75	—	
Emitter-Base Breakdown Voltage (I_E = 10 μAdc, I_C = 0) Non-A Suffix		$V_{(BR)EBO}$	5.0	—	Vdc
A-Suffix, 2N5581, 2N5582			6.0	—	
Collector Cutoff Current (V_{CB} = 60 Vdc, $V_{EB(off)}$ = 3.0 Vdc) A-Suffix, 2N5581, 2N5582		I_{CEX}	—	10	nAdc
Collector Cutoff Current (V_{CB} = 50 Vdc, I_E = 0) Non-A Suffix		I_{CBO}	—	0.01	μAdc
(V_{CB} = 60 Vdc, I_E = 0) A-Suffix, 2N5581, 2N5582			—	0.01	
(V_{CB} = 50 Vdc, I_E = 0, T_A = 150°C) Non-A Suffix			—	10	
(V_{CB} = 60 Vdc, I_E = 0, T_A = 150°C) A-Suffix, 2N5581, 2N5582			—	10	
Emitter Cutoff Current (V_{EB} = 3.0 Vdc, I_C = 0) A-Suffix, 2N5581, 2N5582		I_{EBO}	—	10	nAdc
Base Cutoff Current (V_{CE} = 60 Vdc, $V_{EB(off)}$ = 3.0 Vdc) A-Suffix		I_{BL}	—	20	nAdc
ON CHARACTERISTICS					
DC Current Gain (I_C = 0.1 mAdc, V_{CE} = 10 Vdc)	2N2218,A, 2N2221,A, 2N5581(1)	h_{FE}	20	—	—
	2N2219,A, 2N2222,A, 2N5582(1)		35	—	
(I_C = 1.0 mAdc, V_{CE} = 10 Vdc)	2N2218,A, 2N2221,A, 2N5581		25	—	
	2N2219,A, 2N2222,A, 2N5582		50	—	
(I_C = 10 mAdc, V_{CE} = 10 Vdc)	2N2218,A, 2N2221,A, 2N5581(1)		35	—	
	2N2219,A, 2N2222,A, 2N5582(1)		75	—	
(I_C = 10 mAdc, V_{CE} = 10 Vdc, T_A = −55°C)	2N2218A, 2N2221A, 2N5581		15	—	
	2N2219A, 2N2222A, 2N5582		35	—	
(I_C = 150 mAdc, V_{CE} = 10 Vdc)(1)	2N2218,A, 2N2221,A, 2N5581		40	120	
	2N2219,A, 2N2222,A, 2N5582		100	300	

MOTOROLA SEMICONDUCTORS SMALL-SIGNAL DEVICES

2N2218/19/21/22, A SERIES, 2N5581/82

ELECTRICAL CHARACTERISTICS (continued) (T_A = 25°C unless otherwise noted.)

Characteristic		Symbol	Min	Max	Unit
(I_C = 150 mAdc, V_{CE} = 1.0 Vdc)(1)	2N2218,A, 2N2221,A, 2N5581		20	—	
	2N2219,A, 2N2222,A, 2N5582		50	—	
(I_C = 500 mAdc, V_{CE} = 10 Vdc)(1)	2N2218, 2N2221		20	—	
	2N2219, 2N2222		30	—	
	2N2218A, 2N2221A, 2N5581		25	—	
	2N2219A, 2N2222A, 2N5582		40	—	
Collector-Emitter Saturation Voltage(1)		$V_{CE(sat)}$			Vdc
(I_C = 150 mAdc, I_B = 15 mAdc)	Non-A Suffix		—	0.4	
	A-Suffix, 2N5581, 2N5582		—	0.3	
(I_C = 500 mAdc, I_B = 50 mAdc)	Non-A Suffix		—	1.6	
	A-Suffix, 2N5581, 2N5582		—	1.0	
Base-Emitter Saturation Voltage(1)		$V_{BE(sat)}$			Vdc
(I_C = 150 mAdc, I_B = 15 mAdc)	Non-A Suffix		0.6	1.3	
	A-Suffix, 2N5581, 2N5582		0.6	1.2	
(I_C = 500 mAdc, I_B = 50 mAdc)	Non-A Suffix		—	2.6	
	A-Suffix, 2N5581, 2N5582		—	2.0	

SMALL-SIGNAL CHARACTERISTICS

Characteristic		Symbol	Min	Max	Unit
Current-Gain — Bandwidth Product(2)		f_T			MHz
(I_C = 20 mAdc, V_{CE} = 20 Vdc, f = 100 MHz)	All Types, Except		250	—	
	2N2219A, 2N2222A, 2N5582		300	—	
Output Capacitance(3)		C_{obo}	—	8.0	pF
(V_{CB} = 10 Vdc, I_E = 0, f = 100 kHz)					
Input Capacitance(3)		C_{ibo}			pF
(V_{EB} = 0.5 Vdc, I_C = 0, f = 100 kHz)	Non-A Suffix		—	30	
	A-Suffix, 2N5581, 2N5582		—	25	
Input Impedance		h_{ie}			kohms
(I_C = 1.0 mAdc, V_{CE} = 10 Vdc, f = 1.0 kHz)	2N2218A, 2N2221A		1.0	3.5	
	2N2219A, 2N2222A		2.0	8.0	
(I_C = 10 mAdc, V_{CE} = 10 Vdc, f = 1.0 kHz)	2N2218A, 2N2221A		0.2	1.0	
	2N2219A, 2N2222A		0.25	1.25	
Voltage Feedback Ratio		h_{re}			$\times 10^{-4}$
(I_C = 1.0 mAdc, V_{CE} = 10 Vdc, f = 1.0 kHz)	2N2218A, 2N2221A		—	5.0	
	2N2219A, 2N2222A		—	8.0	
(I_C = 10 mAdc, V_{CE} = 10 Vdc, f = 1.0 kHz)	2N2218A, 2N2221A		—	2.5	
	2N2219A, 2N2222A		—	4.0	
Small-Signal Current Gain		h_{fe}			—
(I_C = 1.0 mAdc, V_{CE} = 10 Vdc, f = 1.0 kHz)	2N2218A, 2N2221A		30	150	
	2N2219A, 2N2222A		50	300	
(I_C = 10 mAdc, V_{CE} = 10 Vdc, f = 1.0 kHz)	2N2218A, 2N2221A		50	300	
	2N2219A, 2N2222A		75	375	
Output Admittance		h_{oe}			μmhos
(I_C = 1.0 mAdc, V_{CE} = 10 Vdc, f = 1.0 kHz)	2N2218A, 2N2221A		3.0	15	
	2N2219A, 2N2222A		5.0	35	
(I_C = 10 mAdc, V_{CE} = 10 Vdc, f = 1.0 kHz)	2N2218A, 2N2221A		10	100	
	2N2219A, 2N2222A		25	200	
Collector Base Time Constant		$rb'C_C$	—	150	ps
(I_E = 20 mAdc, V_{CB} = 20 Vdc, f = 31.8 MHz)	A-Suffix				
Noise Figure		NF	—	4.0	dB
(I_C = 100 μAdc, V_{CE} = 10 Vdc, R_S = 1.0 kohm, f = 1.0 kHz)	2N2219A, 2N2222A				
Real Part of Common-Emitter High Frequency Input Impedance		Re(h_{ie})	—	60	Ohms
(I_C = 20 mAdc, V_{CE} = 20 Vdc, f = 300 MHz)	2N2218A, 2N2219A				
	2N2221A, 2N2222A				

(1) Pulse Test: Pulse Width ≤ 300 μs, Duty Cycle ≤ 2.0%.
(2) f_T is defined as the frequency at which $|h_{fe}|$ extrapolates to unity.
(3) 2N5581 and 2N5582 are Listed C_{cb} and C_{eb} for these conditions and values.

SMALL-SIGNAL DEVICES   MOTOROLA SEMICONDUCTORS

2N2218,A/2N2219,A/2N2221,A/2N2222,A/2N5581/82

ELECTRICAL CHARACTERISTICS (continued) ($T_A = 25°C$ unless otherwise noted.)

Characteristic		Symbol	Min	Max	Unit
SWITCHING CHARACTERISTICS					
Delay Time	($V_{CC} = 30$ Vdc, $V_{BE(off)} = 0.5$ Vdc, $I_C = 150$ mAdc, $I_{B1} = 15$ mAdc) (Figure 14)	t_d	—	10	ns
Rise Time		t_r	—	25	ns
Storage Time	($V_{CC} = 30$ Vdc, $I_C = 150$ mAdc, $I_{B1} = I_{B2} = 15$ mAdc) (Figure 15)	t_s	—	225	ns
Fall Time		t_f	—	60	ns
Active Region Time Constant ($I_C = 150$ mAdc, $V_{CE} = 30$ Vdc) (See Figure 14 for 2N2218A, 2N2219A, 2N2221A, 2N2222A)		T_A	—	2.5	ns

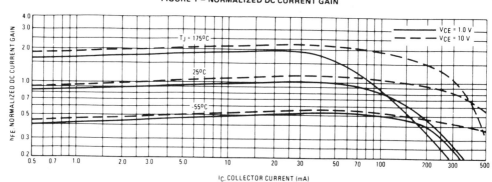

FIGURE 1 — NORMALIZED DC CURRENT GAIN

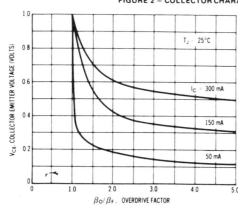

FIGURE 2 — COLLECTOR CHARACTERISTICS IN SATURATION REGION

This graph shows the effect of base current on collector current. β_o (current gain at the edge of saturation) is the current gain of the transistor at 1 volt, and β_s (forced gain) is the ratio of I_c/I_b in a circuit.

EXAMPLE: For type 2N2219, estimate a base current (I_b) to insure saturation at a temperature of 25°C and a collector current of 150 mA.

Observe that at $I_c = 150$ mA an overdrive factor of at least 2.5 is required to drive the transistor well into the saturation region. From Figure 1, it is seen that h_{FE} @ 1 volt is approximately 0.62 of h_{FE} @ 10 volts. Using the guaranteed minimum gain of 100 @ 150 mA and 10 V, $\beta_o = 62$ and substituting values in the overdrive equation, we find:

$$\frac{\beta_o}{\beta_s} = \frac{h_{FE} @ 1.0 V}{I_c/I_b} \qquad 2.5 = \frac{62}{150/I_b} \qquad I_b \approx 6.0 \text{ mA}$$

MOTOROLA SEMICONDUCTORS   SMALL-SIGNAL DEVICES

2N2218,A/2N2219,A/2N2221,A/2N2222,A/2N5581/82

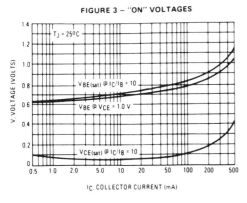

FIGURE 3 – "ON" VOLTAGES

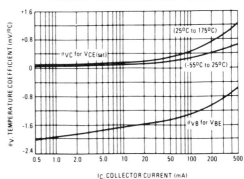

FIGURE 4 – TEMPERATURE COEFFICIENTS

h PARAMETERS

V_{CE} = 10 Vdc, f = 1.0 kHz, T_A = 25°C

This group of graphs illustrates the relationship between h_{fe} and other "h" parameters for this series of transistors. To obtain these curves, a high-gain and a low-gain unit were selected and the same units were used to develop the correspondingly numbered curves on each graph.

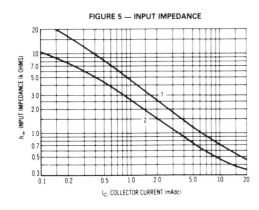

FIGURE 5 — INPUT IMPEDANCE

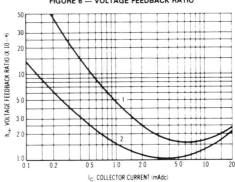

FIGURE 6 — VOLTAGE FEEDBACK RATIO

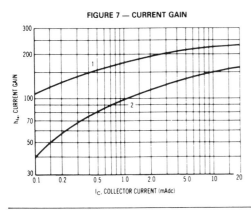

FIGURE 7 — CURRENT GAIN

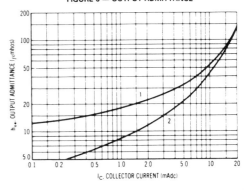

FIGURE 8 — OUTPUT ADMITTANCE

SMALL-SIGNAL DEVICES   MOTOROLA SEMICONDUCTORS

2N2218,A/2N2219,A/2N2221,A/2N2222,A/2N5581/82

SWITCHING TIME CHARACTERISTICS

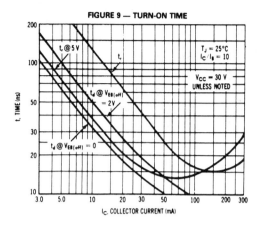

FIGURE 9 — TURN-ON TIME

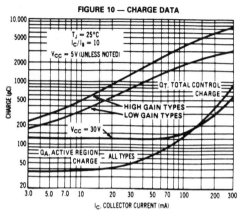

FIGURE 10 — CHARGE DATA

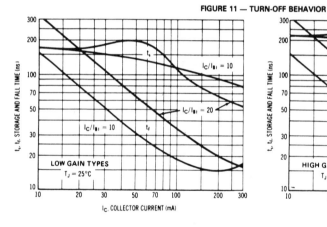

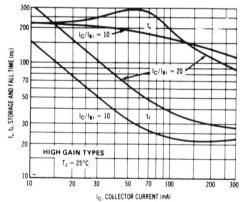

FIGURE 11 — TURN-OFF BEHAVIOR

2N5457 2N5458 2N5459

JUNCTION FIELD-EFFECT TRANSISTORS

SYMMETRICAL SILICON N-CHANNEL

JULY 1968 — DS 5207 R2
(Replaces DS 5207 R1)

Type A

SILICON N-CHANNEL JUNCTION FIELD-EFFECT TRANSISTORS

... depletion mode (Type A) transistors designed for general-purpose audio and switching applications.

- N-Channel for Higher Gain
- Drain and Source Interchangeable
- High AC Input Impedance
- High DC Input Resistance
- Low Transfer and Input Capacitance
- Low Cross-Modulation and Intermodulation Distortion
- Unibloc* Plastic Encapsulated Package

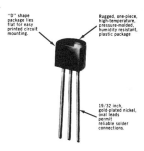

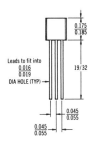

BOTTOM VIEW

TO-92
CASE 29(5)

Drain and Source may be Interchanged.

MAXIMUM RATINGS

Characteristic	Symbol	Rating	Unit
Drain-Source Voltage	V_{DS}	25	Vdc
Drain-Gate Voltage	V_{DG}	25	Vdc
Reverse Gate-Source Voltage	$V_{GS(r)}$	−25	Vdc
Gate Current	I_G	10	mAdc
Total Device Dissipation @ T_A = 25°C Derate above 25°C	P_D	310 2.82	mW mW/°C
Operating Junction Temperature	T_J	135	°C
Storage Temperature Range	T_{stg}	−65 to +150	°C

*Trademark of Motorola Inc.

2N3796
2N3797

CASE 22-03, STYLE 2
TO-18 (TO-206AA)

**MOSFET
LOW-POWER AUDIO**

N-CHANNEL — DEPLETION

MAXIMUM RATINGS

Rating	Symbol	Value	Unit
Drain-Source Voltage	V_{DS}		Vdc
2N3796		25	
2N3797		20	
Gate-Source Voltage	V_{GS}	± 10	Vdc
Drain Current	I_D	20	mAdc
Total Device Dissipation @ T_A = 25°C	P_D	200	mW
Derate above 25°C		1.14	mW/°C
Junction Temperature Range	T_J	+175	°C
Storage Channel Temperature Range	T_{stg}	−65 to +175	°C

ELECTRICAL CHARACTERISTICS (T_A = 25°C unless otherwise noted.)

Characteristic		Symbol	Min	Typ	Max	Unit
OFF CHARACTERISTICS						
Drain-Source Breakdown Voltage		$V_{(BR)DSX}$				Vdc
(V_{GS} = −4.0 V, I_D = 5.0 μA)	2N3796		25	30	—	
(V_{GS} = −7.0 V, I_D = 5.0 μA)	2N3797		20	25	—	
Gate Reverse Current(1)		I_{GSS}				pAdc
(V_{GS} = −10 V, V_{DS} = 0)			—	—	1.0	
(V_{GS} = −10 V, V_{DS} = 0, T_A = 150°C)			—	—	200	
Gate Source Cutoff Voltage		$V_{GS(off)}$				Vdc
(I_D = 0.5 μA, V_{DS} = 10 V)	2N3796		—	−3.0	−4.0	
(I_D = 2.0 μA, V_{DS} = 10 V)	2N3797		—	−5.0	−7.0	
Drain-Gate Reverse Current(1)		I_{DGO}	—	—	1.0	pAdc
(V_{DG} = 10 V, I_S = 0)						
ON CHARACTERISTICS						
Zero-Gate-Voltage Drain Current		I_{DSS}				mAdc
(V_{DS} = 10 V, V_{GS} = 0)	2N3796		0.5	1.5	3.0	
	2N3797		2.0	2.9	6.0	
On-State Drain Current		$I_{D(on)}$				mAdc
(V_{DS} = 10 V, V_{GS} = +3.5 V)	2N3796		7.0	8.3	14	
	2N3797		9.0	14	18	
SMALL-SIGNAL CHARACTERISTICS						
Forward Transfer Admittance		$|Y_{fs}|$				μmhos
(V_{DS} = 10 V, V_{GS} = 0, f = 1.0 kHz)	2N3796		900	1200	1800	
	2N3797		1500	2300	3000	
(V_{DS} = 10 V, V_{GS} = 0, f = 1.0 MHz)	2N3796		900	—	—	
	2N3797		1500	—	—	
Output Admittance		$|Y_{os}|$				μmhos
(V_{DS} = 10 V, V_{GS} = 0, f = 1.0 kHz)	2N3796		—	12	25	
	2N3797		—	27	60	
Input Capacitance		C_{iss}				pF
(V_{DS} = 10 V, V_{GS} = 0, f = 1.0 MHz)	2N3796		—	5.0	7.0	
	2N3797		—	6.0	8.0	
Reverse Transfer Capacitance		C_{rss}	—	0.5	0.8	pF
(V_{DS} = 10 V, V_{GS} = 0, f = 1.0 MHz)						
FUNCTIONAL CHARACTERISTICS						
Noise Figure		NF	—	3.8	—	dB
(V_{DS} = 10 V, V_{GS} = 0, f = 1.0 kHz, R_S = 3 megohms)						

(1) This value of current includes both the FET leakage current as well as the leakage current associated with the test socket and fixture when measured under best attainable conditions.

MOTOROLA SEMICONDUCTORS

SMALL-SIGNAL DEVICES

2N3796, 2N3797

TYPICAL DRAIN CHARACTERISTICS

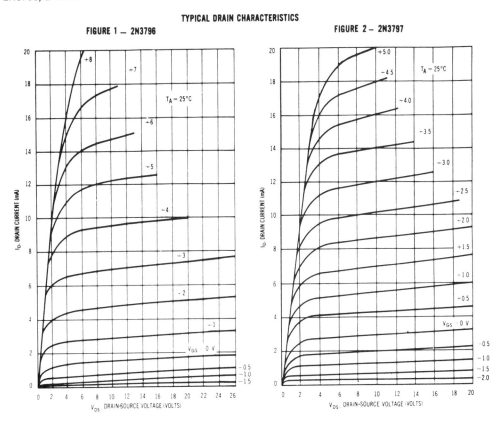

FIGURE 1 — 2N3796
FIGURE 2 — 2N3797

COMMON SOURCE TRANSFER CHARACTERISTICS

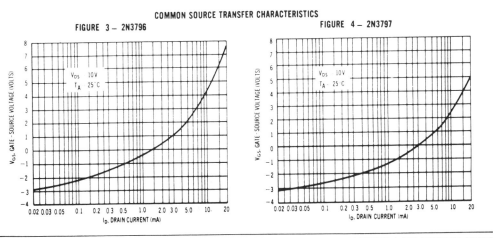

FIGURE 3 — 2N3796
FIGURE 4 — 2N3797

SMALL-SIGNAL DEVICES   MOTOROLA SEMICONDUCTORS

App. A   Manufacturers' Data Sheets

2N3796, 2N3797

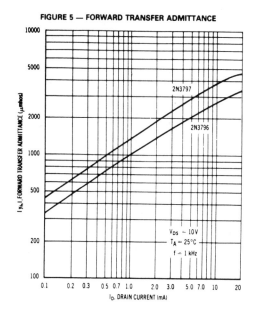

FIGURE 5 — FORWARD TRANSFER ADMITTANCE

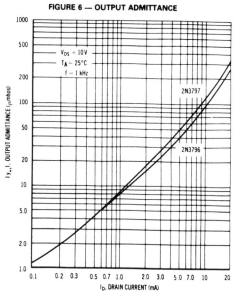

FIGURE 6 — OUTPUT ADMITTANCE

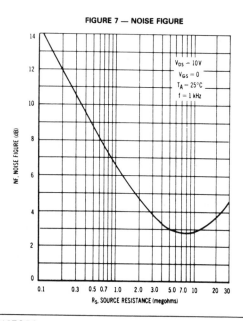

FIGURE 7 — NOISE FIGURE

MOTOROLA SEMICONDUCTORS   SMALL-SIGNAL DEVICES

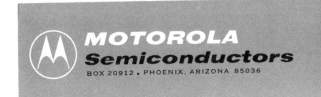

2N4351

SILICON N-CHANNEL
MOS FIELD-EFFECT TRANSISTOR

. . . designed for enhancement-mode operation in low-power switching applications.

- Low Drain-Source Resistance $r_{ds(on)}$ = 300 ohms max
- Low Reverse Transfer Capacitance C_{rss} = 1.3 pF max
- Guaranteed Switching Limits t_{d1}, t_r, t_{d2}, t_f
- Manufactured Using The New Stable Silicon Nitride Process

N-CHANNEL

MOS FIELD-EFFECT
TRANSISTOR

OCTOBER 1970 – DS 5179 R2
(Replaces DS 5179 R1)

MAXIMUM RATINGS (T_A = 25°C unless otherwise noted)

Rating	Symbol	Value	Unit
Drain-Source Voltage	V_{DS}	25	Vdc
Drain-Gate Voltage	V_{DG}	30	Vdc
Gate-Source Voltage*	V_{GS}*	±30	Vdc
Drain Current	I_D	30	mAdc
Power Dissipation at T_A = 25°C Derate above 25°C	P_D	300 1.7	mW mW/°C
Power Dissipation @ T_C = 25°C Derate above 25°C	P_D	800 4.56	mW mW/°C
Operating Junction Temperature	T_J	200	°C
Storage Temperature	T_{stg}	−65 to +200	°C

*Transient potentials of ±75 Volt will not cause gate-oxide failure.

HANDLING CONSIDERATIONS:

MOS field-effect transistors, due to their extremely high input resistance, are subject to potential damage by the accumulation of excess static charge. To avoid possible damage to the devices while handling, testing, or in actual operation, the following procedure should be followed:
1. The leads of the devices should remain wrapped in the shipping foil except when being tested or in actual operation to avoid the build-up of static charge.
2. Avoid unnecessary handling; when handled, the devices should be picked up by the case instead of the leads.
3. The devices should not be inserted or removed from circuits with the power on as transient voltages may cause permanent damage to the devices.

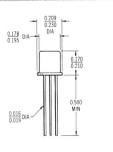

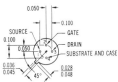

TO-72
CASE 20 (2)

App. A   Manufacturers' Data Sheets   393

2N4351

ELECTRICAL CHARACTERISTICS ($T_A = 25°C$ unless otherwise noted)
Substrate connected to source.

Characteristic	Figure No.	Symbol	Min	Max	Unit
OFF CHARACTERISTICS					
Drain-Source Breakdown Voltage ($I_D = 10~\mu A$, $V_{GS} = 0$)	—	$V_{(BR)DSS}$	25	—	Vdc
Gate Leakage Current ($V_{GS} = \pm 30$ Vdc, $V_{DS} = 0$)	—	I_{GSS}	—	±10	pAdc
Zero-Gate-Voltage Drain Current ($V_{DS} = 10$ V, $V_{GS} = 0$)	—	I_{DSS}	—	10	nAdc
ON CHARACTERISTICS					
Gate-Source Threshold Voltage ($V_{DS} = 10$ V, $I_D = 10~\mu A$)	—	$V_{GS(TH)}$	1.0	5	Vdc
"ON" Drain Current ($V_{GS} = 10$ V, $V_{DS} = 10$ V)	3	$I_{D(on)}$	3	—	mAdc
Drain-Source "ON" Voltage ($I_D = 2$ mA, $V_{GS} = 10$ V)	—	$V_{DS(on)}$	—	1.0	V
SMALL SIGNAL CHARACTERISTICS					
Drain-Source Resistance ($V_{GS} = 10$ V, $I_D = 0$, $f = 1$ kHz)	4	$r_{ds(on)}$	—	300	ohms
Forward Transfer Admittance ($V_{DS} = 10$ V, $I_D = 2$ mA, $f = 1$ kHz)	1	$|y_{fs}|$	1000	—	μmho
Reverse Transfer Capacitance ($V_{DS} = 0$, $V_{GS} = 0$, $f = 140$ kHz)	2	C_{rss}	—	1.3	pF
Input Capacitance ($V_{DS} = 10$ V, $V_{GS} = 0$, $f = 140$ kHz)	2	C_{iss}	—	5.0	pF
Drain-Substrate Capacitance ($V_{D(SUB)} = 10$ V, $f = 140$ kHz)	—	$C_{d(sub)}$	—	5.0	pF
SWITCHING CHARACTERISTICS					
Turn-On Delay	6, 10	t_{d1}	—	45	ns
Rise Time	7, 10	t_r	—	65	ns
Turn-Off Delay	8, 10	t_{d2}	—	60	ns
Fall Time	9, 10	t_f	—	100	ns

For Rise Time / Turn-Off Delay / Fall Time: $I_D = 2.0$ mAdc, $V_{DS} = 10$ Vdc, $V_{GS} = 10$ Vdc (See Figure 10; Times Circuit Determined)

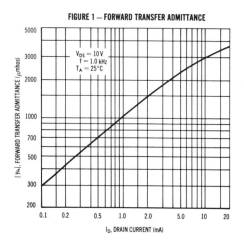

FIGURE 1 — FORWARD TRANSFER ADMITTANCE

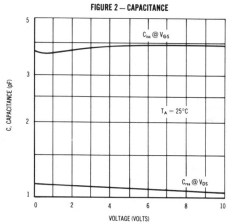

FIGURE 2 — CAPACITANCE

MOTOROLA Semiconductor Products Inc.

1N5758,A thru 1N5762,A

SILICON 3-LAYER BILATERAL TRIGGERS

...annular♦, two terminal devices that exhibit bi-directional negative resistance switching characteristics. These economical, durable devices have been developed for use in thyristor triggering circuits for lamp drivers and universal motor speed controls.

- Switching Voltage Range — 20 to 36 Volts Nominal
- Symmetrical Characteristics
- Passivated Surface for Reliability and Uniformity

SILICON BILATERAL TRIGGERS
DS 6537

*MAXIMUM RATINGS (T_A = 25°C unless otherwise noted)

Rating	Symbol	Value	Unit
Peak Pulse Current (30 µs duration, 120 Hz repetition rate)	I_{pulse}	2.0	Amp
Power Dissipation @ T_A = −40 to +25°C Derate above 25°C	P_D	300 4.0	mW mW/°C
Operating Junction Temperature Range	T_J	−40 to +100	°C
Storage Temperature Range	T_{stg}	−40 to +150	°C

*ELECTRICAL CHARACTERISTICS (T_A = 25°C unless otherwise noted)

Characteristic		Symbol	Min	Max	Unit
Switching Voltage (Both Directions)		V_S			Volts
	1N5758		16	24	
	1N5759		20	28	
	1N5760		24	32	
	1N5761		28	36	
	1N5762		32	40	
	1N5758A		18	22	
	1N5759A		22	26	
	1N5760A		26	30	
	1N5761A		30	34	
	1N5762A		34	38	
Switching Current (Both Directions) (T_A = −40 to +75°C)	1N5758/5762 1N5758A/5762A	I_S	— —	100 25	µA
Switching Voltage Change (Both Directions) ($\Delta I = I_S$ to I = 10 mA)	1N5758,A,1N5759,A 1N5760,A,61,A,62,A	ΔV	5.0 7.0	— —	Volts
Leakage Current (Both Directions), (Applied Voltage = 14 Volts)		I_B	—	10	µA
Switching Voltage Symmetry	1N5758/5762 1N5758A/5762A	$(V_S+)-(V_S-)$	— —	± 4.0 ± 2.0	Volts
Peak Pulse Amplitude (Figure 1) (Both Polarities)	1N5758,A,1N5759,A 1N5760,A,61,A,62,A		3.0 5.0	— —	Volts

*Indicates JEDEC Registered Data.
♦ Annular Semiconductors Patented by Motorola, Inc.

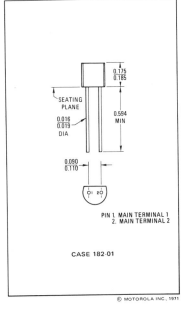

PIN 1. MAIN TERMINAL 1
2. MAIN TERMINAL 2

CASE 182-01

© MOTOROLA INC., 1971

App. A   Manufacturers' Data Sheets

1N5758,A thru 1N5762,A (continued)

TYPICAL ELECTRICAL CHARACTERISTICS

FIGURE 1 – PEAK PULSE AMPLITUDE TEST CIRCUIT

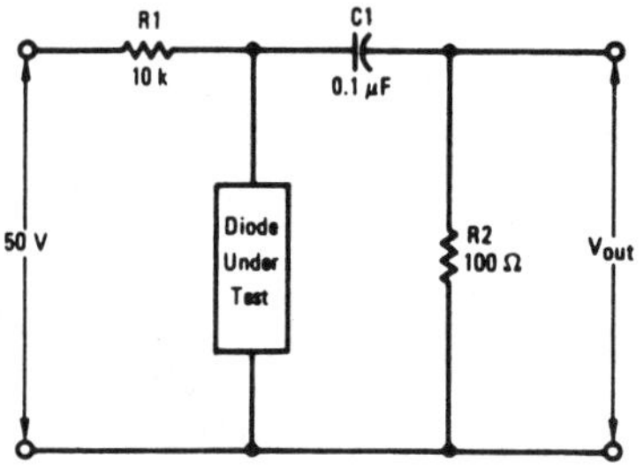

FIGURE 2 – VOLT-AMPERE CHARACTERISTICS

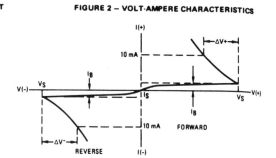

FIGURE 3 – BREAKOVER VOLTAGE BEHAVIOR

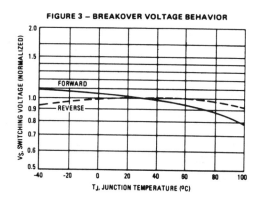

FIGURE 4 – NORMALIZED OUTPUT VOLTAGE BEHAVIOR

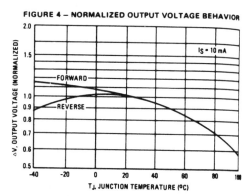

FIGURE 5 – SWITCHING TIMES

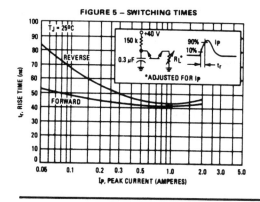

FIGURE 6 – CONTROL CIRCUIT

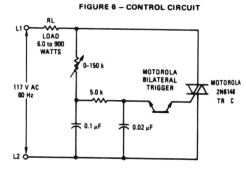

MBS4991
MBS4992

SILICON BIDIRECTIONAL SWITCH

SILICON BIDIRECTIONAL SWITCH (PLASTIC)

6.0–10 VOLTS
500 mW

DS 2510 R2

...designed for full-wave triggering in Triac phase control circuits, half-wave SCR triggering application and as voltage level detectors. Supplied in an inexpensive plastic TO-92 package for high-volume requirements, this low-cost plastic package is readily adaptable for use in automatic insertion equipment.

- Low Switching Voltage — 8.0 Volts Typical
- Uniform Characteristics in Each Direction
- Low On-State Voltage — 1.7 Volts Maximum
- Low Off-State Current — 0.1 μA Maximum
- Low Temperature Coefficient — 0.02 %/°C Typical

MAXIMUM RATINGS

Rating	Symbol	Value	Unit
Power Dissipation	P_D	500	mW
DC Forward Anode Current	I_F	200	mA
DC Gate Current (off-state only)	I_G(off)	5.0	mA
Repetitive Peak Forward Current (1.0% Duty Cycle, 10 μs Pulse Width, T_A = 100°C)	I_{FM}(rep)	2.0	Amp
Non-Repetitive Forward Current 10 μs Pulse Width, T_A = 25°C	I_{FM}(nonrep)	6.0	Amp
Operating Junction Temperature Range	T_J	−55 to +125	°C
Storage Temperature Range	T_{stg}	−65 to +150	°C

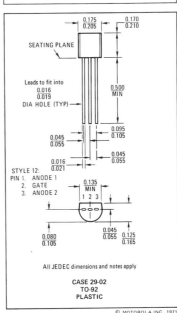

All JEDEC dimensions and notes apply

CASE 29-02
TO-92
PLASTIC

© MOTOROLA INC., 1971

App. A   Manufacturers' Data Sheets

MBS4991 • MBS4992

ELECTRICAL CHARACTERISTICS ($T_A = 25°C$ unless otherwise noted)

Characteristic	Symbol	Min	Typ	Max	Unit
Switching Voltage	V_S				Vdc
MBS4991		6.0	8.0	10	
MBS4992		7.5	8.0	9.0	
Switching Current	I_S				µAdc
MBS4991		–	175	500	
MBS4992		–	90	120	
Switching Voltage Differential	$\|V_{S1}-V_{S2}\|$				Vdc
MBS4991		–	0.3	0.5	
MBS4992		–	0.1	0.2	
Gate Trigger Current	I_{GF}				µAdc
($V_F = 5.0$ Vdc $R_L = 1.0$ K ohm) MBS4992		–	–	100	
Holding Current	I_H				mAdc
MBS4991		–	0.7	1.5	
MBS4992		–	0.2	0.5	
Off-State Blocking Current	I_B				µAdc
($V_F = 5.0$ Vdc, $T_A = 25°C$) MBS4991		–	0.08	1.0	
($V_F = 5.0$ Vdc, $T_A = 85°C$) MBS4991		–	2.0	10	
($V_F = 5.0$ Vdc, $T_A = 25°C$) MBS4992		–	0.08	0.1	
($V_F = 5.0$ Vdc, $T_A = 100°C$) MBS4992		–	6.0	10	
Forward On-State Voltage	V_F				Vdc
($I_F = 175$ mAdc) MBS4991		–	1.4	1.7	
($I_F = 200$ mAdc) MBS4992		–	1.5	1.7	
Peak Output Voltage ($C_C = 0.1$ µF, $R_L = 20$ ohms, (Figure 7)	V_O	3.5	4.8	–	Vdc
Turn-On Time (Figure 8)	t_{on}	–	1.0	–	µs
Turn-Off Time (Figure 9)	t_{off}	–	30	–	µs
Temperature Coefficient of Switching Voltage (–50 to +125°C)	T_C	–	+0.02	–	%/°C

TYPICAL ELECTRICAL CHARACTERISTICS

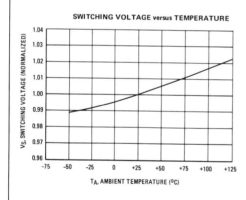

SWITCHING VOLTAGE versus TEMPERATURE

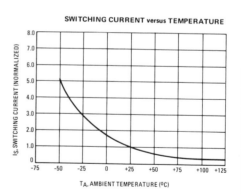

SWITCHING CURRENT versus TEMPERATURE

MOTOROLA Semiconductor Products Inc.

MBS4991 • MBS4992

TURN-ON TIME TEST CIRCUIT

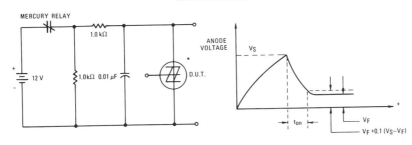

Turn-on time is measured from the time V_S is achieved to the time when the anode voltage drops to within 90% of the difference between V_S and V_F.

TURN-OFF TIME TEST CIRCUIT

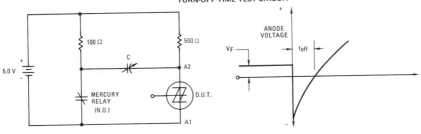

With the SBS in conduction and the relay contacts open, close the contacts to cause anode A2 to be driven negative. Decrease C until the SBS just remains off when anode A2 becomes positive. The turn-off time, t_{off}, is the time from initial contact closure and until anode A2 voltage reaches zero volts.

DEVICE EQUIVALENT CIRCUIT, CHARACTERISTICS AND SYMBOLS

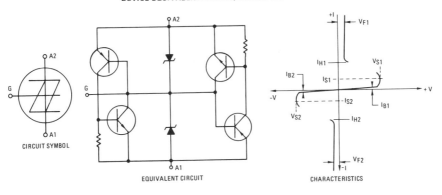

 MOTOROLA Semiconductor Products Inc.

BOX 20912 • PHOENIX, ARIZONA 85036 • A SUBSIDIARY OF MOTOROLA INC.

2N6027
2N6028

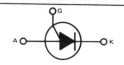

SILICON PROGRAMMABLE UNIJUNCTION TRANSISTORS

40 VOLTS
375 mW

DS 2520

SILICON PROGRAMMABLE UNIJUNCTION TRANSISTORS

... designed to enable the engineer to "program" unijunction characteristics such as R_{BB}, η, I_V, and I_P by merely selecting two resistor values. Application includes thyristor-trigger, oscillator, pulse and timing circuits. These devices may also be used in special thyristor applications due to the availability of an anode gate. Supplied in an inexpensive TO-92 plastic package for high-volume requirements, this package is readily adaptable for use in automatic insertion equipment.

- Programmable — R_{BB}, η, I_V and I_P.
- Low On-State Voltage — 1.5 Volts Maximum @ I_F = 50 mA
- Low Gate to Anode Leakage Current — 10 nA Maximum
- High Peak Output Voltage — 11 Volts Typical
- Low Offset Voltage — 0.35 Volt Typical (R_G = 10 k ohms)

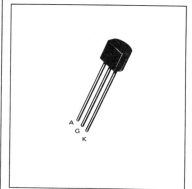

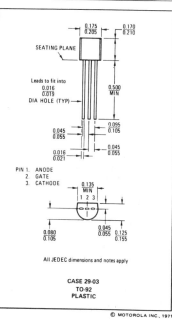

PIN 1. ANODE
2. GATE
3. CATHODE

All JEDEC dimensions and notes apply

CASE 29-03
TO-92
PLASTIC

MAXIMUM RATINGS

Rating	Symbol	Value	Unit
Power Dissipation (1) Derate Above 25°C	P_F $1/\theta_{JA}$	375 5.0	mW mW/°C
DC Forward Anode Current (2) Derate Above 25°C	I_T	200 2.67	mA mA/°C
*DC Gate Current	I_G	±50	mA
Repetitive Peak Forward Current 100 μs Pulse Width, 1.0% Duty Cycle *20 μs Pulse Width, 1.0% Duty Cycle	I_{TRM}	1.0 2.0	Amp Amp
Non-Repetitive Peak Forward Current 10 μs Pulse Width	I_{TSM}	5.0	Amp
*Gate to Cathode Forward Voltage	V_{GKF}	40	Volt
*Gate to Cathode Reverse Voltage	V_{GKR}	−5.0	Volt
*Gate to Anode Reverse Voltage	V_{GAR}	40	Volt
Anode to Cathode Voltage	V_{AK}	±40	Volt
Operating Junction Temperature Range	T_J	−50 to +100	°C
*Storage Temperature Range	T_{stg}	−55 to +150	°C

*Indicates JEDEC Registered Data
(1) JEDEC Registered Data is 300 mW, derating at 4.0 mW/°C.
(2) JEDEC Registered Data is 150 mA.

© MOTOROLA INC., 1971

2N6027 • 2N6028

ELECTRICAL CHARACTERISTICS ($T_A = 25°C$ unless otherwise noted)

Characteristic		Figure	Symbol	Min	Typ	Max	Unit
•Peak Current		2,9,11	I_P				μA
($V_S = 10$ Vdc, $R_G = 1.0$ MΩ)	2N6027			—	1.25	2.0	
	2N6028			—	0.08	0.15	
($V_S = 10$ Vdc, $R_G = 10$ k ohms)	2N6027			—	4.0	5.0	
	2N6028			—	0.70	1.0	
•Offset Voltage		1	V_T				Volts
($V_S = 10$ Vdc, $R_G = 1.0$ MΩ)	2N6027			0.2	0.70	1.6	
	2N6028			0.2	0.50	0.6	
($V_S = 10$ Vdc, $R_G = 10$ k ohms)	(Both Types)			0.2	0.35	0.6	
•Valley Current		1,4,5,	I_V				μA
($V_S = 10$ Vdc, $R_G = 1.0$ MΩ)	2N6027			—	18	50	
	2N6028			—	18	25	
($V_S = 10$ Vdc, $R_G = 10$ k ohms)	2N6027			70	270	—	
	2N6028			25	270	—	
($V_S = 10$ Vdc, $R_G = 200$ Ohms)	2N6027			1.5	—	—	mA
	2N6028			1.0	—	—	
•Gate to Anode Leakage Current		—	I_{GAO}				nAdc
($V_S = 40$ Vdc, $T_A = 25°C$, Cathode Open)				—	1.0	10	
($V_S = 40$ Vdc, $T_A = 75°C$, Cathode Open)				—	3.0	—	
Gate to Cathode Leakage Current		—	I_{GKS}	—	5.0	50	nAdc
($V_S = 40$ Vdc, Anode to Cathode Shorted)							
•Forward Voltage ($I_F = 50$ mA Peak)		1,6	V_F	—	0.8	1.5	Volts
•Peak Output Voltage		3,7	V_O	6.0	11	—	Volts
($V_B = 20$ Vdc, $C_C = 0.2$ μF)							
Pulse Voltage Rise Time		3	t_r	—	40	80	ns
($V_B = 20$ Vdc, $C_C = 0.2$ μF)							

*Indicates JEDEC Registered Data

FIGURE 1 — ELECTRICAL CHARACTERIZATION

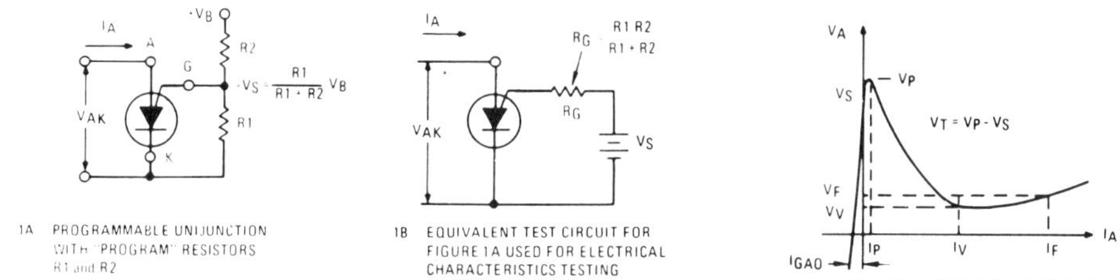

1A PROGRAMMABLE UNIJUNCTION WITH "PROGRAM" RESISTORS R1 and R2

1B EQUIVALENT TEST CIRCUIT FOR FIGURE 1A USED FOR ELECTRICAL CHARACTERISTICS TESTING (ALSO SEE FIGURE 2)

1C — ELECTRICAL CHARACTERISTICS

FIGURE 2 — PEAK CURRENT (I_P) TEST CIRCUIT

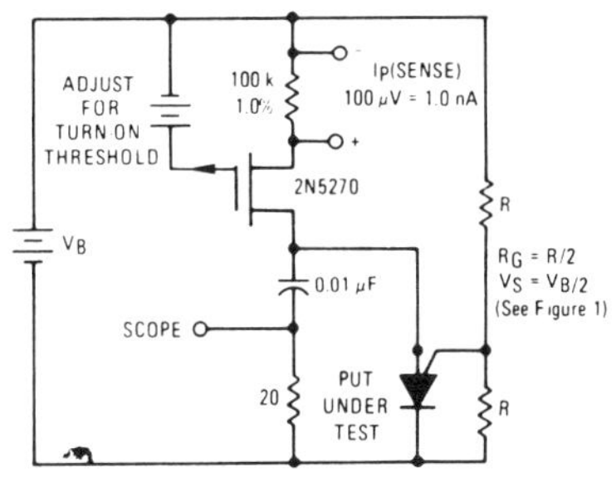

FIGURE 3 — V_O AND t_r TEST CIRCUIT

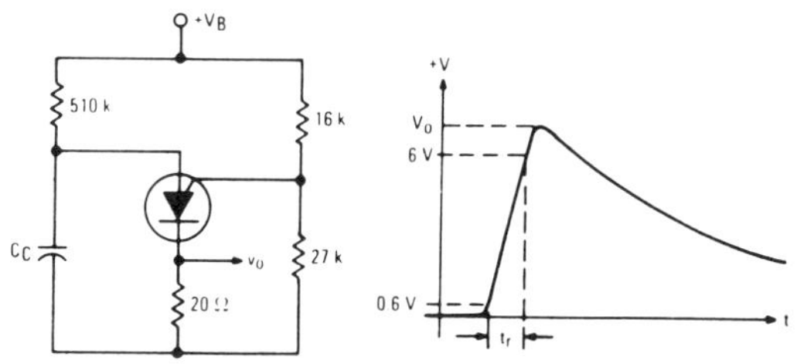

MOTOROLA Semiconductor Products Inc.

App. A   Manufacturers' Data Sheets

2N6027 • 2N6028

TYPICAL VALLEY CURRENT BEHAVIOR

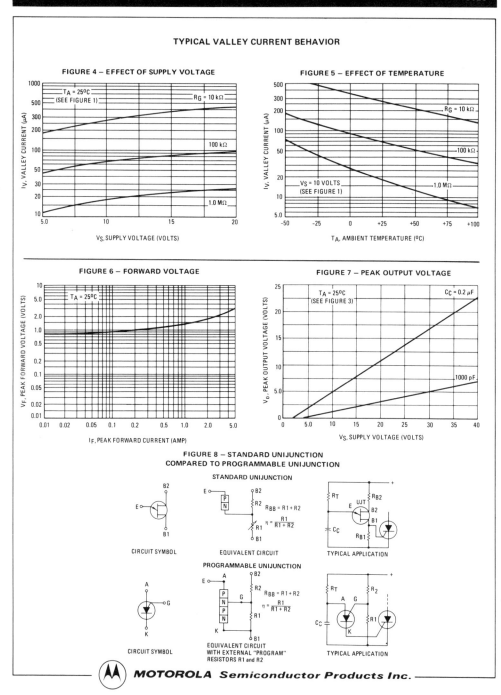

FIGURE 4 — EFFECT OF SUPPLY VOLTAGE

FIGURE 5 — EFFECT OF TEMPERATURE

FIGURE 6 — FORWARD VOLTAGE

FIGURE 7 — PEAK OUTPUT VOLTAGE

FIGURE 8 — STANDARD UNIJUNCTION COMPARED TO PROGRAMMABLE UNIJUNCTION

MOTOROLA Semiconductor Products Inc.

2N5060 thru 2N5064

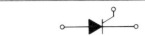

PLASTIC THYRISTORS

.... Annular♦ PNPN devices designed for high volume consumer applications such as relay and lamp drivers, small motor controls, gate drivers for larger thyristors, and sensing and detection circuits. Supplied in an inexpensive plastic TO-92 package which is readily adaptable for use in automatic insertion equipment.

- Sensitive Gate Trigger Current — 200 µA Maximum
- Low Reverse and Forward Blocking Current — 50 µA Maximum, T_C = 125°C
- Low Holding Current — 5.0 mA Maximum
- Passivated Surface for Reliability and Uniformity

PLASTIC SILICON CONTROLLED RECTIFIERS

0.8 AMPERE RMS
30 thru 200 VOLTS

JUNE 1970 — DS 6525
(Replaces PS 41)

MAXIMUM RATINGS(1)

Rating	Symbol	Value	Unit
Peak Reverse Blocking Voltage	V_{RRM}		Volts
2N5060		30*	
2N5061		60*	
2N5062		100*	
2N5063		150*	
2N5064		200*	
Forward Current RMS (See Figures 4 & 5) (All Conduction Angles)	$I_{T(RMS)}$	0.8	Amp
Peak Forward Surge Current, T_A = 25°C (1/2 cycle, Sine Wave, 60 Hz)	I_{TSM}	6.0*	Amp
Circuit Fusing Considerations, T_A = 25°C (t = 1.0 to 8.3 ms)	I^2t	0.15	A^2s
Peak Gate Power — Forward, T_A = 25°C	P_{GM}	0.1*	Watt
Average Gate Power — Forward, T_A = 25°C	$P_{GF(AV)}$	0.01*	Watt
Peak Gate Current — Forward, T_A = 25°C (300 µs, 120 PPS)	I_{GFM}	1.0*	Amp
Peak Gate Voltage — Reverse	V_{GRM}	5.0*	Volts
Operating Junction Temperature Range @ Rated V_{RRM} and V_{DRM}	T_J	−65 to +125*	°C
Storage Temperature Range	T_{stg}	−65 to +150*	°C
Lead Solder Temperature (<1/16" from case, 10 s max)	—	+230*	°C

THERMAL CHARACTERISTICS

Characteristic	Symbol	Max	Unit
Thermal Resistance, Junction to Case	θ_{JC}	75	°C/W
Thermal Resistance, Junction to Ambient	θ_{JA}	200	°C/W

♦Annular Semiconductor Patented by Motorola Inc.
*Indicates JEDEC Registered Data.
(1) Temperature reference point for all case temperatures in center of flat portion of package. (T_C = +125°C unless otherwise noted.)

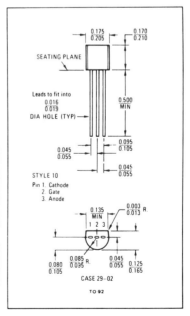

2N5060 thru 2N5064

ELECTRICAL CHARACTERISTICS (R_{GK} = 1000 Ohms)

Characteristic		Symbol	Min	Max	Unit
Peak Forward Blocking Voltage (Note 1)		V_{DRM}			Volts
(T_C = 125°C)	2N5060		30*	–	
	2N5061		60*	–	
	2N5062		100*	–	
	2N5063		150*	–	
	2N5064		200*	–	
Peak Forward Blocking Current (Rated V_{DRM} @ T_C = 125°C)		I_{DRM}	–	50*	µA
Peak Reverse Blocking Current (Rated V_{RRM} @ T_C = 125°C)		I_{RRM}	–	50*	µA
Forward "On" Voltage (Note 2) (I_{TM} = 1.2 A peak @ T_A = 25°C)		V_{TM}	–	1.7*	Volts
Gate Trigger Current (Continuous dc) (Note 3) (Anode Voltage = 7.0 Vdc, R_L = 100 Ohms)	T_C = 25°C T_C = -65°C	I_{GT}	– –	200 350*	µA
Gate Trigger Voltage (Continuous dc) (Anode Voltage = 7.0 Vdc, R_L = 100 Ohms) (Anode Voltage = Rated V_{DRM}, R_L = 100 Ohms)	T_C = 25°C T_C = -65°C T_C = 125°C	V_{GT} V_{GD}	0.8 – 0.1	1.2* –	Volts
Holding Current (Anode Voltage = 7.0 Vdc, initiating current = 20 mA)	T_C = 25°C T_C = -65°C	I_H	– –	5.0 10*	mA
Thermal Resistance, Junction to Case (Note 4)		θ_{JC}	–	75*	°C/W
Thermal Resistance, Junction to Ambient		θ_{JA}	–	200	°C/W

*Indicates JEDEC Registered Data.

1. V_{DRM} and V_{RRM} for all types can be applied on a continuous dc basis without incurring damage. Ratings apply for zero or negative gate voltage but positive gate voltage shall not be applied concurrently with a negative potential on the anode. When checking forward or reverse blocking capability, thyristor devices should not be tested with a constant current source in a manner that the voltage applied exceeds the rated blocking voltage.

2. Forward current applied for 1.0 ms maximum duration, duty cycle ≤ 1.0%.

3. R_{GK} current is not included in measurement.

4. This measurement is made with the case mounted "flat side down" on a heat sink and held in position by means of a metal clamp over the curved surface.

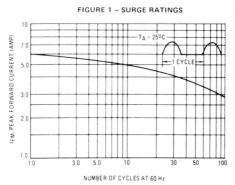

FIGURE 1 – SURGE RATINGS

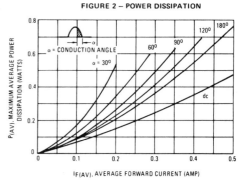

FIGURE 2 – POWER DISSIPATION

MOTOROLA Semiconductor Products Inc.

2N6068, A,B thru 2N6075, A,B

SILICON BIDIRECTIONAL THYRISTORS

... designed primarily for full-wave ac control applications, such as light dimmers, motor controls, heating controls and power supplies; or wherever full-wave silicon gate controlled solid-state devices are needed. Triac type thyristors switch from a blocking to a conducting state for either polarity of applied anode voltage with positive or negative gate triggering.

- Sensitive Gate Triggering (A and B versions) Uniquely Compatible for Direct Coupling to TTL, HTL, CMOS and Operational Amplifier Integrated Circuit Logic Functions.
- Gate Triggering 2 Mode – 2N6068 thru 2N6075
  4 Mode – 2N6068A,B thru 2N6075A,B
- Blocking Voltages to 600 Volts
- All Diffused and Glass Passivated Junctions for Greater Parameter Uniformity and Stability
- Small, Rugged, Thermopad Construction for Low Thermal Resistance, High Heat Dissipation and Durability

SENSITIVE GATE

TRIACS
(THYRISTORS)
4 AMPERES RMS
25 THRU 600 VOLTS

MAXIMUM RATINGS

Rating	Symbol	Value	Unit
*Repetitive Peak Off-State Voltage, Note 1 ($T_J = 110°C$)	V_{DRM}		Volts
2N6068, A,B		25	
2N6069, A,B		50	
2N6070, A,B		100	
2N6071, A,B		200	
2N6072, A,B		300	
2N6073, A,B		400	
2N6074, A,B		500	
2N6075, A,B		600	
*On-State Current RMS ($T_C = 85°C$)	$I_{T(RMS)}$	4.0	Amp
*Peak Surge Current (One Full cycle, 60 Hz, $T_J = -40$ to $+110°C$)	I_{TSM}	30	Amp
Circuit Fusing Considerations ($T_J = -40$ to $+110°C$, t = 1.0 to 8.3 ms)	I^2t	3.6	A^2s
*Peak Gate Power	P_{GM}	10	Watts
*Average Gate Power	$P_{G(AV)}$	0.5	Watt
*Peak Gate Voltage	V_{GM}	5.0	Volts
*Operating Junction Temperature Range	T_J	-40 to $+110$	°C
*Storage Temperature Range	T_{stg}	-40 to $+150$	°C
Mounting Torque (6-32 Screw), Note 2	–	8.0	in. lb.

THERMAL CHARACTERISTICS

Characteristic	Symbol	Max	Unit
*Thermal Resistance, Junction to Case	$R\theta_{JC}$	3.5	°C/W
Thermal Resistance, Case to Ambient	$R\theta_{CA}$	60	°C/W

*Indicates JEDEC Registered Data

NOTES:
1. Ratings apply for open gate conditions. Thyristor devices shall not be tested with a constant current source for blocking capability such that the voltage applied exceeds the rated blocking voltage.
2. Torque rating applies with use of torque washer (Shakeproof WD19523 or equivalent). Mounting torque in excess of 6 in. lb. does not appreciably lower case-to-sink thermal resistance. Main terminal 2 and heatsink contact pad are common.

   For soldering purposes (either terminal connection or device mounting), soldering temperatures shall not exceed +200°C, for 10 seconds. Consult factory for lead bending options.

Thermopad is a Trademark of Motorola Inc.

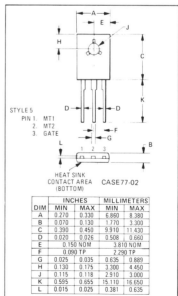

STYLE 5
PIN 1. MT1
2. MT2
3. GATE

HEAT SINK CONTACT AREA (BOTTOM) CASE 77-02

DIM	INCHES		MILLIMETERS	
	MIN	MAX	MIN	MAX
A	0.270	0.330	6.860	8.380
B	0.070	0.130	1.770	3.300
C	0.390	0.450	9.910	11.430
D	0.020	0.026	0.508	0.660
E	0.150 NOM		3.810 NOM	
F	0.090 TP		2.290 TP	
G	0.025	0.035	0.635	0.889
H	0.130	0.175	3.300	4.450
J	0.115	0.118	2.910	3.000
K	0.595	0.655	15.110	16.650
L	0.015	0.025	0.381	0.635

DS 8544 R1

App. A   Manufacturers' Data Sheets

ELECTRICAL CHARACTERISTICS ($T_C = 25°C$ unless otherwise noted)

Characteristic	Symbol	Min	Typ	Max	Unit
*Peak Blocking Current (Either Direction) Rated V_{DRM} @ $T_J = 110°C$, Gate Open	I_{DRM}	—	—	2.0	mA
*On-State Voltage (Either Direction) $I_{TM} = 6.0$ A Peak	V_{TM}	—	—	2.0	Volts

*Peak Gate Trigger Voltage
Main Terminal Voltage = 12 Vdc, R_L = 100 Ohms, $T_J = -40°C$
  MT2 (+), G(+); MT2 (-), G(-) All Types
  MT2 (+), G(-); MT2 (-), G(+) 2N6068A,B thru 2N6075A,B
Main Terminal Voltage = Rated V_{DRM}, R_L = 10 k ohms, $T_J = 110°C$
  MT2 (+), G(+); MT2 (-), G(-) All Types
  MT2 (+), G(-); MT2 (-), G(+) 2N6068A,B thru 2N6075A,B
	V_{GTM}	
		
		—
		—
		
		0.2
		0.2
			
			1.4
			1.4
			
			—
			—
				
				2.5
				2.5
				
				—
				—
					Volts

*Holding Current (Either Direction)
Main Terminal Voltage = 12 Vdc, Gate Open, $T_J = -40°C$
  Initiating Current = 1.0 Adc
                      2N6068 thru 2N6075
                      2N6068A,B thru 2N6075A,B
$T_J = 25°C$ 2N6068 thru 2N6075
               2N6068A,B thru 2N6075A,B
	I_H	
		
		—
		—
		—
		—
			
			—
			—
			—
			—
				
				70
				30
				30
				15
					mA

Turn-On Time (Either Direction) $I_{TM} = 14$ Adc, $I_{GT} = 100$ mAdc	t_{on}	—	1.5	—	μs
Blocking Voltage Application Rate at Commutation @ V_{DRM}, $T_J = 85°C$, Gate Open	dv/dt	—	5.0	—	V/μs

			QUADRANT (See Definition Below)			
	Type	I_{GTM} @ T_J	I mA	II mA	III mA	IV mA
•Peak Gate Trigger Current Main Terminal Voltage = 12 Vdc, R_L = 100 ohms Maximum Value	2N6068 thru 2N6075	+25°C -40°C	30 60	— —	30 60	— —
	2N6068A thru 2N6075A	+25°C -40°C	5.0 20	5.0 20	5.0 20	10 30
	2N6068B thru 2N6075B	+25°C -40°C	3.0 15	3.0 15	3.0 15	5.0 20

* Indicates JEDEC Registered Data.

SAMPLE APPLICATION:
TTL-SENSITIVE GATE 4 AMPERE TRIAC TRIGGERS IN MODES II AND III

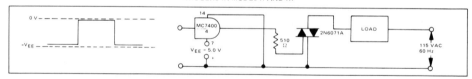

QUADRANT DEFINITIONS

Trigger devices are recommended for gating on Triacs. They provide
1. Consistent predictable turn-on points
2. Simplified circuitry
3. Fast turn-on time for cooler, more efficient and reliable operation

For 2N6068 Thru 2N6075
ELECTRICAL CHARACTERISTICS of RECOMMENDED BIDIRECTIONAL SWITCHES

USAGE	General		Lamp Dimmer
PART NUMBER	MBS4991	MBS4992	MBS100
V_S	6.0 10 V	7.5 9.0 V	3.0 5.0 V
I_S	350 μA Max	120 μA Max	100 400 μA
$V_{S1} - V_{S2}$	0.5 V Max	0.2 V Max	0.35 V Max
Temperature Coefficient		0.02%/°C Typ	

See AN-526 for Theory and Characteristics of Silicon Bidirectional Switches

SENSITIVE GATE LOGIC REFERENCE

IC LOGIC FUNCTIONS	FIRING QUADRANT			
	I	II	III	IV
TTL		2N6068A Series	2N6068A Series	
HTL		2N6068A Series	2N6068A Series	
McMOS (NAND)	2N6068B Series			2N6068B Series
McMOS (Buffer)		2N6068B Series	2N6068B Series	
Operational Amplifier	2N6068A Series			2N6068A Series
Zero Voltage Switch		2N6068A Series	2N6068A Series	

MOTOROLA *Semiconductor Products Inc.*

2N6068,A,B thru 2N6075,A,B

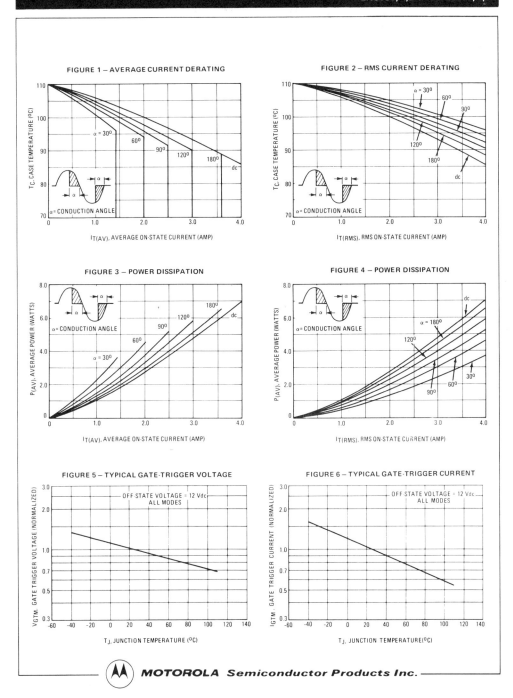

App. A Manufacturers' Data Sheets 407

MFOD2405

INTEGRATED DETECTOR/PREAMPLIFIER FOR FIBER OPTIC SYSTEMS

... designed as a monolithic integrated circuit containing both detector and preamplifier for use in computer, industrial control, and other communications systems.

Packaged in Motorola's hermetic TO-206AC (TO-52) case, the device fits directly into standard fiber optic connectors which also provide excellent RFI immunity. The output of the device is low impedance to provide even less sensitivity to stray interference. The MFOD2405 has a 300 μm (12 mil) optical spot with a high numerical aperture.

- Usable for Data Systems Through 40 Megabaud
- Dynamic Range Greater than 100:1
- Compatible with AMP#228756-1 and Amphenol #905-138-5001 Receptacles Using Motorola Plastic Alignment Bushing #MFOA06 (Included)
- Performance Matched to Motorola Fiber Optics Emitter
- TO-206AC (TO-52) Package — Small, Rugged and Hermetic
- 300 μm (12 mil) Diameter Optical Spot

FIBER OPTICS

INTEGRATED DETECTOR PREAMPLIFIER

MAXIMUM RATINGS (T_A = 25°C unless otherwise noted)

Rating	Symbol	Value	Unit
Supply Voltage	V_{CC}	7.5	Volts
Operating Temperature Range	T_A	−65 to +125	°C
Storage Temperature Range	T_{stg}	−65 to +150	°C

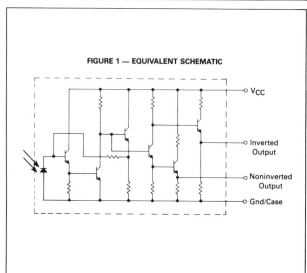

FIGURE 1 — EQUIVALENT SCHEMATIC

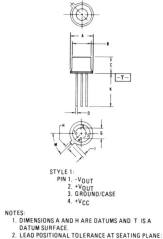

STYLE 1:
PIN 1. $-V_{OUT}$
2. $+V_{OUT}$
3. GROUND/CASE
4. $+V_{CC}$

NOTES:
1. DIMENSIONS A AND H ARE DATUMS AND T IS A DATUM SURFACE.
2. LEAD POSITIONAL TOLERANCE AT SEATING PLANE:
   ⊕ ⌀ 0.036 (0.014) Ⓜ T A Ⓜ H Ⓜ
3. DIMENSIONING AND TOLERANCING PER Y14.5, 1973.

DIM	MILLIMETERS		INCHES	
	MIN	MAX	MIN	MAX
A	5.30	5.38	0.209	0.212
B	4.64	4.69	0.183	0.185
C	3.42	3.60	0.135	0.142
D	0.40	0.48	0.016	0.019
G	2.54 BSC		0.100 BSC	
H	0.91	1.16	0.036	0.046
J	0.83	1.21	0.033	0.048
K	12.70	—	0.500	—
M	45° BSC		45° BSC	

CASE 210D-02

© MOTOROLA INC., 1984   DS2690R1

ELECTRICAL CHARACTERISTICS ($V_{CC} = 5.0$ V, $T_A = 25°C$)

Characteristics	Symbol	Conditions	Min	Typ	Max	Units
Power Supply Current	I_{CC}	Circuit A	3.0	4.5	6.0	mA
Quiescent dc Output Voltage (Non-Inverting Output)	V_q	Circuit A	0.6	0.7	0.8	Volts
Quiescent dc Output Voltage (Inverting Output)	V_q	Circuit A	2.7	3.0	3.3	Volts
Output Impedance	z_o		—	200	—	Ohms
RMS Noise Output	V_{NO}	Circuit A	—	0.5	1.0	mV

OPTICAL CHARACTERISTICS ($T_A = 25°C$)

Characteristics	Symbol	Conditions	Min	Typ	Max	Units
Responsivity ($V_{CC} = 5.0$ V, $\lambda = 820$ nm, $P = 10$ μW*)	R	Circuit B	3.0	4.5	7.0	mV/μW
Pulse Response	t_r, t_f	Circuit B	—	10	15	ns
Numerical Aperture of Input Core (200 μm [8 mil] diameter core)	NA		—	0.70	—	—
Signal-to-Noise Ratio @ $P_{in} = 2.0$ μW peak*	S/N		—	24	—	dB
Maximum Input Power for Negligible Distortion in Output Pulse*		Circuit B	—	—	120	μW

RECOMMENDED OPERATING CONDITIONS

	Symbol	Conditions	Min	Typ	Max	Units
Supply Voltage	V_{CC}		4.0	5.0	6.0	Volts
Capacitive Load (Either Output)	C_L		—	—	100	pF
Input Wavelength	λ		—	820	—	nm

*Power launched into Optical Input Port. The designer must account for interface coupling losses as discussed in AN-804.

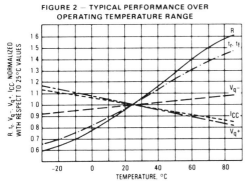

FIGURE 2 — TYPICAL PERFORMANCE OVER OPERATING TEMPERATURE RANGE

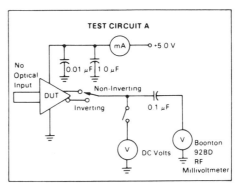

TEST CIRCUIT A

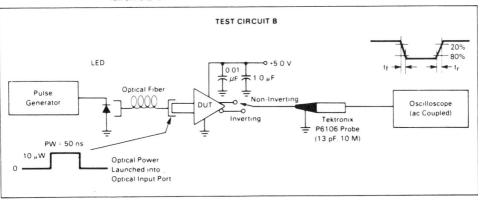

TEST CIRCUIT B

App. A Manufacturers' Data Sheets

4N29, 4N29A
4N30
4N31
4N32, 4N32A
4N33

NPN PHOTO DARLINGTON AND PN INFRARED EMITTING DIODE

... Gallium Arsenide LED optically coupled to a Silicon Photo Darlington Transistor designed for applications requiring electrical isolation, high-current transfer ratios, small package size and low cost; such as: interfacing and coupling systems, phase and feedback controls, solid-state relays and general-purpose switching circuits.

- High Isolation Voltage V_{ISO} = 7500 V (Min)
- High Collector Output Current @ I_F = 10 mA —
  - I_C = 50 mA (Min) — 4N32,33
  - 10 mA (Min) — 4N29,30
  - 5.0 mA (Min) — 4N31
- Economical, Compact, Dual-In-Line Package
- Excellent Frequency Response — 30 kHz (Typ)
- Fast Switching Times @ I_C = 50 mA
  - t_{on} = 2.0 μs (Typ)
  - t_{off} = 25 μs (Typ) — 4N29,30,31
  - 60 μs (Typ) — 4N32,33
- 4N29A, 4N32A are UL Recognized — File Number E54915

OPTO COUPLER/ISOLATOR
DARLINGTON OUTPUT

MAXIMUM RATINGS (T_A = 25°C unless otherwise noted)

Rating	Symbol	Value	Unit
INFRARED-EMITTING DIODE MAXIMUM RATINGS			
Reverse Voltage	V_R	3.0	Volts
Forward Current — Continuous	I_F	80	mA
Forward Current — Peak (Pulse Width = 300 μs, 2.0% Duty Cycle)	I_F	3.0	Amp
Total Power Dissipation @ T_A = 25°C Negligible Power in Transistor Derate above 25°C	P_D	150 2.0	mW mW/°C
PHOTOTRANSISTOR MAXIMUM RATINGS			
Collector-Emitter Voltage	V_{CEO}	30	Volts
Emitter-Collector Voltage	V_{ECO}	5.0	Volts
Collector-Base Voltage	V_{CBO}	30	Volts
Total Power Dissipation @ T_A = 25°C Negligible Power in Diode Derate above 25°C	P_D	150 2.0	mW mW/°C
TOTAL DEVICE RATINGS			
Total Device Dissipation @ T_A = 25°C Equal Power Dissipation in Each Element Derate above 25°C	P_D	250 3.3	mW mW/°C
Operating Junction Temperature Range	T_J	-55 to +100	°C
Storage Temperature Range	T_{stg}	-55 to +150	°C
Soldering Temperature (10 s)	—	260	°C

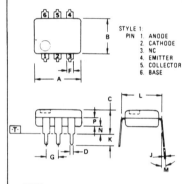

NOTES:
1. DIMENSIONS A AND B ARE DATUMS.
2. T IS SEATING PLANE.
3. POSITIONAL TOLERANCES FOR LEADS:
   ⌖ ⌀0.13 (0.005) Ⓜ T A Ⓜ B Ⓜ
4. DIMENSION L TO CENTER OF LEADS WHEN FORMED PARALLEL.
5. DIMENSIONING AND TOLERANCING PER ANSI Y14.5, 1973.

DIM	MILLIMETERS		INCHES	
	MIN	MAX	MIN	MAX
A	8.13	8.89	0.320	0.350
B	6.10	6.60	0.240	0.260
C	2.92	5.08	0.115	0.200
D	0.41	0.51	0.016	0.020
F	1.02	1.78	0.040	0.070
G	2.54 BSC		0.100 BSC	
J	0.20	0.30	0.008	0.012
K	2.54	3.81	0.100	0.150
L	7.62 BSC		0.300 BSC	
M	0°	15°	0°	15°
N	0.38	2.54	0.015	0.100
P	1.27	2.03	0.050	0.080

CASE 730A-01

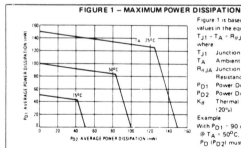

FIGURE 1 — MAXIMUM POWER DISSIPATION

Figure 1 is based upon using limit values in the equation:
$T_{J1} - T_A = R_{\theta JA}(P_{D1} + K_\theta P_{D2})$
where
- T_{J1} Junction Temperature (100°C)
- T_A Ambient Temperature
- $R_{\theta JA}$ Junction to Ambient Thermal Resistance (500°C/W)
- P_{D1} Power Dissipation in One Chip
- P_{D2} Power Dissipation in Other Chip
- K_θ Thermal Coupling Coefficient (20%)

Example
With P_{D1} = 90 mW in the LED @ T_A = 50°C, the Darlington P_D (P_{D2}) must be less than 50 mW.

4N29, 4N29A, 4N30, 4N31, 4N32, 4N32A, 4N33

LED CHARACTERISTICS ($T_A = 25°C$ unless otherwise noted)

Characteristic	Symbol	Min	Typ	Max	Unit
*Reverse Leakage Current ($V_R = 3.0$ V, $R_L = 1.0$ M ohms)	I_R	–	0.005	100	µA
*Forward Voltage ($I_F = 50$ mA)	V_F	–	1.2	1.5	Volts
Capacitance ($V_R = 0$ V, $f = 1.0$ MHz)	C	–	150	–	pF

PHOTOTRANSISTOR CHARACTERISTICS ($T_A = 25°C$ and $I_F = 0$ unless otherwise noted)

Characteristic	Symbol	Min	Typ	Max	Unit
*Collector-Emitter Dark Current ($V_{CE} = 10$ V, Base Open)	I_{CEO}	–	8.0	100	nA
*Collector-Base Breakdown Voltage ($I_C = 100$ µA, $I_E = 0$)	$V_{(BR)CBO}$	30	110	–	Volts
*Collector-Emitter Breakdown Voltage ($I_C = 100$ µA, $I_B = 0$)	$V_{(BR)CEO}$	30	75	–	Volts
*Emitter-Collector Breakdown Voltage ($I_E = 100$ µA, $I_B = 0$)	$V_{(BR)ECO}$	5.0	8.0	–	Volts
DC Current Gain ($V_{CE} = 5.0$ V, $I_C = 500$ µA)	h_{FE}	–	15 K	–	–

COUPLED CHARACTERISTICS ($T_A = 25°C$ unless otherwise noted)

Characteristic		Symbol	Min	Typ	Max	Unit
*Collector Output Current (1) ($V_{CE} = 10$ V, $I_F = $ mA, $I_B = 0$)	4N32, 4N33	I_C	50	80	–	mA
	4N29, 4N30		10	40	–	
	4N31		5.0	–	–	
Isolation Surge Voltage (2, 5) (60 Hz ac Peak, 5 Seconds)		V_{ISO}	7500	–	–	Volts
	*4N29, 4N32		2500	–	–	
	*4N30, 4N31, 4N33		1500	–	–	
Isolation Resistance (2) ($V = 500$ V)		–	–	10^{11}	–	Ohms
*Collector-Emitter Saturation Voltage (1) ($I_C = 2.0$ mA, $I_F = 8.0$ mA)	4N31	$V_{CE(sat)}$	–	0.8	1.2	Volts
	4N29, 4N39, 4N32, 4N33		–	0.8	1.0	
Isolation Capacitance (2) ($V = 0$, $f = 1.0$ MHz)		–	–	0.8	–	pF
Bandwidth (3) ($I_C = 2.0$ mA, $R_L = 100$ ohms, Figures 6 and 8)		–	–	30	–	kHz

SWITCHING CHARACTERISTICS (Figures 7 and 9), (4)

Characteristic		Symbol	Min	Typ	Max	Unit
Turn-On Time ($I_C = 50$ mA, $I_F = 200$ mA, $V_{CC} = 10$ V)		t_{on}	–	2.0	5.0	µs
Turn-Off Time ($I_C = 50$ mA, $I_F = 200$ mA, $V_{CC} = 10$ V)	4N29, 30, 31	t_{off}	–	25	40	µs
	4N32, 33		–	60	100	

*Indicates JEDEC Registered Data.
(1) Pulse Test: Pulse Width = 300 µs, Duty Cycle ≤ 2.0%.
(2) For this test, LED pins 1 and 2 are common and phototransistor pins 4, 5, and 6 are common.
(3) I_F adjusted to yield $I_C = 2.0$ mA and $i_c = 2.0$ mA P-P at 10 kHz.
(4) t_d and t_r are inversely proportional to the amplitude of I_F; t_s and t_f are not significantly affected by I_F.
(5) Isolation Surge Voltage, V_{ISO}, is an internal device dielectric breakdown rating.

DC CURRENT TRANSFER CHARACTERISTICS

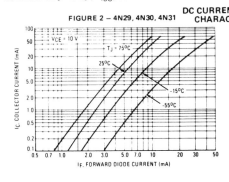

FIGURE 2 – 4N29, 4N30, 4N31

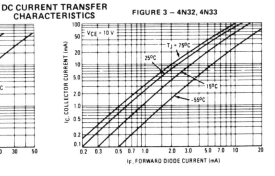

FIGURE 3 – 4N32, 4N33

App. A   Manufacturers' Data Sheets

4N29, 4N29A, 4N30, 4N31, 4N32, 4N32A, 4N33

TYPICAL ELECTRICAL CHARACTERISTICS
(Printed Circuit Board Mounting)

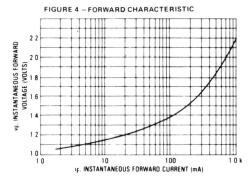

FIGURE 4 – FORWARD CHARACTERISTIC

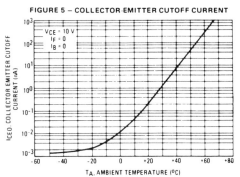

FIGURE 5 – COLLECTOR-EMITTER CUTOFF CURRENT

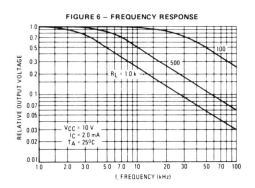

FIGURE 6 – FREQUENCY RESPONSE

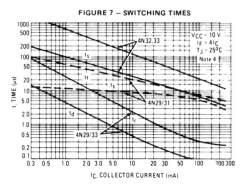

FIGURE 7 – SWITCHING TIMES

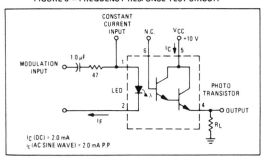

FIGURE 8 – FREQUENCY RESPONSE TEST CIRCUIT

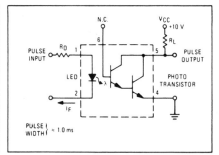

FIGURE 9 – SWITCHING TIME TEST CIRCUIT

4N29, 4N29A, 4N30, 4N31, 4N32, 4N32A, 4N33

TYPICAL APPLICATIONS

FIGURE 10 – VOLTAGE CONTROLLED TRIAC

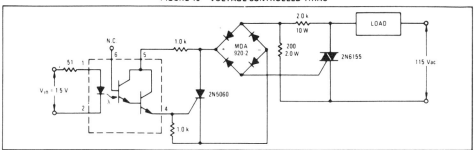

FIGURE 11 – AC SOLID STATE RELAY

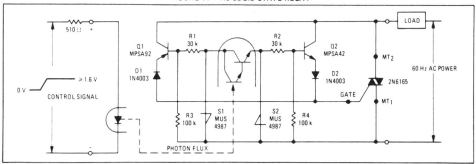

FIGURE 12 – OPTICALLY COUPLED ONE SHOT

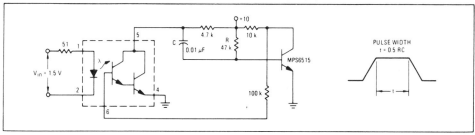

FIGURE 13 – ZERO VOLTAGE SWITCH

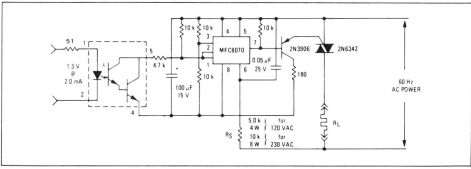

App. A   Manufacturers' Data Sheets

Operational Amplifiers/Buffers

LM741/LM741A/LM741C/LM741E Operational Amplifier

General Description

The LM741 series are general purpose operational amplifiers which feature improved performance over industry standards like the LM709. They are direct, plug-in replacements for the 709C, LM201, MC1439 and 748 in most applications.

The amplifiers offer many features which make their application nearly foolproof: overload protection on the input and output, no latch-up when the common mode range is exceeded, as well as freedom from oscillations.

The LM741C/LM741E are identical to the LM741/LM741A except that the LM741C/LM741E have their performance guaranteed over a 0°C to +70°C temperature range, instead of −55°C to +125°C.

Schematic and Connection Diagrams (Top Views)

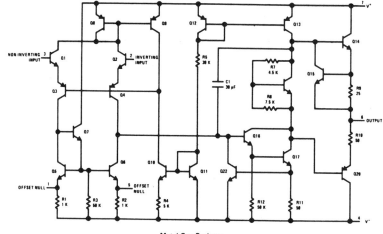

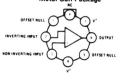

Metal Can Package

Order Number LM741H, LM741AH, LM741CH or LM741EH
See NS Package H08C

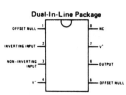

Dual-In-Line Package

Order Number LM741CN or LM741EN
See NS Package N08B
Order Number LM741CJ
See NS Package J08A

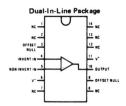

Dual-In-Line Package

Order Number LM741CN-14
See NS Package N14A
Order Number LM741J-14, LM741AJ-14
or LM741CJ-14
See NS Package J14A

Absolute Maximum Ratings

	LM741A	LM741E	LM741	LM741C
Supply Voltage	±22V	±22V	±22V	±18V
Power Dissipation (Note 1)	500 mW	500 mW	500 mW	500 mW
Differential Input Voltage	±30V	±30V	±30V	±30V
Input Voltage (Note 2)	±15V	±15V	±15V	±15V
Output Short Circuit Duration	Indefinite	Indefinite	Indefinite	Indefinite
Operating Temperature Range	−55°C to +125°C	0°C to +70°C	−55°C to +125°C	0°C to +70°C
Storage Temperature Range	−65°C to +150°C	−65°C to +150°C	−65°C to +150°C	−65°C to +150°C
Lead Temperature (Soldering, 10 seconds)	300°C	300°C	300°C	300°C

Electrical Characteristics (Note 3)

PARAMETER	CONDITIONS	LM741A/LM741E MIN	TYP	MAX	LM741 MIN	TYP	MAX	LM741C MIN	TYP	MAX	UNITS
Input Offset Voltage	$T_A = 25°C$										
	$R_S \leq 10\ k\Omega$					1.0	5.0		2.0	6.0	mV
	$R_S \leq 50\ \Omega$		0.8	3.0							mV
	$T_{AMIN} \leq T_A \leq T_{AMAX}$										
	$R_S \leq 50\ \Omega$			4.0							mV
	$R_S \leq 10\ k\Omega$						6.0			7.5	mV
Average Input Offset Voltage Drift				15							µV/°C
Input Offset Voltage Adjustment Range	$T_A = 25°C$, $V_S = ±20V$	±10				±15			±15		mV
Input Offset Current	$T_A = 25°C$		3.0	30		20	200		20	200	nA
	$T_{AMIN} \leq T_A \leq T_{AMAX}$			70		85	500			300	nA
Average Input Offset Current Drift				0.5							nA/°C
Input Bias Current	$T_A = 25°C$		30	80		80	500		80	500	nA
	$T_{AMIN} \leq T_A \leq T_{AMAX}$			0.210			1.5			0.8	µA
Input Resistance	$T_A = 25°C$, $V_S = ±20V$	1.0	6.0		0.3	2.0		0.3	2.0		MΩ
	$T_{AMIN} \leq T_A \leq T_{AMAX}$, $V_S = ±20V$	0.5									MΩ
Input Voltage Range	$T_A = 25°C$							±12	±13		V
	$T_{AMIN} \leq T_A \leq T_{AMAX}$				±12	±13					V
Large Signal Voltage Gain	$T_A = 25°C$, $R_L \geq 2\ k\Omega$										
	$V_S = ±20V$, $V_O = ±15V$	50									V/mV
	$V_S = ±15V$, $V_O = ±10V$				50	200		20	200		V/mV
	$T_{AMIN} \leq T_A \leq T_{AMAX}$, $R_L \geq 2\ k\Omega$,										
	$V_S = ±20V$, $V_O = ±15V$	32									V/mV
	$V_S = ±15V$, $V_O = ±10V$				25			15			V/mV
	$V_S = ±5V$, $V_O = ±2V$	10									V/mV
Output Voltage Swing	$V_S = ±20V$										
	$R_L \geq 10\ k\Omega$	±16									V
	$R_L \geq 2\ k\Omega$	±15									V
	$V_S = ±15V$										
	$R_L \geq 10\ k\Omega$				±12	±14		±12	±14		V
	$R_L \geq 2\ k\Omega$				±10	±13		±10	±13		V
Output Short Circuit Current	$T_A = 25°C$	10	25	35		25			25		mA
	$T_{AMIN} \leq T_A \leq T_{AMAX}$	10		40							mA
Common-Mode Rejection Ratio	$T_{AMIN} \leq T_A \leq T_{AMAX}$										
	$R_S \leq 10\ k\Omega$, $V_{CM} = ±12V$				70	90		70	90		dB
	$R_S \leq 50\ k\Omega$, $V_{CM} = ±12V$	80	95								dB

App. A   Manufacturers' Data Sheets

Electrical Characteristics (Continued)

PARAMETER	CONDITIONS	LM741A/LM741E MIN	TYP	MAX	LM741 MIN	TYP	MAX	LM741C MIN	TYP	MAX	UNITS
Supply Voltage Rejection Ratio	$T_{AMIN} \leq T_A \leq T_{AMAX}$: $V_S = \pm 20V$ to $V_S = \pm 5V$										
	$R_S \leq 50\Omega$	86	96								dB
	$R_S \leq 10\,k\Omega$				77	96		77	96		dB
Transient Response	$T_A = 25°C$, Unity Gain										
Rise Time			0.25	0.8		0.3			0.3		μs
Overshoot			6.0	20		5			5		%
Bandwidth (Note 4)	$T_A = 25°C$	0.437	1.5								MHz
Slew Rate	$T_A = 25°C$, Unity Gain	0.3	0.7			0.5			0.5		V/μs
Supply Current	$T_A = 25°C$					1.7	2.8		1.7	2.8	mA
Power Consumption	$T_A = 25°C$										
	$V_S = \pm 20V$		80	150							mW
	$V_S = \pm 15V$					50	85		50	85	mW
LM741A	$V_S = \pm 20V$										
	$T_A = T_{AMIN}$			165							mW
	$T_A = T_{AMAX}$			135							mW
LM741E	$V_S = \pm 20V$			150							mW
	$T_A = T_{AMIN}$			150							mW
	$T_A = T_{AMAX}$			150							mW
LM741	$V_S = \pm 15V$										
	$T_A = T_{AMIN}$					60	100				mW
	$T_A = T_{AMAX}$					45	75				mW

Note 1: The maximum junction temperature of the LM741/LM741A is 150°C, while that of the LM741C/LM741E is 100°C. For operation at elevated temperatures, devices in the TO-5 package must be derated based on a thermal resistance of 150°C/W junction to ambient, or 45°C/W junction to case. The thermal resistance of the dual-in-line package is 100°C/W junction to ambient.

Note 2: For supply voltages less than ±15V, the absolute maximum input voltage is equal to the supply voltage.

Note 3: Unless otherwise specified, these specifications apply for $V_S = \pm 15V$, $-55°C \leq T_A \leq +125°C$ (LM741/LM741A). For the LM741C/LM741E, these specifications are limited to $0°C \leq T_A \leq +70°C$.

Note 4: Calculated value from: BW (MHz) = 0.35/Rise Time(μs).

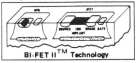

LF351 Wide Bandwidth JFET Input Operational Amplifier

Operational Amplifiers/Buffers

BI-FET II™ Technology

General Description

The LF351 is a low cost high speed JFET input operational amplifier with an internally trimmed input offset voltage (BI-FET II™ technology). The device requires a low supply current and yet maintains a large gain bandwidth product and a fast slew rate. In addition, well matched high voltage JFET input devices provide very low input bias and offset currents. The LF351 is pin compatible with the standard LM741 and uses the same offset voltage adjustment circuitry. This feature allows designers to immediately upgrade the overall performance of existing LM741 designs.

The LF351 may be used in applications such as high speed integrators, fast D/A converters, sample-and-hold circuits and many other circuits requiring low input offset voltage, low input bias current, high input impedance, high slew rate and wide bandwidth. The device has low noise and offset voltage drift, but for applications where these requirements are critical, the LF356 is recommended. If maximum supply current is important, however, the LF351 is the better choice.

Features

- Internally trimmed offset voltage — 10 mV
- Low input bias current — 50 pA
- Low input noise voltage — 16 nV/√Hz
- Low input noise current — 0.01 pA/√Hz
- Wide gain bandwidth — 4 MHz
- High slew rate — 13 V/μs
- Low supply current — 1.8 mA
- High input impedance — $10^{12}\,\Omega$
- Low total harmonic distortion $A_V = 10$, $R_L = 10k$, $V_O = 20$ Vp-p, BW = 20 Hz-20 kHz — <0.02%
- Low 1/f noise corner — 50 Hz
- Fast settling time to 0.01% — 2 μs

Typical Connection

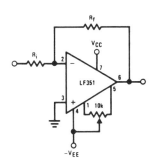

Simplified Schematic

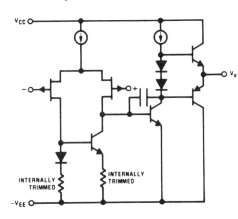

Connection Diagrams (Top Views)

Metal Can Package

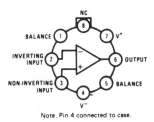

Note. Pin 4 connected to case.

Order Number LF351H
See NS Package H08C

Dual-In-Line Package

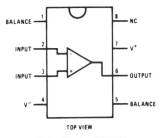

TOP VIEW

Order Number LF351N
See NS Package N08A

App. A Manufacturers' Data Sheets

LF351

Absolute Maximum Ratings

Supply Voltage	±18V
Power Dissipation (Note 1)	500 mW
Operating Temperature Range	0°C to +70°C
T_j(MAX)	115°C
Differential Input Voltage	±30V
Input Voltage Range (Note 2)	±15V
Output Short Circuit Duration	Continuous
Storage Temperature Range	−65°C to +150°C
Lead Temperature (Soldering, 10 seconds)	300°C

DC Electrical Characteristics (Note 3)

SYMBOL	PARAMETER	CONDITIONS	LF351 MIN	LF351 TYP	LF351 MAX	UNITS
V_{OS}	Input Offset Voltage	$R_S = 10 k\Omega$, $T_A = 25°C$		5	10	mV
		Over Temperature			13	mV
$\Delta V_{OS}/\Delta T$	Average TC of Input Offset Voltage	$R_S = 10 k\Omega$		10		$\mu V/°C$
I_{OS}	Input Offset Current	$T_j = 25°C$, (Notes 3, 4)		25	100	pA
		$T_j \leq 70°C$			4	nA
I_B	Input Bias Current	$T_j = 25°C$, (Notes 3, 4)		50	200	pA
		$T_j \leq 70°C$			8	nA
R_{IN}	Input Resistance	$T_j = 25°C$		10^{12}		Ω
A_{VOL}	Large Signal Voltage Gain	$V_S = \pm 15V$, $T_A = 25°C$ $V_O = \pm 10V$, $R_L = 2 k\Omega$	25	100		V/mV
		Over Temperature	15			V/mV
V_O	Output Voltage Swing	$V_S = \pm 15V$, $R_L = 10 k\Omega$	±12	±13.5		V
V_{CM}	Input Common-Mode Voltage Range	$V_S = \pm 15V$	±11	+15 −12		V
CMRR	Common-Mode Rejection Ratio	$R_S \leq 10 k\Omega$	70	100		dB
PSRR	Supply Voltage Rejection Ratio	(Note 5)	70	100		dB
I_S	Supply Current			1.8	3.4	mA

AC Electrical Characteristics (Note 3)

SYMBOL	PARAMETER	CONDITIONS	LF351 MIN	LF351 TYP	LF351 MAX	UNITS
SR	Slew Rate	$V_S = \pm 15V$, $T_A = 25°C$		13		V/μs
GBW	Gain Bandwidth Product	$V_S = \pm 15V$, $T_A = 25°C$		4		MHz
e_n	Equivalent Input Noise Voltage	$T_A = 25°C$, $R_S = 100\Omega$, f = 1000 Hz		16		nV/$\sqrt{Hz}$
i_n	Equivalent Input Noise Current	$T_j = 25°C$, f = 1000 Hz		0.01		pA/$\sqrt{Hz}$

Note 1: For operating at elevated temperature, the device must be derated based on a thermal resistance of 150°C/W junction to ambient or 45°C/W junction to case.
Note 2: Unless otherwise specified the absolute maximum negative input voltage is equal to the negative power supply voltage.
Note 3: These specifications apply for $V_S = \pm 15V$ and $0°C \leq T_A \leq +70°C$. V_{OS}, I_B and I_{OS} are measured at $V_{CM} = 0$.
Note 4: The input bias currents are junction leakage currents which approximately double for every 10°C increase in the junction temperature, T_j. Due to the limited production test time, the input bias currents measured are correlated to junction temperature. In normal operation the junction temperature rises above the ambient temperature as a result of internal power dissipation, P_D. $T_j = T_A + \theta_{jA} P_D$ where θ_{jA} is the thermal resistance from junction to ambient. Use of a heat sink is recommended if input bias current is to be kept to a minimum.
Note 5.: Supply voltage rejection ratio is measured for both supply magnitudes increasing or decreasing simultaneously in accordance with common practice.

Typical Applications

Supply Current Indicator/Limiter

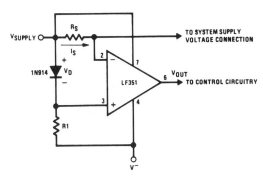

- V_{OUT} switches high when $R_S I_S > V_D$

Hi-Z_{IN} Inverting Amplifier

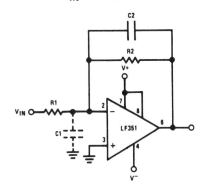

Parasitic input capacitance C1 ≅ (3 pF for LF351 plus any additional layout capacitance) interacts with feedback elements and creates undesirable high frequency pole. To compensate, add C2 such that: R2C2 ≅ R1C1.

Ultra-Low (or High) Duty Cycle Pulse Generator

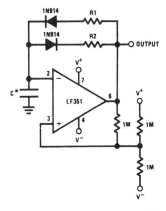

- $t_{OUTPUT\ HIGH} \approx R1C \ln \dfrac{4.8 - 2V_S}{4.8 - V_S}$
- $t_{OUTPUT\ LOW} \approx R2C \ln \dfrac{2V_S - 7.8}{V_S - 7.8}$

where $V_S = V^+ + |V^-|$

*low leakage capacitor

Long Time Integrator

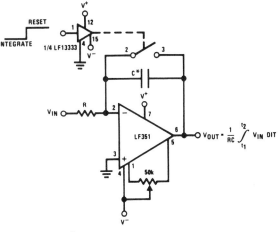

* Low leakage capacitor
* 50k pot used for less sensitive V_{OS} adjust

App. A   Manufacturers' Data Sheets

MC1723
MC1723C

MONOLITHIC VOLTAGE REGULATOR

The MC1723 is a positive or negative voltage regulator designed to deliver load current to 150 mAdc. Output current capability can be increased to several amperes through use of one or more external pass transistors. MC1723 is specified for operation over the military temperature range (-55°C to +125°C) and the MC1723C over the commercial temperature range (0 to +70°C).

- Output Voltage Adjustable from 2 Vdc to 37 Vdc
- Output Current to 150 mAdc Without External Pass Transistors
- 0.01% Line and 0.03% Load Regulation
- Adjustable Short-Circuit Protection

VOLTAGE REGULATOR

SILICON
MONOLITHIC
INTEGRATED CIRCUIT

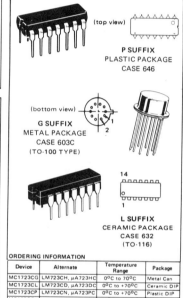

(top view)

P SUFFIX
PLASTIC PACKAGE
CASE 646

(bottom view)

G SUFFIX
METAL PACKAGE
CASE 603C
(TO-100 TYPE)

L SUFFIX
CERAMIC PACKAGE
CASE 632
(TO-116)

ORDERING INFORMATION

Device	Alternate	Temperature Range	Package
MC1723CG	LM723CH, µA723HC	0°C to 70°C	Metal Can
MC1723CL	LM723CD, µA723DC	0°C to +70°C	Ceramic DIP
MC1723CP	LM723CN, µA723PC	0°C to +70°C	Plastic DIP
MC1723G	—	-55°C to +125°C	Metal Can
MC1723L	—	-55°C to +125°C	Ceramic DIP

FIGURE 1 — CIRCUIT SCHEMATIC

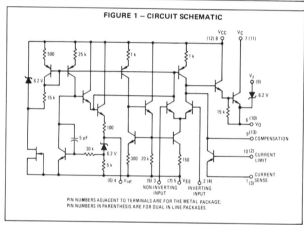

PIN NUMBERS ADJACENT TO TERMINALS ARE FOR THE METAL PACKAGE.
PIN NUMBERS IN PARENTHESIS ARE FOR DUAL IN LINE PACKAGES.

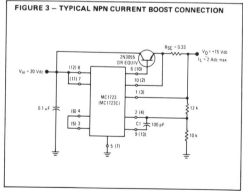

FIGURE 2 — TYPICAL CIRCUIT CONNECTION

$V_O \cong 7 \left(\frac{R1 + R2}{R2} \right)$ $I_{SC} = \frac{V_{sense}}{R_{SC}} = \frac{0.66}{R_{SC}}$ at $T_J = +25°C$

For best results 10 k < R2 < 100 k
For minimum drift R3 = R1||R2

FIGURE 3 — TYPICAL NPN CURRENT BOOST CONNECTION

©MOTOROLA INC., 1976

MAXIMUM RATINGS (T_A = +25°C unless otherwise noted.)

Rating	Symbol	Value	Unit
Pulse Voltage from V_{CC} to V_{EE} (50 ms)	$V_{in(p)}$	50	V_{peak}
Continuous Voltage from V_{CC} to V_{EE}	V_{in}	40	Vdc
Input-Output Voltage Differential	$V_{in} - V_O$	40	Vdc
Maximum Output Current	I_L	150	mAdc
Current from V_{ref}	I_{ref}	15	mAdc
Current from V_z	I_z	25	mA
Voltage Between Non-Inverting Input and V_{EE}	V_{ie}	8.0	Vdc
Differential Input Voltage	V_{id}	±5.0	Vdc

Power Dissipation and Thermal Characteristics
Plastic Package
  T_A = +25°C
    Derate above T_A = +25°C
    Thermal Resistance, Junction to Air
Metal Package
  T_A = +25°C
    Derate above T_A = +25°C
    Thermal Resistance, Junction to Air
  T_C = +25°C
    Derate above T_A = +25°C
    Thermal Resistance, Junction to Case
Dual In-Line Ceramic Package
    Derate above T_A = +25°C
    Thermal Resistance, Junction to Air
	P_D
	$1/\theta_{JA}$
	θ_{JA}
	
	P_D
	$1/\theta_{JA}$
	θ_{JA}
	P_D
	$1/\theta_{JA}$
	θ_{JC}
	P_D
	$1/\theta_{JA}$
	θ_{JA}
		1.25
		10
		100
		
		1.0
		6.6
		150
		2.1
		14
		35
		1.5
		10
		100
			W
			mW/°C
			°C/W
			
			Watt
			mW/°C
			°C/W
			Watts
			mW/°C
			°C/W
			Watt
			mW/°C
			°C/W

Operating and Storage Junction Temperature Range
  Metal Package
  Dual In-Line Ceramic and Ceramic Flat Packages
	T_J, T_{stg}	
		−65 to +150
		−65 to +175
			°C

Operating Ambient Temperature Range
  MC1723C
  MC1723
	T_A	
		0 to +70
		−55 to +125
			°C

ELECTRICAL CHARACTERISTICS (Unless otherwise noted: T_A = +25°C, V_{in} 12 Vdc, V_O = 5.0 Vdc, I_L = 1.0 mAdc, R_{SC} = 0, C1 = 100 pF, C_{ref} = 0 and divider impedance as seen by the error amplifier ≤ 10 kΩ connected as shown in Figure 1)

Characteristic	Symbol	MC1723 Min	MC1723 Typ	MC1723 Max	MC1723C Min	MC1723C Typ	MC1723C Max	Unit
Input Voltage Range	V_{in}	9.5	−	40	9.5	−	40	Vdc
Output Voltage Range	V_O	2.0	−	37	2.0	−	37	Vdc
Input-Output Voltage Differential	$V_{in} - V_O$	3.0	−	38	3.0	−	38	Vdc
Reference Voltage	V_{ref}	6.95	7.15	7.35	6.80	7.15	7.50	Vdc
Standby Current Drain (I_L = 0, V_{in} = 30 V)	I_{IB}	−	2.3	3.5	−	2.3	4.0	mAdc

Output Noise Voltage (f = 100 Hz to 10 kHz)
  C_{ref} = 0
  C_{ref} = 5.0 μF
	V_N	
		−
		−
			
			20
			2.5
				
				−
				−
					
					−
					−
						
						20
						2.5
							
							−
							−
								μV(RMS)

Average Temperature Coefficient of Output Voltage (T_{low} ① < T_A < T_{high} ②)	TCV_O	−	0.002	0.015	−	0.003	0.015	%/°C

Line Regulation
  (T_A = +25°C) $\{$12 V < V_{in} < 15 V
                           12 V < V_{in} < 40 V
  (T_{low} ① < T_A < T_{high} ②)
    12 V < V_{in} < 15 V
	Reg_{in}	
		−
		−
		
		−
			
			0.01
			0.02
			
			−
				
				0.1
				0.2
				
				0.3
					
					−
					−
					
					−
						
						0.01
						0.1
						
						−
							
							0.1
							0.5
							
							0.3
								%V_O

Load Regulation (1.0 mA < I_L < 50 mA)
  T_A = +25°C
  T_{low} ① < T_A < T_{high} ②
	Reg_{load}	
		−
		−
			
			0.03
			−
				
				0.15
				0.6
					
					−
					−
						
						0.03
						−
							
							0.2
							0.6
								%V_O

Ripple Rejection (f = 50 Hz to 10 kHz)
  C_{ref} = 0
  C_{ref} = 5.0 μF
	Rej_R	
		−
		−
			
			74
			86
				
				−
				−
					
					−
					−
						
						74
						86
							
							−
							−
								dB

Short Circuit Current Limit (R_{SC} = 10 Ω, V_O = 0)	I_{SC}	−	65	−	−	65	−	mAdc
Long Term Stability	$\Delta V_O / \Delta t$	−	0.1	−	−	0.1	−	%/1000 Hr

① T_{low} = 0°C for MC1723C
       = −55°C for MC1723

② T_{high} = +70°C for MC1723C
         = +125°C for MC1723

MOTOROLA Semiconductor Products Inc.

MC1723, MC1723C

TYPICAL APPLICATIONS

Pin numbers adjacent to terminals are for the metal package;
pin numbers in parenthesis are for the dual in-line packages.

FIGURE 16 – TYPICAL CONNECTION FOR $2 < V_O < 7$

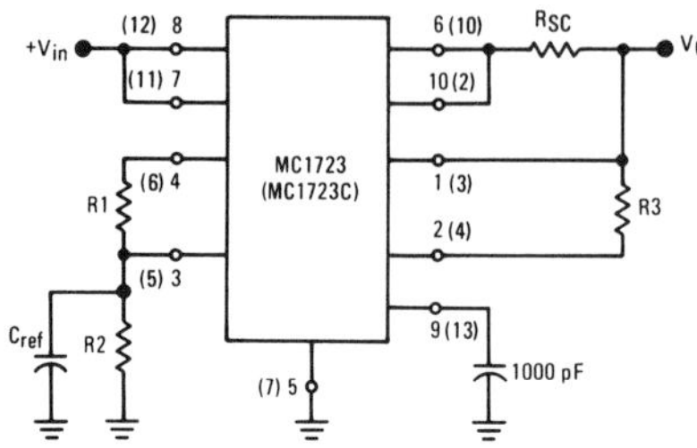

$$V_O \cong 7 \left[\frac{R2}{R1 + R2} \right] \quad I_{SC} = \frac{V_{sense}}{R_{SC}} \cong \frac{0.66}{R_{SC}} \text{ at } T_J = +25^\circ C$$

For best results $10\,k < R1 + R2 < 100\,k$.
For minimum drift $R3 = R1 \| R2$.

FIGURE 17 – MC1723,C FOLDBACK CONNECTION

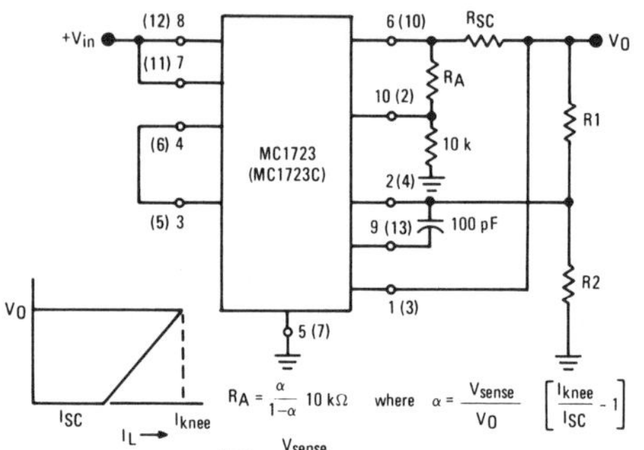

$$R_A = \frac{\alpha}{1-\alpha} \, 10\,k\Omega \quad \text{where} \quad \alpha = \frac{V_{sense}}{V_O} \left[\frac{I_{knee}}{I_{SC}} - 1 \right]$$

$$R_{SC} = \frac{V_{sense}}{(1-\alpha) \, I_{SC}}$$

FIGURE 18 – +5 V, 1-AMPERE SWITCHING REGULATOR

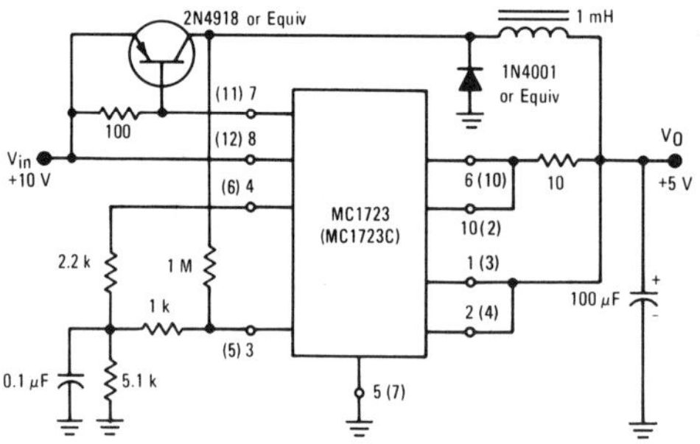

FIGURE 19 – +5 V, 1-AMPERE HIGH EFFICIENCY REGULATOR

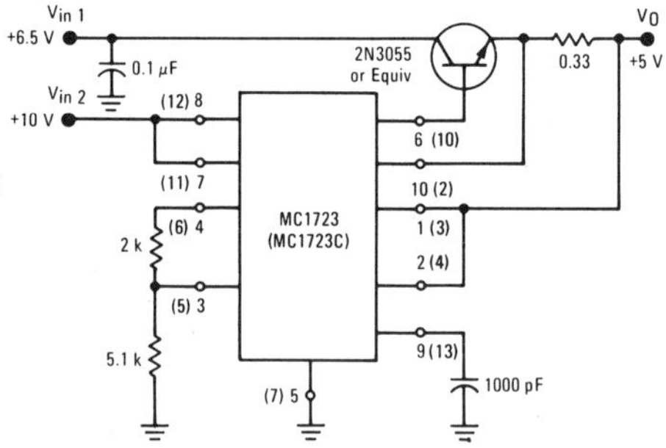

FIGURE 20 – +15 V, 1-AMPERE REGULATOR WITH REMOTE SENSE

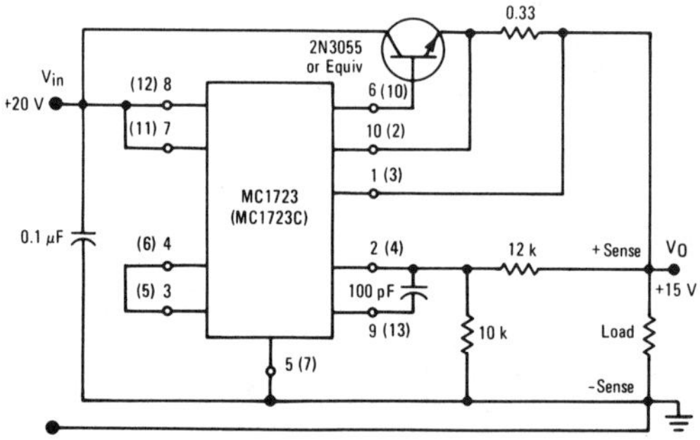

FIGURE 21 – –15 V NEGATIVE REGULATOR

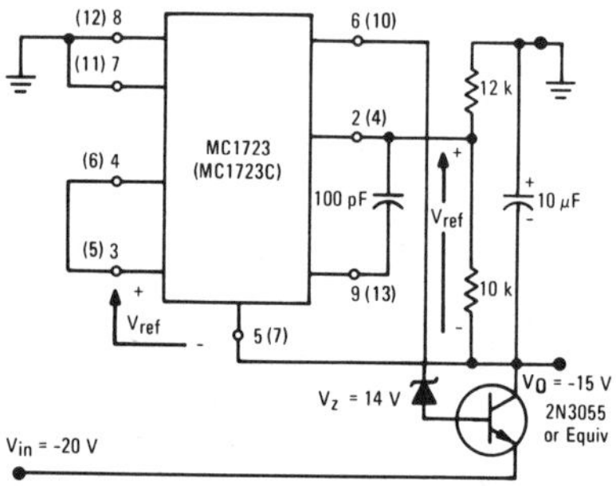

MOTOROLA Semiconductor Products Inc.

LM117
LM217
LM317

3-TERMINAL ADJUSTABLE OUTPUT POSITIVE VOLTAGE REGULATOR

The LM117/217/317 are adjustable 3-terminal positive voltage regulators capable of supplying in excess of 1.5 A over an output voltage range of 1.2 V to 37 V. These voltage regulators are exceptionally easy to use and require only two external resistors to set the output voltage. Further, they employ internal current limiting, thermal shutdown and safe area compensation, making them essentially blow-out proof.

The LM117 series serve a wide variety of applications including local, on card regulation. This device also makes an especially simple adjustable switching regulator, a programmable output regulator, or by connecting a fixed resistor between the adjustment and output, the LM117 series can be used as a precision current regulator.

- Output Current in Excess of 1.5 Ampere in TO-3 and TO-220 Packages
- Output Current in Excess of 0.5 Ampere in TO-39 Package
- Output Adjustable between 1.2 V and 37 V
- Internal Thermal Overload Protection
- Internal Short-Circuit Current Limiting Constant with Temperature
- Output Transistor Safe-area Compensation
- Floating Operation for High Voltage Applications
- Standard 3-lead Transistor Packages
- Eliminates Stocking Many Fixed Voltages

3-TERMINAL ADJUSTABLE POSITIVE VOLTAGE REGULATOR

SILICON MONOLITHIC INTEGRATED CIRCUIT

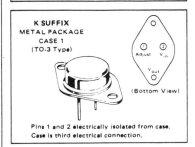

K SUFFIX
METAL PACKAGE
CASE 1
(TO-3 Type)

(Bottom View)

Pins 1 and 2 electrically isolated from case.
Case is third electrical connection.

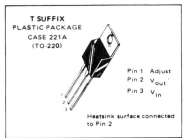

T SUFFIX
PLASTIC PACKAGE
CASE 221A
(TO-220)

Pin 1 Adjust
Pin 2 V_{out}
Pin 3 V_{in}

Heatsink surface connected to Pin 2

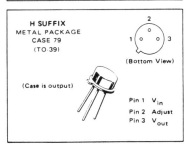

H SUFFIX
METAL PACKAGE
CASE 79
(TO-39)

(Bottom View)

(Case is output)

Pin 1 V_{in}
Pin 2 Adjust
Pin 3 V_{out}

STANDARD APPLICATION

* = C_{in} is required if regulator is located an appreciable distance from power supply filter.

** = C_o is not needed for stability, however it does improve transient response.

$$V_{out} = 1.25\ V\left(1 + \frac{R_2}{R_1}\right) + I_{Adj} R_2$$

Since I_{Adj} is controlled to less than 100 µA, the error associated with this term is negligible in most applications

ORDERING INFORMATION

Device	Temperature Range	Package
LM117H	$T_J = -55^\circ C$ to $+150^\circ C$	Metal Can
LM117K	$T_J = -55^\circ C$ to $+150^\circ C$	Metal Power
LM217H	$T_J = -25^\circ C$ to $+150^\circ C$	Metal Can
LM217K	$T_J = -25^\circ C$ to $+150^\circ C$	Metal Power
LM317H	$T_J = 0^\circ C$ to $+125^\circ C$	Metal Can
LM317K	$T_J = 0^\circ C$ to $+125^\circ C$	Metal Power
LM317T	$T_J = 0^\circ C$ to $+125^\circ C$	Plastic Power

LM117, LM217, LM317

MAXIMUM RATINGS

Rating	Symbol	Value	Unit
Input-Output Voltage Differential	V_I-V_O	40	Vdc
Power Dissipation	P_D	Internally Limited	
Operating Junction Temperature Range LM117 LM217 LM317	T_J	 –55 to +150 –25 to +150 0 to +125	°C
Storage Temperature Range	T_{stg}	–65 to +150	°C

ELECTRICAL CHARACTERISTICS ($V_I - V_O$ = 5 V; I_O = 0.5 A for K and T packages; I_O = 0.1 A for H package; T_J = T_{low} to T_{high} [see Note 1]; I_{max} and P_{max} per Note 2; unless otherwise specified.)

Characteristic	Figure	Symbol	LM117/217 Min	LM117/217 Typ	LM117/217 Max	LM317 Min	LM317 Typ	LM317 Max	Unit
Line Regulation (Note 3) T_A = 25°C, 3 V ≤ $V_I - V_O$ ≤ 40 V	1	Reg$_{line}$	–	0.01	0.02	–	0.01	0.04	%/V
Load Regulation (Note 3) T_A = 25°C, 10 mA ≤ I_O ≤ I_{max} V_O ≤ 5 V V_O ≥ 5 V	2	Reg$_{load}$	 – –	 5 0.1	 15 0.3	 – –	 5 0.1	 25 0.5	 mV % V_O
Adjustment Pin Current	3	I_{Adj}	–	50	100	–	50	100	μA
Adjustment Pin Current Change 2.5 V ≤ $V_I - V_O$ ≤ 40 V 10 mA ≤ I_L ≤ I_{max}, P_D ≤ P_{max}	1, 2	ΔI_{Adj}	–	0.2	5	–	0.2	5	μA
Reference Voltage (Note 4) 3 V ≤ $V_I - V_O$ ≤ 40 V 10 mA ≤ I_O ≤ I_{max}, P_D ≤ P_{max}	3	V_{ref}	1.20	1.25	1.30	1.20	1.25	1.30	V
Line Regulation (Note 3) 3 V ≤ $V_I - V_O$ ≤ 40 V	1	Reg$_{line}$	–	0.02	0.05	–	0.02	0.07	%/V
Load Regulation (Note 3) 10 mA ≤ I_O ≤ I_{max} V_O ≤ 5 V V_O ≥ 5 V	2	Reg$_{load}$	 – –	 20 0.3	 50 1	 – –	 20 0.3	 70 1.5	 mV %V_O
Temperature Stability (T_{low} ≤ T_J ≤ T_{high})	3	T_S	–	0.7	–	–	0.7	–	%V_O
Minimum Load Current to Maintain Regulation ($V_I - V_O$ = 40 V)	3	I_{Lmin}	–	3.5	5	–	3.5	10	mA
Maximum Output Current $V_I - V_O$ ≤ 15 V, P_D ≤ P_{max} K and T Packages H Package $V_I - V_O$ = 40 V, P_D ≤ P_{max}, T_A = 25°C K and T Packages H Package	3	I_{max}	 1.5 0.5 0.25 –	 2.2 0.8 0.4 0.07	 – – – –	 1.5 0.5 0.15 –	 2.2 0.8 0.4 0.07	 – – – –	A
RMS Noise, % of V_O T_A = 25°C, 10 Hz ≤ f ≤ 10 KHz	–	N	–	0.003	–	–	0.003	–	%V_O
Ripple Rejection, V_O = 10 V, f = 120 Hz (Note 5) Without C_{ADJ} C_{ADJ} = 10 μF	4	RR	 – 66	 65 80	 – –	 – 66	 65 80	 – –	dB
Long Term Stability, T_J = T_{high} (Note 6) T_A = 25°C for Endpoint Measurements	3	S	–	0.3	1	–	0.3	1	%/1.0k Hrs
Thermal Resistance Junction to Case H Package (TO-39) K Package (TO-3) T Package (TO-220)	–	$R_{\theta JC}$	 – – –	 12 2.3 –	 15 3 –	 – – –	 12 2.3 5	 15 3 –	°C/W

NOTES: (1) T_{low} = –55°C for LM117 T_{high} = +150°C for LM117
= –25°C for LM217   = +150°C for LM217
= 0°C for LM317   = +125°C for LM317
(2) I_{max} = 1.5 A for K (TO-3) and T (TO-220) Packages
= 0.5 A for H (TO-39) Package
P_{max} = 20 W for K (TO-3) and T (TO-220) Packages
= 2 W for H (TO-39) Package
(3) Load and line regulation are specified at constant junction temperature. Changes in V_O due to heating effects must be taken into account separately. Pulse testing with low duty cycle is used.
(4) Selected devices with tightened tolerance reference voltage available.
(5) C_{ADJ}, when used, is connected between the adjustment pin and ground.
(6) Since Long Term Stability cannot be measured on each device before shipment, this specification is an engineering estimate of average stability from lot to lot.

Manufacturers' Data Sheets   App. A

LM117, LM217, LM317

APPLICATIONS INFORMATION

BASIC CIRCUIT OPERATION

The LM117 is a 3-terminal floating regulator. In operation, the LM117 develops and maintains a nominal 1.25 volt reference (V_{ref}) between its output and adjustment terminals. This reference voltage is converted to a programming current (I_{PROG}) by R1 (see Figure 17), and this constant current flows through R2 to ground. The regulated output voltage is given by:

$$V_{out} = V_{ref}(1 + \frac{R2}{R1}) + I_{Adj} R2$$

Since the current from the adjustment terminal (I_{Adj}) represents an error term in the equation, the LM117 was designed to control I_{Adj} to less than 100 μA and keep it constant. To do this, all quiescent operating current is returned to the output terminal. This imposes the requirement for a minimum load current. If the load current is less than this minimum, the output voltage will rise.

Since the LM117 is a floating regulator, it is only the voltage differential across the circuit which is important to performance, and operation at high voltages with respect to ground is possible.

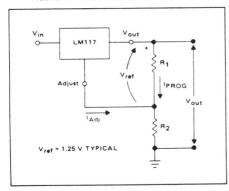

FIGURE 17 – BASIC CIRCUIT CONFIGURATION

LOAD REGULATION

The LM117 is capable of providing extremely good load regulation, but a few precautions are needed to obtain maximum performance. For best performance, the programming resistor (R1) should be connected as close to the regulator as possible to minimize line drops which effectively appear in series with the reference, thereby degrading regulation. The ground end of R2 can be returned near the load ground to provide remote ground sensing and improve load regulation.

EXTERNAL CAPACITORS

A 0.1 μF disc or 1 μF tantalum input bypass capacitor (C_{in}) is recommended to reduce the sensitivity to input line impedance.

The adjustment terminal may be bypassed to ground to improve ripple rejection. This capacitor (C_{ADJ}) prevents ripple from being amplified as the output voltage is increased. A 10 μF capacitor should improve ripple rejection about 15dB at 120 Hz in a 10 volt application.

Although the LM117 is stable with no output capacitance, like any feedback circuit, certain values of external capacitance can cause excessive ringing. An output capacitance (C_O) in the form of a 1 μF tantalum or 25 μF aluminum electrolytic capacitor on the output swamps this effect and insures stability.

PROTECTION DIODES

When external capacitors are used with any I.C. regulator it is sometimes necessary to add protection diodes to prevent the capacitors from discharging through low current points into the regulator.

Figure 18 shows the LM117 with the recommended protection diodes for output voltages in excess of 25 V or high capacitance values ($C_O > 25$ μF, $C_{ADJ} > 10$ μF). Diode D_1 prevents C_O from discharging thru the I.C. during an input short circuit. Diode D_2 protects against capacitor C_{ADJ} discharging through the I.C. during an output short circuit. The combination of diodes D1 and D2 prevents C_{ADJ} from discharging through the I.C. during an input short circuit.

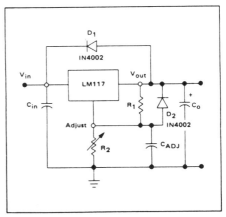

FIGURE 18 – VOLTAGE REGULATOR WITH PROTECTION DIODES

LM117, LM217, LM317

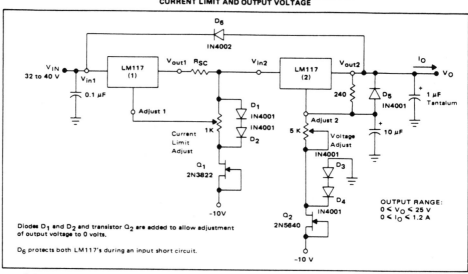

FIGURE 19 – "LABORATORY" POWER SUPPLY WITH ADJUSTABLE CURRENT LIMIT AND OUTPUT VOLTAGE

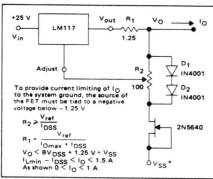

FIGURE 20 – ADJUSTABLE CURRENT LIMITER

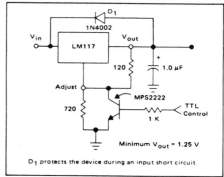

FIGURE 21 – 5 V ELECTRONIC SHUT DOWN REGULATOR

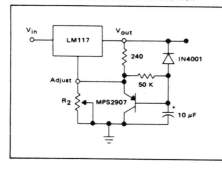

FIGURE 22 – SLOW TURN-ON REGULATOR

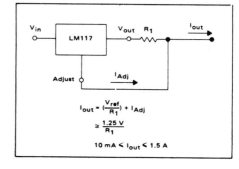

FIGURE 23 – CURRENT REGULATOR

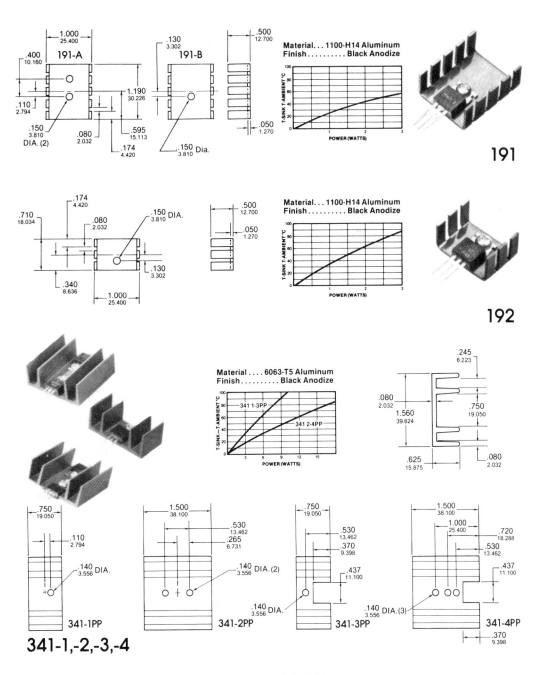

AHAM TOR INC

App. A   Manufacturers' Data Sheets

Appendix B

Answers to Selected Problems

2-4
$$I = I_o(e^{V/0.025} - 1)$$

From graph: $V = 0.3$ V
$I = 10$ mA

$$I_o = \frac{I}{e^{V/0.025} - 1} = 61.4 \text{ nA}$$

2-6 Fit a tangent to curve at $V = 0.3$ V:

$$r = \frac{\Delta V}{\Delta I} = \frac{0.35 - 0.25}{20 - 0} = 5 \,\Omega$$

$$A_A \approx 0.25 \text{ V}$$

2-10 The diodes are in parallel—for a given voltage, the net current is exactly twice that of one diode. Cut-in voltage is equal to that of a single diode and r is one-half of the dynamic resistance of a single diode.

3-1
$$I_C = 2.0 \text{ mA} \qquad I_E = 2.01 \text{ mA}$$

$$\alpha \approx \frac{I_C}{I_E} = 0.995$$

$$I_E = I_C + I_B \quad I_B = (2.01 - 2.0) \text{ mA} = 10 \text{ }\mu\text{A}$$

$$\beta = \frac{\alpha}{1 - \alpha} = 200 \quad \text{also } \beta = \frac{I_C}{I_B}$$

3-7 At $V_{CE} = 7$ V and $I_B = 200$ μA then $I_C = 18$ mA

$$\beta = \frac{I_C}{I_B} = \frac{18 \text{ mA}}{200 \text{ }\mu\text{A}} = 90$$

4-3
$$R_{C(\max)} = \frac{V_{CC}}{I_C} = \frac{20}{5 \text{ m}} = 4 \text{ k}\Omega$$

4-8
$$S = \beta + 1 = 51$$

4-13 $$h_{ie} = \frac{\Delta V_{be}}{\Delta I_B} = \frac{0.6 - 0.5 \text{ V}}{20 - 0 \text{ μA}} = 5 \text{ kΩ}$$

4-17 $$r_{be} \approx 800 \text{ Ω} \quad r_{ce} \approx \frac{1}{50 \times 10^{-6}} = 20 \text{ kΩ}$$

$$gm = \frac{65}{800} = 81.3 \text{ mS}$$

$$f_\beta = \frac{100 \times 10^6}{65} = 1.54 \text{ MHz}$$

$$C_{be} + C_{bc} = \frac{1}{2\pi(1.54 \times 10^6)800} = 120 \text{ pF}$$

$$C_{be} = 129 - 10 = 119 \text{ pF}$$

5-2 $$R_D + R_S = \frac{V_{DD} - V_{DSQ}}{I_{DQ}} = 2.38 \text{ kΩ}$$

$$V_{GSQ} = -3\left(1 - \sqrt{\frac{I_{DQ}}{I_{DSS}}}\right) = -3(1 - \sqrt{0.7}) = -0.49 \text{ V}$$

$$R_S = \frac{V_{GSQ}}{I_{DQ}} = \frac{0.49}{7 \text{ m}} = 70 \text{ Ω}$$

$$R_D = (2.38 - 0.07)\text{k} = 2.31 \text{ kΩ}$$

5-7 $$gm = \frac{\Delta I_D}{\Delta V_{GS}} = \frac{3 - 2}{-0.9 + 1.5} = 1.67 \text{ mS}$$

5-12 From Figure 5.17(b) for $I_D = 4$ mA, $V_{GS} = 3.2$ V, and $V_{DD} = 20$ V

$$V_{DD}/V_{GS} = 3.2/20 = 0.16$$

$$\frac{R_{G2}}{R_{G1} + R_{G2}} = 0.16 \quad \text{Let } R_{G2} = 100 \text{ kΩ}$$

$$R_{G1} = 525 \text{ kΩ}$$

$$R_D = \frac{V_{DD} - V_{DS}}{I_D} = \frac{20 - 10}{4 \text{ m}} = 2.5 \text{ kΩ}$$

5-16 V_p Defined as V_{GS} when $I_D = 0$

$$\therefore V_p \approx -4 \text{ V}$$

8-2 $$V_m = \frac{50}{0.707} = 70.72 \text{ V}$$

$$I_m = \frac{V_m}{R_L} = 2.36 \text{ A}$$

$$I_{dc} = \frac{2(2.36)}{\pi} = 1.5 \text{ A}$$

$$r = 0.483 \text{ (From Eq. 8.11)}$$

8-9

$$C_1 C_2 L = \frac{3300}{0.05(100)} = 660$$

when $R_L = 50\ \Omega$

$$r = \frac{3300}{50(660)} = 0.1$$

when $R_L = 200\ \Omega$

$$r = \frac{3300}{200(660)} = 0.025$$

9-2

$$V_o = -h_{fe} i_b R_c = -i_b(80)(10)$$
$$V_i = -i_b h_{ie} = -i_b(1)$$
$$A_V = \frac{V_o}{V_i} = \frac{800}{1} = 800$$

9-8

$$v_i = i_b h_{ie} + v_o$$
$$v_o = (h_{fe} i_b + i_b) R_E = (h_{fe} + 1) R_E i_b$$
$$A_V = \frac{v_o}{v_i} = \frac{(h_{fe} + 1) R_E}{h_{ie} + (h_{fe} + 1) R_E} = \frac{(50 + 1)1}{1 + (50 + 1)1} = 0.98$$

10-1

$$R_i = R_1 \| R_2 \| h_{ie} \approx 1.78\ \text{k}\Omega$$
$$R_o = \frac{1}{h_{oe}} = 40\ \text{k}\Omega$$
$$A_{ISC} = -\frac{h_{fe} R_B}{R_B + h_{ie}} = -66.6$$
$$A_{VOC} = A_{ISC} \frac{R_o}{R_i} = -1497$$

10-6

$$V_o = V_i \frac{R_i}{R_i + R_s} A_V$$

where

$$R_i = 688\ \text{k}\Omega\ \text{and}\ A_V = -30.5$$
$$R_s = 100\ \text{k}\Omega\ \text{and}\ V_i = 100\ \mu\text{V}$$
$$V_o = -2.66\ \text{mV}$$

10-11

$$20 \log(23) = 27.23\ dB$$
$$\text{Net gain} = 3(27.2) = 81.7\ dB$$

From Table 10-1 (when $n = 3$)

$$f_2 = 0.51 f_2 = 561\ \text{kHz}$$
$$f_1 = 1.96 f_1 = 78.4\ \text{Hz}$$

11-2 At 910 kHz $Q \approx 5718$

11-6
$$R_p = \frac{r_{ds}R_{res}}{r_{ds} + R_{res}} = 50 \text{ k}\Omega$$

$$Q_e = \tfrac{1}{2}Q_o = 14.15$$

$$A_{V(\text{resonant})} = A_{VOC}\frac{R_p}{r_{ds}} = -\mu\frac{R_p}{r_{ds}}$$

$$A_{V(\text{resonant})} = -\text{gm } R_p = -25$$

$$BW = \frac{f_o}{Q_e} = \frac{225}{14.15} \text{ kHz} = 15.9 \text{ kHz}$$

11-11
$$f_{o1} = 60 - 0.35(6) = 57.9 \text{ MHz}$$
$$f_{o2} = 60 + 0.35(6) = 62.1 \text{ MHz}$$
$$BW_1 = BW_2 = 0.7(6) = 4.2 \text{ MHz}$$

$$Q_{e1} = \frac{57.9}{4.2} \approx 13.8$$

$$Q_{e2} = \frac{62.1}{4.2} = 14.8$$

12-4
$$\theta_T = \theta_{JC} + \theta_{CA}$$
$$\theta_{CA} = \theta_T - \theta_{JC}$$
$$\theta_{CA} = 8.33 - 4.2 = 4.13°\text{C/W}$$

12-9 Same Q-point as Problem 8; however, a new load line is needed.

$$P_1 \approx 12.4 \text{ V}, P_2 = 4.4 \text{ V (Figure 12.3)}$$
$$M_0 \approx M_3 \approx M_4 \approx 0, M_1 = 157 \text{ mA}, M_2 \approx -20 \text{ mA}$$
$$P_0 = M_1^2 R_L = 197 \text{ mW}$$
$$P_{dc} = V_{CC}I_Q = 6.5 \text{ W}$$

$$\eta = \frac{197 \times 10^{-3}}{6.5} \times 100 = 3\%$$

$$D_T \approx D_2 = 12\%$$

13-2 Current feedback decreases R_i

$$R_{if} = \frac{R_i}{1 + \beta A_I} = 0.5 \text{ k}\Omega$$

$$1 + \beta A_I = 1.8/0.5 = 3.6$$

$$A_{If_{SC}} = \frac{A_I}{1 + \beta A_I} = \frac{80}{3.6} = 22.2$$

$$R_{of} = R_o(1 + \beta A_I) = 40 \text{ k}(3.6) = 144 \text{ k}\Omega$$

$$A_I = A_{If_{SC}}\frac{R_{of}}{R_{of} + R_2} = 22.2\left(\frac{144}{144 + 1}\right) = 22$$

13-9

$$A_v = 100 \pm 5$$

$$A_{vf} = \frac{A_v}{1 + \beta A_v} = \frac{105}{1 + \frac{105}{23}} = 18.9$$

$$A_{vf} = \frac{95}{1 + \frac{95}{23}} = 18.5$$

$$\Delta A_{vf} = \frac{18.9 - 18.5}{2(18.7)} \times 100 = 1.05\%$$

$$A_{vf} = 18.7 \pm 1.05\% \text{ with feedback}$$

The percent change in gain is approximately $\frac{1}{5}$ of that without feedback. Note that it is reduced by roughly $(1 + \beta A_v)$. The change could be minimized further by increasing β at the cost of decreasing the gain.

14-1

$$R_i = \frac{(50 + 1)^2(0.1)}{1 + (50)(0.1)(0.0125)} \approx 245 \text{ k}\Omega$$

$$A_I = \frac{(50 + 1)^2(50)(12.5 \times 10^{-3})(0.1)}{1 + (50)(12.5 \times 10^{-3})(0.1)} \approx 153$$

14-4

$$I = \frac{5.6 - 0.6}{0.4} \text{ mA} = 12.5 \text{ mA}$$

$$V_{E1} = V_{E2} = -0.6 \text{ V } (V_1 = V_2 = 0)$$

$$I_{C1} = I_{C2} = I/2 = 6.25 \text{ mA}$$

$$V_{C1} = V_{C2} = 6 - (6.25)(0.56) = 2.5 \text{ V}$$

$$I_Z = \frac{12 - 5}{1.5} = 4.67 \text{ mA}$$

15-4

$$|A_v| = g_m R_T \qquad R_T = r_{ds} \| R_D$$

$$\text{needed gain} = 10 = g_m R_T$$

$$R_T = 10 \text{ k}\Omega$$

$$R_D = \frac{r_{ds} R_T}{r_{ds} - R_T} = 12.5 \text{ k}\Omega$$

15-9

$$R_a = 4R_b = R$$
$$C_a = \tfrac{1}{2}C_b = C = 1000 \text{ pF}$$

then:

$$R = \frac{1}{2\pi f_o C} \approx 3.18 \text{ k}\Omega$$

$$R_a = 3.18 \text{ k}\Omega$$

$$R_b = 12.7 \text{ k}\Omega$$

$$R_f = 40 \text{ k}\Omega \text{ and } R_1 = 1 \text{ k}\Omega$$

16-4
$$V_1 = 60 - \frac{30}{2} = 45 \text{ V}$$
$$V_2 = 60 + \frac{30}{2} = 75 \text{ V}$$

16-9
$$I_{max} = \frac{20 - 6.2}{50} \text{ A} = 276 \text{ mA}$$
$$P_{max} = V_Z I_{max} = 1.55 \text{ W}$$

16-13
$$I_{in} = \frac{9.3 - 0.7}{10} = 0.86 \text{ mA}$$
$$I = \frac{20 - 9.3}{0.2} = 53.5 \text{ mA}$$

Allow a minimum of 5 mA for the conducting diode
$$I_o = I - 5 = 48.5 \text{ mA}$$
$$N = \frac{I_o}{I_{in}} = \frac{48.5}{0.86} = 56$$

17-2(a)
$$I_Z = \frac{15 - 12}{0.1} \text{ mA} = 30 \text{ mA}$$
$$P_Z = (12)(0.03) = 0.36 \text{ W}$$

(b)
$$I_{Z\text{min}} = 5 \text{ mA}$$
$$I_o = I - I_Z$$
$$I_{o\text{max}} = 30 - 5 = 25 \text{ mA}$$

17-5 No load;
$$I_{in} = \frac{20 - 15 - 0.6}{50} = 88 \text{ mA}$$
$$V_o \approx 15.6 \text{ V}$$
$$P_D = I_{in} V_o = 1.37 \text{ W}$$

With 250 Ω load:
$$I_L = \frac{15.6}{250} = 62.4 \text{ mA}$$
$$I_B + I_C = 88 \text{ mA} - 62.4 \text{ mA} = 25.6 \text{ mA}$$
$$P_D = V_o(I_C + I_B) = (15.6)(25.6 \times 10^{-3}) = 399 \text{ mW}$$

17-9
$$I_{B\text{max}} = 20 \text{ mA}$$
$$I_L = I_E = \frac{15 \text{ V}}{5 \text{ Ω}} = 3 \text{ A}$$
$$\beta = \frac{3}{0.02} = 150$$
$$V_{CE} = V_i - V_L = 20 - 15 = 5 \text{ V}$$
$$I_C \approx I_L = 3 \text{ A}$$
$$P_D = V_{CE} I_C = 15 \text{ W}$$

Index

A

Ac bilateral trigger diode, 113–14
  manufacturer's data sheet on, 395–96
Ac load line, 63
Ac ripple, 145
Acceptor atoms, 13
Aluminum, 4–5, 17
Amplifiers
  amplifier loading, 182–84
    effect of source and load impedances in a current amplifier, 182–83
  cascading of, 186–90
    bandwidth, 188–90
    current gain, 186–87
    voltage gain, 188
  classifications of, 160
  current amplifiers (*See* Current amplifiers)
  graphical analysis of, 63–65
  impedance matching, 184–85
  input and output impedance, 178–79, 181–82
  integrated circuit (IC) (*See* Integrated circuit (IC) amplifiers)
  large-signal (*See* Power amplifiers)
  power amplifiers (*See* Power amplifiers)
  properties common to all, 160
  real and apparent gain, 179–82
  selectivity of, 198–99
  single-stage BJT amplifier, 163–66
  single-stage FET amplifier, 166–69
  small-signal (*See* Small-signal amplifiers)
  systematic analysis for gain calculations, 162
  terminology used with, 160–61
  tuned amplifiers (*See* Tuned amplifiers)
  voltage amplifiers (*See* Voltage amplifiers)

Amplitude-stabilized oscillators, 294–95
Analog systems, 359–74
  analog circuits, 368–70
    current-to-voltage converter, 369
    voltage-to-current, 368–69
    voltage follower (buffer), 369–70
  application of, 370–74
  principles of analog computation, 359–68
    difference amplifier (subtractor), 363, 364
    differentiator, 367–68
    integrator, 364, 365–67
    operational amplifier as chief building block in, 359–60
    scaler circuit, 360, 361, 362
    sign changer (inverter), 360, 361
    summing amplifier (adder), 361, 362–63
Anode, 19
Antimony, 11–12
Arsenic, 11
Avalanche breakdown, 23, 24, 115

B

Bandwidth, 192
  adjustments in the single-tuned amplifier, 199–200
  definition of, 161
  effect of cascading amplifiers on, 188–90
Barkhausen criterion, 282–83
Bias curve equation, 53
Biasing, 51–63
  calculation of the operating point, 52–55
  drawing the dc load line, 62–63
  fixed-bias circuit (*See* Fixed-bias circuit)

Biasing (*Contd.*)
   self-bias circuit (*See* Self-bias circuit)
   thermal stability
      for a fixed-bias circuit, 54–55
      for a self-bias circuit, 58–59
Bilateral diode switch (diac), 116–17
Bilateral triode (*See* Triac)
Bipolar junction transistors, 23, 34–48, 113, 114, 134
   amplification of time-varying signals by, 63–65
   biasing (*See* Biasing)
   contrasted to JFETs and MOSFETs, 82
   currents in the, 35–41
      BJT operation, 36–37
      current relationships, 37–41
   fabrication processes, 34–35
   high-frequency model, 76–79
      cutoff frequency, 78–79
      determination of the ac short-circuit current gain, 77–78
      development of the *CE* hybrid-π model, 76–77, 105, 164
   low-frequency model, 63–76, 79
      CB h-parameter model, 72, 73
      CC h-parameter model, 72–73, 74
      CE h-parameter model, 66–70, 163–64
      CE small-signal models, 70–71
      h-parameter conversions, 73, 74–76, 387
      hybrid or *h*-parameters, 65–71
      normalized *h*-parameters, 71–73
   manufacturer's data sheet on, 382–83
   permissible region of operation, 47–48
   static characteristics of, 41–46
      common-base configuration, 41–42, 43
      common-collector configuration, 41, 46
      common-emitter configuration, 41, 42–46
      contained in two curves, 41
   structure of, 34
   temperature sensitivity of, 45–46
   three regions of, 34
   transistor ratings, 46–48
Boron, 12, 17
Bound electron, 9, 10
Breakdown voltage, 23–24
   mechanisms causing, 23
      avalanche breakdown, 23, 24
      Zener breakdown, 23, 24
   special purpose diodes for exhibiting, 24
Breakover voltage, 115, 120, 121

C

Cadmium oxide, 131
Cadmium selenide (CdSe), 128
   spectral response of, 130
Cadmium sulphide (CdS), 128
   spectral response of, 129, 130
Capacitive, definition of, 76
Capacitors, 153, 154, 155
   bypass, 163
   coupling, 163
   tapped, 202
Cathode, 19
Center-tapped bridge rectifiers, 152–53
Channel resistance, 84
Charge carriers, 9–11
   majority, 12, 13, 14, 20
   minority, 12, 13, 14, 21, 27–28
   motion within an NPN transistor, 38
Clamping circuits, 303–8
   dc restorer circuits, 303–6
   latching circuits, 306–8
Clipping circuits, 298–303, 308, 310
   single-level, 298–301
      negative, 299–301
      positive, 298–99
   two-level, 298, 301–3
Collector-base reverse saturation current, 37
Colpitts oscillators, 286–88
Complementary-symmetry amplifiers, 233–35
Complementary UJT devices, 111
Conduction band, 5
Conductors, definition of, 3
Contact potential, 19
Covalent bonding, 6–8, 12
   breaking up of, 8–9
   definition of, 6
Crossover distortion, 227
Crystal oscillators, 295–96
Current amplifiers, 160, 182–83
   characteristics of, 183
Current-feedback amplifiers, 247–51
   current gain, 247–48
   determining the parameters of, 250–51
   equivalent circuit, 249–51
   input resistance, 249
   output resistance, 249
Current limiting circuits (*See* Power supply, current limiting circuits)

Current transfer curve, 212
Curve tracer, 21
Cut-in voltage, 24, 25
Cutoff frequencies, 161

D

Dark resistance, definition of, 129
Darlington emitter-follower circuits, 261–64
  determining the parameters of, 264
  equivalent circuit, 263
Dc load line equation
  drawing the dc load line, 62–63
  for the fixed-bias circuit, 53, 62
  for the self-bias circuit, 57, 62
Dc restorer circuits, 303–6
Dc short-circuit current gain
  in the common-base configuration, 39
  in the common-emitter configuration, 40–41, 45
    designing a self-bias circuit with a spread of, 60–62
    as a function of the operating point, 45
    range of value for, 54, 384
Decade, definition of, 172
Depletion region, formation of, 18–19
Depletion-region capacitance, definition of, 27
Derating curve, 217
Diac, 113
  definition of, 116
Die, 17
Differential amplifiers, 261, 264–71
  basic, 264–65
  characteristics of ideal, 264–65
  improved circuits, 265–71
    common-mode gain, 268–69, 271
    common-mode rejection ratio, 269–70, 271
    difference gain, 268, 269, 271
    integrated circuit, 271–72
Differentiator circuit, 367–68
Diffusion, 16, 17, 19
  explanation of, 13–14
Diffusion capacitance, definition of, 27, 76
Donor atoms, 12
Dot convention, 145
Double-tuned amplifiers (*See* Tuned amplifiers, double-tuned)
Drain current, 87
  equation for, 86

Drift, 19
  explanation of, 13
Dynamic resistance, 24–25

E

Electron-hole pair, 8–9
Electronic crowbar, 356–57
Electronic filters, 153, 154
Emitter-follower circuits, 261–64
Energy bands, 5–6
Energy gap, 5
Enhancement, explanation of, 96
Epitaxial growth, 35
Epitaxial mesa transistor, 35

F

Fiber optics emitters and detectors, 139–41
  manufacturer's data sheet on, 408–9
Field effect transistors (FETs)
  high-frequency model, 105, 167
  JFET (*See* Junction field-effect transistors)
  MOSFET (*See* MOSFETs)
  small-signal model, 102–5
Filters, 153–58
  capacitors (*See* Capacitors)
  definition of, 145
  electronic filters, 153, 154
  inductors, 153, 154, 156
  multiple-element filters (*See* Multiple-element filters)
Fixed-bias circuit, 52–55, 63
  calculation of the operating point, 52–54
  thermal stability, 54–55
Four-layer (Shockley) diode, 114–16
Free electrons, 3, 5, 8, 11, 12
  definition of, 8
Frequency response, 160–61
  (*See also* Small-signal amplifiers, frequency response)
Full-wave rectifiers, 148–53, 154, 156
  center-tapped bridge rectifier, 152–53
  full-wave bridge rectifiers, 149–52, 155
    advantages of, 149
  full-wave center-tapped rectifier, 148–49, 155

G

Gain
   definition of, 160
   determining real and apparent, 179–82
   importance of knowing magnitude of short-circuit current gain and open-circuit voltage gain, 180–81
   measurement of, 171–72
   systematic analysis for gain calculations, 161–62
Gain-bandwidth product, 79
Gallium, 12
Gallium arsenide (GaAs) infrared emitting diode, 132, 133, 137
Gallium arsenide phosphide light-emitting diode, 133, 137, 138
Gate turnoff switches (GTOs), 122, 123
Germanium, 6, 19
   broken bonds in, 9
   core for, 8
   crystal structure for, 6–7
   diode characteristics, 21, 22, 23

H

Half-wave rectifiers, 145–48, 154
   average current, 147
   half-wave operation, 145–47
   ripple factor, 147–48
Harmonic distortion, 214, 220–23
   methods of analysis, 223
   in the output waveshape for a series-fed class-A amplifier, 221–23
Hartley oscillators, 285–86, 288
Heat sinks, 216, 217–19
HEXFET, 98–99, 100
Holding current, 121

I

IGFET, 94
Impedance
   of a BJT tuned amplifier, 201
   definition of, 160
   effect of source and load impedances in a current amplifier, 182–83
   of FET tuned amplifiers, 202

Impedance (*Contd.*)
   input, 178, 181–82
   matching, 184–85
      with an interstage transformer, 201
      with tapped capacitors, 202
      with tapped inductors, 201, 202
   output, 178–79, 182
Indium, 12
Inductors, 153, 154, 156
   tapped, 201, 202
Inert, definition of, 5
Infrared-emitting diodes (IREDs), 133, 137
Insulated-gate field effect transistor (*See* IGFET)
Insulators, 5
   definition of, 3
Integrated circuit (IC) amplifiers, 235–38
   dual IC power amplifiers, 235–36
   mono IC power amplifiers, 235, 236, 237–38
Integrated circuit (IC) differential amplifiers, 271–72
Integrated circuit (IC) regulators, 334–39
   applications of, 335, 336, 337
   features of, 334–35
   manufacturer's data sheet on, 420–22
   terminology used in specification sheets, 335, 336
   three-terminal IC power regulators, 337–39
      manufacturer's data sheet on, 423–27
Integrated circuits, definition of, 1
Integrator circuit, 364, 365–67
Intrinsic standoff ratio, equation for, 108–9
Inverting operational amplifers, 277–78
   equation for closed-loop gain in, 278
Ionized, definition of, 4
Irradiance, 133

J

JFETs (*See* Junction field-effect transistors)
Junction diodes (*See* Semiconductor diodes)
Junction field-effect transistors, 82–94
   biasing, 87–94
      self-bias, 87–89
      voltage-divider bias, 89–94
   channel resistance, 84
   constant resistance and constant current regions, 85–86, 87
   depletion region, 83–84

Junction field-effect transistors (*Contd.*)
   drain characteristics, 86–87
   equations for determining the operating
      point for, 88–89
   manufacturer's data sheet on, 389
   pinchoff voltage, 84–85
   saturation drain current, 86
   small-signal parameters, 104
   summary of operation of, 86–87
   transfer characteristic curve, 86–87

L

Latching circuits, 306–8
Latching current, 121
Light-actuated silicon-controlled rectifiers
      (LASCRs), 132, 136
Light-emitting diodes (LEDs), 133, 137, 138
Load line equation (*See* Dc load line equation)

M

Metal-oxide-semiconductor field-effect
      transistors (*See* MOSFETs)
Monolithic operational amplifiers (*See*
      Operational amplifiers)
MOSFETs, 82, 94–98
   biasing, 101–2
   depletion-mode, 94–95
   dual-gate, 97–98, 105
   electrostatic discharge, 100
   enhancement-mode, 96–97
   HEXFET, 98–99, 100
   manufacturer's data sheets on, 390–94
   small-signal parameters, 104
   VFET (vertical enhancement-mode
      MOSFET), 98, 99
Multiple-element filters, 154–58
   L-section, 155, 156
   π-section, 155, 156

N

Negative feedback, 230
   definition of, 242
Negative feedback amplifiers, 242–58
   classified according to action on gain, 243

Negative feedback amplifiers (*Contd.*)
   current-feedback amplifiers (*See* Current-
      feedback amplifiers)
   effect of feedback on frequency response,
      251–53
   effect of feedback on nonlinear distortion
      and noise, 258
   general feedback concepts, 242–43
   series-feedback amplifiers (*See* Series-
      feedback amplifiers)
   seven uses of negative feedback in an
      amplifier, 242–43
   shunt-feedback amplifiers (*See* Shunt-
      feedback amplifiers)
   voltage-feedback amplifiers (*See* Voltage-
      feedback amplifiers)
Noninverting operational amplifiers, 278–79
Notch filter, 292
NPN transistors, 34, 134
   calculating the operating point of (*See*
      Biasing, calculation of the operating
      point)
   common-base characteristics for, 42, 43
   common-base configuration using, 41
   common-emitter characteristics for, 43–44
   common-emitter configuration using, 42, 43
   currents in, 35–39, 40–41

O

Operational amplifiers, 261, 272–76
   applications of, 272
   basic amplifier configurations, 276–80
      inverting amplifiers (*See* Inverting
         operational amplifiers)
      noninverting amplifiers, 278–79
      trimming input offset voltage, 280
   characteristics of, 273
   common-mode rejection ratio (CMRR),
      269–70, 275
   definition of terms, 274–76
   internal operation, 273–74
   manufacturer's data sheets on, 414–19
   table on ideal vs. practical, 273
   (*See also* Analog systems, principles of
      analog computation)
Optical couplers/isolators, 139
   manufacturer's data sheet on, 410–13
   table of, 139

Index   **439**

Optically-triggered triac driver, 132, 136–37
Oscillators (*See* Sinusoidal oscillators)

P

Peak voltage, equation for, 110
Pentavalent impurity atoms, 11–12
Phosphorus, 11
Photodetectors, 132–37
  light-actuated SCR, 132, 136
  optically-triggered triac driver, 132, 136–37
  photodarlington, 132, 135–36, 141
  photodiodes, 132, 133–34, 135, 141
  phototransistors, 132, 134–35, 141
  spectral characteristics of, 132, 133
Photoelectric devices, 128–41
  fiber optics emitters and detectors (*See* Fiber optics emitters and detectors)
  optical couplers/isolators (*See* Optical couplers/isolators)
  photoconductive cells, 128–31, 133
  photodetectors (*See* Photodetectors)
  photoemitters, 137–38
    infrared-emitting diode (IRED), 133, 137
    light-emitting diode (LED), 133, 137, 138
  photovoltaic cells, 131–32, 133
Photoresist, definition of, 16
Pinchoff voltage, 84, 85
PNP transistors, 34, 35–36
  calculating the operating point for, 52
  currents in, 40
Positive feedback, definition of, 242
Power amplifiers, 160, 161, 209–39
  classes of, 209, 210
  complementary-symmetry amplifiers, 233–35
  integrated circuit (IC) amplifiers (*See* Integrated circuit (IC) amplifiers)
  power efficiency and dissipation in, 214–20
    derating curve, 217
    equations for ac output power, 214, 215
    equation for dc input power, 214
    heat sink application, 216, 217–19
    power efficiency calculation, 215
    ratio for, 214
    transistor maximum power dissipation, 215–17
    using manufacturer's data sheets to establish a transistor's permissible region of operation, 215–20, 382–83
  push-pull amplifiers (*See* Push-pull amplifiers)

Power amplifiers (*Contd.*)
  schemes for eliminating the need for transformers at both the input and output, 231
    complementary-symmetry amplifiers, 231, 233–35
    transformerless push-pull configuration with NPN and PNP transistors, 231–32
  series-fed class-A amplifiers (*See* Series-fed class-A amplifiers)
  summary on, 238–39
  transformer-coupled push-pull amplifiers (*See* Transformer-coupled push-pull amplifiers)
  transformer-coupled single-ended class-A amplifiers, 223–24
Power control systems, 346–57
  electronic crowbar, 356–57
  principles of power control, 346–52
    firing of an SCR, 350–51
    full-wave bridge with one SCR, 348
    full-wave bridge with two SCRs, 348, 349
    full-wave power control as a function of the conduction angle, 350
    half-wave power control circuit, 348, 349–50
    a single SCR in series, 346–47
    triac power control circuit, 347–48
    two SCRs connected in inverse parallel, 347
    using the UJT to fire an SCR, 351–52
  SCR universal motor speed and direction control, 354–55
  SCS (silicon-controlled switch) alarm circuit, 353–54
  triac light-intensity control, 352–53
  12-volt battery charger, 355–56
Power ratings, 216
Power supply, 145
  circuits of the, 319
  complete, 339–44
    an adjustable 0 to 22 V power supply, 342
    a dual low-current power supply, 343–44
    five components in, 340
    using manufacturer's data sheet to determine, 340–42, 377, 423–27
  current limiting circuits, 332–33
    diode overcurrent protection, 332–33
    principle of foldback, 335
    transistor overcurrent protection, 333
  regulated, 319–44
Precision rectifier circuits, 310–11

Programmable unijunction transistors (PUTs), 111–12, 117–18, 119
   manufacturer's data sheet on, 400–402
Pulsating direct current, 147
Push-pull amplifiers
   other, 231–32
   transformer-coupled (*See* Transformer-coupled push-pull amplifiers)

Q

Quality factor (Q), 195–97
Quiescent point, 52
   determining (*See* Biasing, calculation of the operating point)

R

RC phase-shift oscillators, 288–91
Rectifiers
   definition of, 145
   full-wave (*See* Full-wave rectifiers)
   half-wave (*See* Half-wave rectifiers)
Regulators, 319–32
   definition of, 319
   integrated circuit regulators (*See* Integrated circuit regulators)
   series regulators (*See* Series regulators)
   shunt regulators, 325
   Zener regulator circuits (*See* Zener regulator circuits)
Resistance, equation for, 13
Resistivity, 13
Reverse blocking thyristors (*See* Silicon-controlled rectifiers (SCRs))
Reverse recovery time, definition of, 28
Reverse saturation current, 22–23, 28
   collector-base, 37
   factors determining its value at room temperature, 23
   importance of temperature in the magnitude of, 23
Ripple factor, 147–48
Rise time, definition of, 131

S

Saturation drain current, 86
Scaler circuit, 360, 361, 362
Schottky-barrier diode, 29, 30
   manufacturer's data sheet on, 381
Selenium, 131
Self-bias circuit, 55–59
   advantages over fixed-bias, 55, 59
   designing a, 59–62
      designing for a β spread, 60–62
   determining the operating point for, 56–58
   thermal stability, 58–59
Semiconductor devices, advantages of, 3
Semiconductor diodes, 16–30
   with bias, 19–21
      forward, 19–20
      reverse, 19, 20–21
   contact potential, 19, 20
   depletion region, 18–19, 20
   diode characteristics, 21–24
      breakdown voltage (*See* Breakdown voltage)
      reverse saturation current (*See* Reverse saturation current)
   diode equation, 21
   manufacture of a PN junction diode, 16–18
   manufacturer's data sheet, 376–77
   multilayered, 112–19
      ac bilateral trigger diode, 113–114, 395–96
      bilateral diode switch (diac), 116–17
      four-layer (Shockley) diode, 114–16, 119, 120
      silicon bilateral switches (SBSs) (*See* Silicon bilateral switches (SBS))
      silicon unilateral (SUS) switches, 117–19
   with no bias, 18–19, 20
   special purpose diodes (*See* Special purpose diodes)
   use of piecewise linear equivalent circuit, 24–28
      capacitative effects in, 26–27
      recovery time, 27–28
      regions, 1–3, 24–26
Semiconductor physics, 3–14
   classifying matter, 3–6
      on the basis of conductivity, 5–6
      energy bands, 5–6
      shell structure, 3–5
Semiconductors, 6–14
   charge carriers (*See* Charge carriers)
   conduction in, 13–14
      diffusion, 13–14
      drift, 13
   covalent bonding (*See* Covalent bonding)

Semiconductors (*Contd.*)
  definition of, 6
  doped, 11–13
    N-type, 11–12, 13, 14, 16
    P-type, 11, 12–13, 14, 16
  electron-hole pair, 8–9
  free electrons, 3, 5, 8
  intrinsic (pure), 9, 11
    drawbacks to the use of, 11
  schematic representation of a crystal, 7
Series-fed class-A amplifiers, 209, 210–14, 223, 224
  determining ac and dc power and power efficiency for, 215
  determining the collector current waveshape, 212–14
  determining the dc Q-point, 211
  determining the harmonic content and distortion in the output waveshape, 221–23
Series-feedback amplifiers, 253, 254, 255
  determining the parameters of, 254–55, 256
  effects of feedback on circuit performance, 258
  equations for, 255
Series regulators, 323–32
  basic, 323–25
    operation of, 324–25
  with differential amplifier feedback, 330–31
  with operational amplifier feedback, 331–32
  with transistor feedback, 325–30
Shell structure, 3–5
Shockley diode, 114–16, 119, 120
Shunt-feedback amplifiers, 253, 254, 255, 256–57, 258
  determining the parameters of, 256–57, 258
  effects of feedback on circuit performance, 258
  equations for, 255
Silicon, 6, 12, 13, 14, 19, 131
  broken bonds in, 9
  core for, 8
  crystal structure of, 6, 7
  diode characteristics, 21, 22, 23
  growing a crystal of, 16, 17
Silicon bilateral switches (SBSs), 117–19
  manufacturer's data sheet on, 397–99
Silicon-controlled rectifiers (SCRs), 118, 119–22, 123, 124, 125

Silicon-controlled rectifiers (*Contd.*)
  characteristics of, 120–22
  complementary, 123
  light-actuated, 136
  manufacturer's data sheet on, 403–4
  operation of, 119–20
Silicon-controlled switches (SCSs), 123
  alarm circuit, 353–54
Silicon unilateral switches (SUSs), 117–19
Single-stage BJT amplifiers, 163–66
Single-stage FET amplifiers, 166–69
Single-tuned amplifiers (*See* Tuned amplifiers, single-tuned)
Sinusoidal oscillators, 282–96
  amplitude-stabilized oscillators, 294–95
  Colpitts oscillators, 286–88
  criteria for oscillation, 282–84
    Barkhausen criterion, 282–83
  crystal oscillators, 295–96
  Hartley oscillators, 285–86, 288
  RC phase-shift oscillators, 288–91
  tuned-output oscillators, 291–92
  twin-T oscillators, 292
  Wien-bridge oscillators, 293–94, 295
Small-signal amplifiers, 161
  difference between large-signal and, 161
  frequency response, 169–75
    complete, 175
    determining upper cutoff frequency and current gain magnitude response, 171–73
    high frequency, 170–73
    low frequency, 173–75
    midband frequency, 163–66, 167–69
  systematic analysis for gain calculations, 162–69
    four-step procedure outline, 162
    for single-stage BJT amplifier, 163–66
    for single-stage FET amplifier, 166–69
Special purpose diodes, 28–30
  varactor diode (*See* Varactor diode)
  Zener diode (*See* Zener diode)

T

Tetrode thyristor (*See* Silicon-controlled switches (SCSs))
Thermal runaway, 54–55

Thermal stability factor S, 54–55
  guidelines governing choice of, 60
Threshold voltage, 96–97
Thyristors, 117, 118, 119–26, 346
  bilateral triode (*See* Triac)
  gate turnoff switch (GTO), 122, 123
  silicon-controlled rectifier (*See* Silicon-
    controlled rectifiers (SCRs))
  silicon-controlled switch (*See* Silicon-
    controlled switches (SCSs))
Transconductance
  equation for, 77
Transformer-coupled push-pull amplifiers,
    225–30
  advantage of the push-pull amplifier, 227
  class AB operation, 225, 228–30
  class B operation, 225, 226–28
  differences between class-B and class-AB,
    228, 229–30
  drawback of, 230
Transformer-coupled single-ended class-A
    amplifiers, 223–24
  performance comparisons of the series-fed
    vs., 224
Transformers
  as impedance-matching devices, 224
Transistors
  bipolar junction (*See* Bipolar junction
    transistors)
  determining suitability of, 215–20
  diffused planar, 35
  epitaxial mesa, 35
  establishing proper operating point for (*See*
    Biasing, calculating the operating point)
  germanium, 52
  maximum power dissipation, 215–17
  most important and useful equations for, 40
  NPN (*See* NPN transistors)
  permissible region of operation, 47–48
  PNP (*See* PNP transistors)
  programmable unijunction (*See*
    Programmable unijunction transistors)
  ratings, 46–48
  silicon, 42, 43, 44, 46, 52, 53, 58, 60, 63, 79,
    211, 215, 382–83
  temperature sensitivity of, 46, 54–55
Triac, 118, 123–26
  light-intensity control, 352–53
  manufacturer's data sheet on, 405–7

Triac (*Contd.*)
  optically-triggered triac driver, 132, 136–37
  possible configurations, 125
  remote gate and junction gate, 123–24
Trivalent impurity atoms, 12–13, 17
Tuned amplifiers, 192–206
  characteristics of an ideal, 192, 193
  coupling of, 201–2, 203–4
  definition of, 192
  double-tuned, 202–6
    selectivity of, 202, 203, 206
    stagger-tuned amplifiers, 203, 204, 205–6
    synchronously-tuned amplifiers, 202–3,
      204
  single-tuned, 192, 193–200
    adjusting, 199–200
    quality factor (Q), 195–97
    resonant amplifier performance, 197–99
    selectivity of, 198–99
Tuned-output oscillators, 291–92
Tungsten lamps, 132
Turnoff current gain, 122
Twin-T oscillators, 292

U

Unijunction transistors (UJTs), 108–111, 117
  compared to the JFET, 108
  compared to the PUT, 111
  complementary, 111
  input and output characteristics, 110–11

V

Valence band, 5
Valence shell, 4, 5
Varactor diode, 28, 29
  manufacturer's data sheet on, 380
VFET, 98, 99
Virtual ground, 277–78
  definition of, 277
Voltage amplifiers, 160, 183–84
  cascading, 188
  characteristics of, 184
Voltage-feedback amplifiers, 243–47
  determining the parameters of, 246–47
  equivalent circuit, 246–47
  input resistance, 245

Index 443

Voltage-feedback amplifiers (*Contd.*)
  output resistance, 245
  voltage gain, 244–45
Voltage gain, definition of, 64

W

Wafers, 16
Wave-shaping circuits, 308–14
  peak and average detecting circuits, 312, 314
  spike-forming circuits (differentiators), 308–12, 313
Wien-bridge oscillators, 293–94, 295

Z

Zener breakdown, 23, 24
Zener diode, 28, 29, 266, 267, 303, 321, 333
  manufacturer's data sheet on, 378–79
Zener regulator circuits, 321–23
  determining the regulation and voltage for, 321–22
  manufacturer's data sheet on diode used, 378–79
  two important facts in designing, 322–23